Use of Microbes for Control and Eradication of Invasive Arthropods

Progress in Biological Control

Volume 6

Published:

Volume 1
H.M.T. Hokkanen and A.E. Hajek (eds.):
Environmental Impacts of Microbial Insecticides – Need and Methods for Risk Assessment. 2004   ISBN 978-1-4020-0813-9

Volume 2
J. Eilenberg and H.M.T. Hokkanen (eds.):
An Ecological and Societal Approach to Biological Control. 2007
ISBN 978-1-4020-4320-8

Volume 3
J. Brodeur and G. Boivin (eds.):
Trophic and Guild Interactions in Biological Control. 2006
ISBN 978-1-4020-4766-4

Volume 4
J. Gould, K. Hoelmer and J. Goolsby (eds.):
Classical Biological Control of *Bemisia tabaci* in the United States. 2008
ISBN 978-1-4020-6739-6

Volume 5
J. Romeis, A.M. Shelton and G. Kennedy (eds.):
Integration of Insect-Resistant Genetically Modified Crops within IPM Programs. 2008
HB ISBN 978-1-4020-8372-3; PB ISBN 978-1-4020-8459-1

Forthcoming:

Ecological & Evolutionary Relationships among Entomophagous Arthropods and Non-prey Foods
By J. Lundgren

Biocontrol-based Integrated Management of Oilseed Rape Pests
Edited by I.H. Williams and H.M.T. Hokkanen

Biological Control of Plant-Parasitic Nematodes: Building Coherence between Microbial Ecology and Molecular Mechanisms
Edited by Y. Spiegel and K. Davies

Egg Parasitoids in Agroecosystems with Emphasis on *Trichogramma*
Edited by J. Parra, F. Consoli and R. Zucchi

For other titles published in this series, go to
www.springer.com/series/6417

Ann E. Hajek · Travis R. Glare ·
Maureen O'Callaghan
Editors

Use of Microbes for Control and Eradication of Invasive Arthropods

Editors
Dr. Ann E. Hajek
Cornell University
Dept. Entomology
Ithaca NY 14853-2601
USA

Dr. Travis R. Glare
AgResearch Ltd.
Biocontrol, Biosecurity & Bioprocessing
Lincoln Research Centre
Christchurch 8140
New Zealand

Dr. Maureen O'Callaghan
AgResearch Ltd.
Biocontrol, Biosecurity & Bioprocessing
Lincoln Research Centre
Christchurch 8140
New Zealand

Cover: The cover photo collage was created by Kent Loeffler, Cornell University. The background map was created by NASA (http://visibleearth.nasa.gov/). Inset photos are as follows (clockwise starting at the top): Sporulating cadaver of early instar gypsy moth, *Lymantria dispar*, killed by the fungal pathogen *Entomophaga maimaiga* (Photo courtesy of Ann E Hajek); Octospores and free spores of *Thelohania solenopsae*, the microsporidian pathogen infecting *Solenopsis* spp. (Photo courtesy of USDA, ARS); Electron micrograph of *Bacillus thuringiensis* sporangium (Photo courtesy of José Bresciani and Jørgen Eilenberg); Adult rhinoceros beetle, *Oryctes rhinoceros* (Photo courtesy of Sada Nand Lal).

ISBN: 978-1-4020-8559-8   e-ISBN: 978-1-4020-8560-4

Library of Congress Control Number: 2008931200

© Springer Science+Business Media B.V. 2009
No part of this work may be reproduced, stored in a retrieval system, or transmitted
in any form or by any means, electronic, mechanical, photocopying, microfilming, recording
or otherwise, without written permission from the publisher, with the exception
of any material supplied specifically for the purpose of being entered
and executed on a computer system, for exclusive use by the purchaser of the work.

Printed on acid-free paper

9 8 7 6 5 4 3 2 1

springer.com

We would like to dedicate this book to the many devoted insect pathologists over many years who have studied and developed entomopathogens for control of invasive arthropods.

Progress in Biological Control

Series Preface

Biological control of pests, weeds, and plant and animal diseases utilising their natural antagonists is a well-established and rapidly evolving field of science. Despite its stunning successes world-wide and a steadily growing number of applications, biological control has remained grossly underexploited. Its untapped potential, however, represents the best hope to providing lasting, environmentally sound, and socially acceptable pest management. Such techniques are urgently needed for the control of an increasing number of problem pests affecting agriculture and forestry, and to suppress invasive organisms which threaten natural habitats and global biodiversity.

Based on the positive features of biological control, such as its target specificity and the lack of negative impacts on humans, it is the prime candidate in the search for reducing dependency on chemical pesticides. Replacement of chemical control by biological control – even partially as in many IPM programs – has important positive but so far neglected socio-economic, humanitarian, environmental and ethical implications. Change from chemical to biological control substantially con- tributes to the conservation of natural resources, and results in a considerable reduc- tion of environmental pollution. It eliminates human exposure to toxic pesticides, improves sustainability of production systems, and enhances biodiversity. Public demand for finding solutions based on biological control is the main driving force in the increasing utilisation of natural enemies for controlling noxious organisms.

This book series is intended to accelerate these developments through exploring the progress made within the various aspects of biological control, and via documenting these advances to the benefit of fellow scientists, students, public officials, policymakers, and the public at large. Each of the books in this series is expected to provide a comprehensive, authoritative synthesis of the topic, likely to stand the test of time.

Heikki M.T. Hokkanen, Series Editor

Preface

One of the main reasons that we organized this edited volume is to increase international awareness of the growing use of invertebrate pathogens for control and eradication of invasive arthropods. As the numbers of invasive species continues to rise, more insect pathologists have been involved with work on their control using entomopathogens. In fact, this is not a new area of focus for insect pathologists; work on microbes against invasive arthropods began more than a century ago with classical biological control introductions of entomopathogenic fungi against invasive species in the 1890s. Chapters in this book cover entomopathogens that have been developed for control of invasive species over many decades (e.g. a nematode against *Sirex noctilio* and *Bacillus thuringiensis* against gypsy moth) while other chapters focus on development of control measures for very recent invasives (e.g. emerald ash borer first found in the US in 2002). Since both the United States and New Zealand are countries with abundant trade, which is a key pathway for invasives, we have been very aware of the growing numbers of invasive pests arriving in our own countries and the need for control strategies. We have been closely involved with their control using microbes, at varying levels (from laboratory bench to field studies to national committees evaluating eradication programs using the entomopathogen *B. thuringiensis*).

Within the past few years, symposia on use of microbes for invasive control have been organized twice at the annual meetings of the Society of Invertebrate Pathology (2005 – Anchorage, Alaska, and 2007 – Quebec City, Quebec, Canada), demonstrating interest in this subject across the international community of invertebrate pathologists. However, no written summaries, covering the different types of pathogens being studied, developed and used for control, have previously addressed this subject. This could be due to the fact that the subject is very diverse, including programs using very different microbes (viruses, bacteria, fungi, protists and nematodes) in a diversity of contexts: from eradication of new populations of invasive species, to control of established populations of invasives as well as basic studies of host/pathogen interactions and epizootiology. Especially for eradication programs, the lack of written summaries may also relate to the practical focus of these programs, which are about applied pest control rather than research. We hope that those working with invasive arthropods will find this book useful as a resource and that it will serve to support further work on this subject as well as, eventually, increased use of entomopathogens for control of invasives.

We would like to thank Cornell University for sabbatical funding for Dr. Hajek while at AgResearch, Lincoln, New Zealand and support from AgResearch, where this book was mostly organized. We would also very much like to thank the many authors who have contributed excellent chapters to this edited book. The final preparation of this book was facilitated by Sue Zydenbos and Lois McKay. We also thank Heikki Hokkanen for organizing this book series and Springer for their support of this subject.

Lincoln, New Zealand   Ann E. Hajek, Travis R. Glare &
                                                              Maureen O'Callaghan

Contents

Part I Introduction

1. **Invasive Arthropods and Approaches for Their Microbial Control** 3
   Ann E. Hajek

Part II Ecological Considerations

2. **Naturally Occurring Pathogens and Invasive Arthropods** 19
   Ted E. Cottrell and David I. Shapiro-Ilan

3. **Population Ecology of Managing Insect Invasions** 33
   Andrew M. Liebhold and Patrick C. Tobin

Part III Eradication

4. **Use of Pathogens for Eradication of Exotic Lepidopteran Pests in New Zealand** 49
   Travis R. Glare

5. **North American Eradications of Asian and European Gypsy Moth** .. 71
   Ann E. Hajek and Patrick C. Tobin

Part IV Control

6. **Exotic Aphid Control with Pathogens** 93
   Charlotte Nielsen and Stephen P. Wraight

7. *Steinernema scapterisci* **as a Biological Control Agent of** *Scapteriscus* **Mole Crickets** 115
   J. Howard Frank

8 **The Use of *Oryctes* Virus for Control of Rhinoceros Beetle in the Pacific Islands** 133
Trevor A. Jackson

9 **Use of Microbes for Control of *Monochamus alternatus*, Vector of the Invasive Pinewood Nematode** 141
Mitsuaki Shimazu

10 **Use of Entomopathogens against Invasive Wood Boring Beetles in North America** 159
Ann E. Hajek and Leah S. Bauer

11 **Control of Gypsy Moth, *Lymantria dispar*, in North America since 1878** 181
Leellen F. Solter and Ann E. Hajek

12 **Controlling the Pine-Killing Woodwasp, *Sirex noctilio*, with Nematodes** 213
Robin A. Bedding

13 **Fire Ant Control with Entomopathogens in the USA** 237
David H. Oi and Steven M. Valles

14 **Biological Control of the Cassava Green Mite in Africa with Brazilian Isolates of the Fungal Pathogen *Neozygites tanajoae*** 259
Italo Delalibera Júnior

15 **Microbial Control for Invasive Arthropod Pests of Honey Bees** 271
Rosalind R. James

Part V Safety and Public Issues

16 **Human Health Effects Resulting from Exposure to *Bacillus thuringiensis* Applied during Insect Control Programmes** 291
David B. Levin

17 **Environmental Impacts of Microbial Control Agents Used for Control of Invasive Pests** 305
Maureen O'Callaghan and Michael Brownbridge

Part VI   Conclusions

18 Considerations for the Practical Use of Pathogens for Control and Eradication of Arthropod Invasive Pests 331
Travis R. Glare, Maureen O'Callaghan and Ann E. Hajek

Index 351

Contributors

Leah S. Bauer USDA, Forest Service, Northern Research Station, 1407 S. Harrison Rd., East Lansing, Michigan 48823, USA, lbauer@fs.fed.us

Robin A. Bedding CSIRO Entomology, P.O. Box 1700, Canberra ACT 2601, Australia, robin.bedding@csiro.au

Michael Brownbridge AgResearch Ltd., Biocontrol, Biosecurity & Bioprocessing, Lincoln Science Centre, Christchurch 8140, New Zealand, michael.brownbridge@agresearch.co.nz

Ted E. Cottrell USDA, Agricultural Research Service, SE Fruit and Tree Nut Research Lab, 21 Dunbar Rd., Byron, Georgia 31008, USA, ted.cottrell@ars.usda.gov

Italo Delalibera Júnior Department of Entomology, Plant Pathology and Agricultural Zoology, ESALQ/University of São Paulo, Av. Pádua Dias 11, C.P. 9, 13418–900 Piracicaba, São Paulo, Brasil, italo@esalq.usp.br

J. Howard Frank Department of Entomology and Nematology, University of Florida, Gainesville, Florida 32611-0630, USA, jhfrank@ufl.edu

Travis R. Glare AgResearch Ltd., Biocontrol, Biosecurity & Bioprocessing, Lincoln Science Centre, Christchurch 8140, New Zealand, travis.glare@agresearch.co.nz

Ann E. Hajek Department of Entomology, Comstock Hall, Cornell University, Ithaca, New York 14853–2601, USA, aeh4@cornell.edu

Trevor A. Jackson AgResearch Ltd., Biocontrol, Biosecurity & Bioprocessing, Lincoln Science Centre, Christchurch 8140, New Zealand, trevor.jackson@agresearch.co.nz

Rosalind R. James USDA, Agricultural Research Service, Pollinating Insects—Biology, Management, Systematics Research Unit, 5310 Old Main Hill, Logan, Utah 84322–5310, USA, Rosalind.James@ars.usda.gov

David B. Levin Department of Biosystems Engineering, E2-376 EITC, University of Manitoba, Winnipeg, Manitoba R3T 5V6 Canada, levindb@cc.umanitoba.ca

Andrew M. Liebhold USDA, Forest Service, Northern Research Station, 180 Canfield St., Morgantown, West Virginia 26505, USA, aliebhold@fs.fed.us

Charlotte Nielsen Department of Entomology, Comstock Hall, Cornell University, Ithaca, New York 14853-2601, USA; Department of Ecology, University of Copenhagen, Faculty of Life Sciences, Thorvaldsensvej 40, DK-1871 Frederiksberg C, Denmark, chni@life.ku.dk

Maureen O'Callaghan AgResearch Ltd., Biocontrol, Biosecurity & Bioprocessing, Lincoln Science Centre, Christchurch 8140, New Zealand, maureen.ocallaghan@agresearch.co.nz

David H. Oi USDA, Agricultural Research Service, Center for Medical, Agricultural and Veterinary Entomology, 1600 SW 23rd Dr., Gainesville, Florida 32608, USA, david.oi@ars.usda.gov

David I. Shapiro-Ilan USDA, Agricultural Research Service, SE Fruit and Tree Nut Research Lab, 21 Dunbar Rd., Byron, Georgia 31008, USA, david.shapiro@ars.usda.gov

Mitsuaki Shimazu Forestry and Forest Products Research Institute, Department of Forest Entomology, Matsunosato 1, Tsukuba, Ibaraki Pref. 305-8687 Japan, shimazu@ffpri.affrc.go.jp

Leellen F. Solter Division of Biodiversity and Ecological Entomology, Illinois Natural History Survey, 1816 S. Oak Street, Champaign, Illinois 61801, USA, lsolter@uiuc.edu

Patrick C. Tobin USDA, Forest Service, Northern Research Station, 180 Canfield St., Morgantown, West Virginia 26505, USA, ptobin@fs.fed.us

Steven M. Valles USDA, Agricultural Research Service, Center for Medical, Agricultural and Veterinary Entomology, 1600 SW 23rd Dr., Gainesville, Florida 32608, USA, steven.valles@ars.usda.gov

Stephen P. Wraight USDA, Agricultural Research Service, Biological Integrated Pest Management Research Unit, Robert W. Holley Center for Agriculture and Health, Tower Road, Ithaca, New York 14853-2901, USA, stephen.wraight@ars.usda.gov

Part I
Introduction

Chapter 1
Invasive Arthropods and Approaches for Their Microbial Control

Ann E. Hajek

Abstract Invasive arthropod species cause ever-increasing economic, environmental and public health problems. Microbes (i.e. viruses, bacteria, fungi, nematodes and protists) have been used very successfully for eradicating and controlling a range of invasive arthropods in diverse ecosystems worldwide. Many eradication and control programs using microbes have used inundative augmentation (widespread application) approaches while some control programs have instead focused on classical biological control (point release and natural spread). This chapter provides a short history of past use of microbes for control of invasive arthropods as well as an introduction to the subjects that will be covered in this book.

1.1 Globalization and Invasive Species

Throughout history, arthropods have always competed with humans for managed and unmanaged resources as well as causing problems to public health. Initially, arthropods that became pestiferous were native species, but as humans began dispersing to new areas, the arthropods associated with humans, their domesticated animals and crop plants accompanied them. A few of the movements of arthropods have been purposeful and beneficial to humans, such as introducing honeybees for pollination and honey production or introducing arthropod biological control agents. However, the vast majority of introductions have been accidental and usually this dispersal is made possible through human means. (Of course, natural dispersal occurs too but this is slow and takes place over evolutionary times, doesn't usually cross biogeographic borders and usually occurs in only one direction (Nentwig 2007b)). While some introduced species have had little impact, failing to establish or not competing strongly for local resources, a low percentage become established and populations have grown, resulting in crop damage, displacement of indigenous species or adversely impacting animal or public health.

A.E. Hajek
Department of Entomology, Comstock Hall, Cornell University, Ithaca,
New York 14853-2601 USA
e-mail: aeh4@cornell.edu

Given the difficulties in detection of small arthropods, many of which exhibit secretive behaviors, along with the constantly increasing movement of items and people around the world, the numbers of arthropods being introduced to new areas are increasing at an alarming rate. It has been estimated that in the United States alone (including Hawaii), 4,500 arthropod species have been introduced (Pimentel *et al.* 2005). We regularly hear of the detection of new invasive species partly due to the ever-increasing rate of new introductions and partly due to improved surveillance methods, reflecting the recognition of the serious risks associated with invasive arthropods. While agriculture began about 10,000 years ago, probably leading to the first major movements of pests, the date that Columbus discovered America is often accepted as an arbitrary timing for the beginning of what we now call 'biological invasions' (Nentwig 2007a). Thus, after Columbus' voyages to America approximately 500 years ago, faster movement of people and goods was possible and, with improvements in transportation methods and availability, the age of globalization began. To aid in globalization, global regulatory organizations work toward facilitating exchanges of goods between all nations, eliminating obstructions to free trade. As global trade increases, so does accidental movements of organisms. Unfortunately, global regulations have not been put in place to prevent the increasing numbers of invasions. For example, in the United States alone, in 1800 only 50 alien arthropods were recorded as established; by 1990, 2000 invasive arthropods had become established (OTA 1993).

Improved transportation speed, access and availability have facilitated dispersal of hitchhiking arthropods from areas where they are native to areas where they are exotic, or alien. Various methods have led to arthropods breaching the biogeographic barriers that defined the original distributions of species. Unintentionally introduced species have been transported as tramps in vehicles, ships and planes; faster means of transportation now allow introductions of species that in past years would not have survived slower voyages over long distances (Nentwig 2007b). For example, cockroach and ant species that are native to more tropical areas would not be able to survive months of transport across colder areas but, when moved quickly by protected means (e.g. within an airplane), they are able to survive and thus some have now been introduced globally as tramps. Some species are transported by waterways and through shipping. The alfalfa snout beetle, *Otiorhynchus ligustici*, was most probably transported in ballast from England to northern New York State (Shields *et al.* 2004). Other species have been introduced with living plants, with harvested plant material including various types of wood or with soil associated with plants (Nentwig 2007b). The global growth in business travel, tourism and immigration from one continent to another adds to possibilities for transporting arthropods that might become introduced to new areas.

In fact, only a small fraction of species that are introduced become established over the long-term (Simberloff & Gibbons 2004) and only a small fraction of those becoming established cause serious problems (Williamson 1996). The alien species that become established and have potential to or actually cause economic or environmental harm, or harm to human health are termed 'invasive species' (National Invasive Species Council 2000). This is a subjective and broad definition.

The necessity to address problems due to invasive species is viewed differently by those mostly interested in preserving native biodiversity versus those purely interested in whether non-native species that have been introduced are having an economic impact.

We have seen that it does not take many new species well-suited to exploiting a new habitat to create catastrophic problems; sometimes only one species can monopolize or take over a previously diverse and balanced ecosystem and change it irreversibly. Ecologists have been very intrigued by the question of what allows a species to become a damaging invasive, especially because in many cases these pests are not pests in their areas of origin. Among the ecological hypotheses that have been proposed, some ecologists think that species that become invasives are pre-adapted to the new environment or are superior in some way when compared with native species (Hufbauer & Torchin 2007). Another hypothesis that has long received support is that when new species enter new environments, they are not accompanied by the natural enemies controlling their populations in their area of origin (= enemy release hypothesis); this hypothesis is supported by the fact that many invasives do not cause problems in their areas of origin and can, in fact, be uncommon endemics. The evidence that introduced species have fewer natural enemies in their new environments compared with their native environments supports this hypothesis (Torchin & Mitchell 2004). It has been hypothesized that the natural enemies attacking invasive species in a new environment would be generalists rather than specialists since generalists would be more likely to shift to new host species (Torchin & Mitchell 2004, van der Putten *et al.* 2005).

Ecosystems that have been altered in some way may be more susceptible to being overcome by an invasive species. For example, when exotic pines were planted as monocultures over large areas in Australia, the system was ripe for an invasive species, the European woodwasp *Sirex noctilio*, to easily spread and kill trees (e.g. Haugen & Underdown 1990). Likewise, agricultural monocultures are highly susceptible to being overcome when invasive species that can utilize the crops being grown are introduced. In most cases, agriculture relies on exotic plant and animal species, which can be susceptible to exotic arthropods, and, in particular, those arthropods from the areas where the agricultural crop or animal originated.

Many books have chronicled the impacts of invasive species, including arthropods (e.g. Van Driesche & Van Driesche 2004). Introduced species that become invasive can have severe impacts on entire ecosystems. For example, over a 20 year period, the European balsam woolly adelgid, *Adelges piceae*, spread and killed over 95% of mature Fraser firs (*Abies fraseri*) in the southern Appalachian Mountains in the eastern US (as cited in Pimentel *et al.* 2002). The extensive tree mortality has been associated with regional loss of two native bird species and invasion by three other bird species due to substantial changes in the forest (Alsop & Laughlin 1991). Although Fraser fir regeneration is now extensive (younger trees are less susceptible to *A. piceae*), stand characteristics such as age and distribution of Fraser fir have been changed. The long-term impacts to the new generation of trees remains to be seen but it seems possible that the overall effect will be that these infested forests will have fewer Fraser firs (Ragenovich & Mitchell

2006). Beginning in the 19th century and through 1979, four species of wasps in the family Vespidae were introduced to New Zealand (Cook *et al.* 2002); New Zealand is now recorded as having the highest density of such wasps in the world. While these wasps cause abundant stings each year, they also have a strong impact on the native beech (*Nothofagus* spp.) forests. In particular, *Vespula vulgaris* and *Vespula germanica* reduce the crop of homopteran-secreted honeydew by > 90%, thus impacting food resources of native birds (Beggs & Rees 1999). In addition, the wasps also directly impact native invertebrate biodiversity and compete with native birds for invertebrate prey. The red imported fire ant, *Solenopsis invicta*, has spread throughout the southeastern United States where it has reduced biodiversity and harmed wildlife (Wojcik *et al.* 2001). In these instances, entire communities associated with invasives are affected, often leading to permanent changes in ecosystems.

In general, the impacts of invasive species on native ecosystems and biodiversity are not well documented whereas the economic impacts of invasive species have received more attention. Costs incurred as a result of invasive species take three general forms: direct costs of damage caused by invasives (e.g. productivity losses), costs of control efforts and costs of preventing new introductions. The latter is the concern of quarantine (or biosecurity) agencies and will not be covered in this book. It has been estimated that damage and control costs in the US for the imported fire ant alone equal US$1 billion per year (Pimentel *et al.* 2005). Structural damage caused by the Formosan termite (*Coptotermes formosanus*) in the southern United States has been estimated at US$1 billion per year (Pimentel *et al.* 2005), and this estimate was made before the 2005 hurricanes that resulted in heavy infestations in ravaged areas and increased potential for this destructive termite to be dispersed by humans. It is much more difficult or even impossible to put a cost to environmental changes caused by invasive species. Red imported fire ants kill poultry chicks, lizards, snakes, ground-nesting birds and many native invertebrates and, while some of these can be associated with a monetary loss, it is not so easy to assign a monetary value to loss of native snakes, lizards and many invertebrates. It is likewise difficult to assign a monetary value to loss of native Fraser firs (a species that is not logged) and associated communities in the forests of the Appalachian Mountains of the US.

1.2 Managing Invasive Arthropods

Exotic arthropod pests have been problematic since agriculture began and pest management strategies have been developed for these purposes. So how does controlling the ever-increasing numbers of invasive arthropods differ from controlling native arthropod pests? This book focuses on eradication and control of arthropods that are invasive because this group of pests presents unique challenges.

First, the goal of eradication is never an objective for native pests so this type of control is unique to invasive pests. To confound the difficulty of undertaking an eradication campaign, newly introduced species usually initially exist at low densities and often first invade areas with large human populations (i.e. airports and ports of entry are generally in urban centers), making control efforts more difficult. Detection can be particularly difficult and often invasive species are only discovered by chance. Even where routine surveys are conducted to detect invasives, the chance of successful detection is often dependent on timing of the survey and training of the operators.

Pests of agricultural crops are a mixture of native and introduced arthropods; it has been estimated that 40% of crop pests in the US have been introduced (Pimentel 1993). In New Zealand, it is estimated that 90% of invertebrate pests are aliens (Barlow & Goldson 2002). In cases where the invasive that has become established is an agricultural pest and cannot be targeted for eradication because it is already well-established and has spread, pest management can become more similar to control measures used for native pests. However, many invasives are pests in native ecosystems and, in these cases, standard management practices that have been developed to protect crops in agriculture and forestry are not appropriate, making control efforts more challenging. Invasive arthropods can increase to huge populations that spread like wildfire, leaving behind decimated native ecosystems that will never be the same. For example, the invasive emerald ash borer (*Agrilus planipennis*), first found in the US in 2002 in southeastern Michigan, is currently spreading, already leaving behind > 20 million dead ash (*Fraxinus* spp.) trees (USDA Forest Service, 2007).

Spread of these new pests can be very fast and their population dynamics once they invade new areas can be chaotic; these are unique characteristics of populations of invasive arthropods and, once again, require more creative control strategies. Commonly, there is very rapid population growth of invasive pests once they are established, whereas established populations of these species in their places of origin exist at lower population densities. This 'outbreak' phenomenon means there can be large populations at the expanding front of a pest invasion. A newer approach for controlling invasive species that have become established and are spreading (so eradication is no longer possible) is "slow the spread'. This strategy involves aggressively targeting the leading edge of a new invasion to limit the rate of colonization of new areas (see Chapter 5). In addition, this method reduces the outbreak impact at the leading edge.

One type of control that is frequently exploited for combatting invasive arthropods is classical biological control; this strategy is defined as 'the intentional introduction of an exotic biological control agent for permanent establishment and long-term pest control' (Eilenberg *et al.* 2001). In fact, historically, practitioners predominantly have focused on invasive pests with classical biological programs (Brewer & Charlet 1999). This strategy is more appropriate for targeting pests occurring in habitats with some degree of permanence (e.g. wetlands, forests, orchards) where effective, environmentally safe natural enemies of invasive pests are not present or effective. It is also a strategy in keeping with the theory that invasive pests succeed due to lack of natural enemies in their new area; this strategy is based on seeking to introduce to the invaded ecosystem the natural enemies of the pest that control its populations in its area of origin.

The other control strategy used against invasive pests is inundative augmentation: 'the use of living organisms to control pests when control is achieved exclusively by the organisms themselves that have been released' (Eilenberg *et al.* 2001). This control method is similar to strategies using synthetic chemical pesticides as it seeks to kill pests by direct mass application. Thus, when living organisms are used in this way, they are often referred to as biopesticides.

Invasions occur in several phases (Liebhold & Tobin 2008) and different methods for mitigation are appropriate for each phase. First, the initial arrival of the pest can be prevented by international quarantines that restrict movement of exotics and by inspections at ports of entry. This first phase is outside of the subject of this book but is covered in a recent review (Follett & Neven 2006). Subsequently, exotics become established, forming initial reproducing populations. Many insects fail to find hosts that can maintain a reproductive population, or they find conditions (i.e. winter) unsuitable for survival. Those that successfully form an initial population can be targeted for eradication if found before significant spread has occurred. Once the population of the invasive has developed and spread, becoming part of new ecosystems, invasives can only be controlled.

1.2.1 Preventing Establishment

Often no one knows when and where invasives were first introduced and only sometimes scientific detective work after the fact can help to trace the source and determine when and how the initial introduction occurred. When invasive pests are initially introduced, the numbers of individuals establishing are frequently very few. If the invasive is a species with a track record elsewhere, then monitoring methods have sometimes already been developed and can be used to aid in detection; in fact, customs and airport authorities are constantly searching for species that are known to be serious pests elsewhere and are also easily transported. However, methods for detection are not always available, either because the new invasive is not a pest elsewhere or simply because the species is not easy to detect, e.g. adults of wood borers that do not rely on sex pheromones can be difficult to detect because a standard detection tool, pheromone traps, is thus not appropriate and, in addition, immature stages of these pests live inside wood and can escape visual inspections. The period while populations of new invasive pests are at very low densities (i.e. just after establishment) is the optimal time when eradication should be undertaken for greatest chance of success; it becomes increasingly difficult or impossible to eradicate when the target is already at moderate densities and/or has already spread very far. Decisions on whether or not to undertake an eradication campaign are often based on costs and benefits of the program as well as the prediction of whether success is possible (Myers *et al.* 2000), although sometimes politics or public attitude can influence whether or not eradication programs are conducted regardless of the economics (see Chapter 18). Of course, availability or development of effective methods for detecting and monitoring low density populations are critically important to success in eradication. Eradication programs require blanketing the areas known to be infested with control agents to drive populations of the invasive pest below levels where they can reproduce; generally highly effective inundative augmentation methods are used. Thus, eradication programs are very expensive, they require large amounts of biopesticide and a very large crew, often needing to be available on short notice, in order to eliminate populations of the invasive pest before the population reaches higher density and spreads so that the species cannot be eradicated. In addition, it is important that authorities do not terminate eradication programs as soon as populations cannot be detected; they must continue surveillance for a time to ensure that the invasive species is truly eradicated (see Chapter 18).

1.2.2 Preventing Increase and Slowing Spread

After an invasive species has become established and is increasing in numbers and spreading, the next step often taken is implementation of a domestic quarantine to prevent movement of the pest (usually via humans) outside of the area of establishment. Alternately, barrier zones are created to prevent spread of the organism on its own. Control is easier for organisms that spread along a continuous population front. However, for organisms with the ability to travel long distances, either on their own or aided by humans, it is more difficult or even impossible to halt spread.

Once a population of an invasive is established in a new area, the focus of the program changes to control. Suppressing the population should aid in decreasing problems locally as well as decreasing the chance of further long distance dispersal into more areas. Suppression of spreading populations is often through inundative augmentation, including the release of large quantities of a pathogen to decrease populations of the invasive. Inundative augmentation is also used for control of established populations of invasives, especially when their populations increase to damaging levels. A more long-term strategy for control of established invasives is introduction of exotic pathogens for permanent long term control (= classical biological control). This strategy has been used extensively through the release of herbivorous arthropods and plant pathogens for control of invasive weeds, and releases of parasitoids and predators for control of invasive arthropods. Although classical biological control of arthropods has not utilized pathogens as frequently, this is not due to lack of success but perhaps more likely due to difficulties in finding and working with microorganisms during foreign exploration (i.e. searching for natural enemies in the area of origin) and possibly due to less knowledge of microbiology among entomologists involved in invasive responses and classical biological control. The numerous classical biological control programs using entomopathogens, including successes, have recently been reviewed (Hajek *et al.* 2007) and some of those pertaining to invasive species are covered in this book.

1.3 Use of Microbes Against Invasive Arthropods

1.3.1 Advantages of Using Microbes

Bacteria, viruses, fungi, nematodes and protists comprise the major groups of arthropod-pathogens used for eradication and control. Along with this diversity in microorganisms comes a diversity in pathogenicity (ability to infect) and virulence (speed of kill). However, in general the pathogens being used only affect arthropods and often have narrow host ranges, so they are more acceptable for use in the urban/suburban areas where invasives are often found first. Due to differing pathogenicity and ecology, different types of arthropod pathogens are appropriate for different pests in different circumstances. For example, strains of the bacterial pathogen *Bacillus thuringiensis* (*Bt*) can kill lepidopteran hosts within a day or two (Glare & O'Callaghan 2000). In addition, *Bt* is relatively easy to mass produce outside of insects and numerous companies sell different strains for pest control. Therefore, *Bt* has been used numerous times in eradication programs in urban areas (see Chapters 4 and 5). However, although *Bt* is a commonly occurring soil bacterium, it has rarely been known to cause epizootics and thus *Bt* is not considered for classical biological control programs. In contrast, the fungal pathogen *Entomophaga maimaiga* presently cannot be mass-produced but persists in the northeastern US, frequently causing epizootics in gypsy moth (*Lymantria dispar*) populations. This species therefore would never be considered for an eradication campaign but is very appropriate for classical biological control as it is capable of maintaining itself in the host population. For both eradication and control programs, entomopathogens offer a natural alternative in comparison with synthetic chemical pesticides. Introductions of invasive pests often occur around ports or airports, areas with large human populations, and extensive use of synthetic chemical pesticides is not possible or is not accepted by the public. Many invasive pests attack plants growing near human habitation or they attack plants in environmentally sensitive natural areas; in either case, the public frequently prefers an environmentally benign pest control option. In many of these environments, non-target impacts on other arthropods may be of concern, so use of host specific entomopathogens is also advantageous.

1.3.2 History of Use of Pathogens for Classical Biological Control of Invasive Arthropods

Classical biological control using pathogens against arthropods was first recorded in 1894–1895 but this strategy was used relatively little until the 1950s (Hajek *et al.* 2005, 2007). For introductions between 1894 and 1950 (about which there is adequate documentation) 81.8% targeted invasive insect hosts instead of native hosts. For all except one of the introductions before 1950, fungal pathogens were introduced and the principal targets were hemipterans and soil-dwelling scarabs.

For the majority (63.6%) of the 131 total release programs for which the area of endemism of the pest(s) and success in establishment could be determined, the targeted arthropod pest was an invasive and not native. The percentage of programs yielding successful establishment did not differ between programs targeting native versus invasive pests (71.4–72.4% establishment). Among the five pathogens released most commonly, the *Oryctes rhinoceros* virus (see Chapter 8), *Entomophaga maimaiga* (the fungal pathogen infecting gypsy moth) (see Chapter 11) and the nematode targeting *Sirex noctilio* (see Chapter 12) all targeted invasive hosts. These pathogens were used frequently because of their success in control of hosts.

Several major successes with classical biological control of invasive conifer-feeding sawflies in North America using viruses occurred in the mid-1900s; there are currently no on-going programs with the viruses so they are not covered elsewhere in this book and are therefore mentioned briefly here. European pine sawfly (*Neodiprion sertifer*) in North America (Hajek *et al.* 2005) was first reported in New Jersey in 1925 and it then spread in eastern North America (Cunningham & Entwistle 1981). A nucleopolyhedrovirus (NPV) was obtained from Sweden in 1949 and was subsequently released in Canada and the USA in the early 1950s and in the UK in the early 1960s. Methods for mass production were developed and this virus was also extensively used for inundative releases by Forest Service personnel, Christmas tree growers and private individuals until 1970 when registration became necessary for further use. Outbreak populations of the invasive European spruce sawfly (*Gilpinia hercyniae*) in eastern Canada and the northeastern US were controlled by an NPV and introduced parasites; the virus was first noted in 1936 but by 1952 it had been transferred or had spread through most infested areas (Cunningham & Entwistle 1981). This NPV had not been purposefully introduced and must have accidentally accompanied sawflies or parasites from Europe. The virus, in combination with parasitoids, appears to have permanently solved problems due to *G. hercyniae* in eastern North America.

1.3.3 History of Use of Pathogens for Inundative Augmentation of Invasive Arthropods

Metchnikoff is generally credited with being the first to conduct experimental work on the application of entomopathogens against economically important pests (Cameron 1973, Zimmermann *et al.* 1995). He worked with *Metarhizium anisopliae* and species of crop pests that were native to eastern Europe and Russia: *Anisoplia austriaca* and *Bothynoderes* (= *Cleonus*) *punctiventris*. Subsequent use of mass-produced entomopathogens for inundative control focused primarily on arthropod pests that were native or which had been moved extensively through agriculture for many years (e.g. scale insects and whitefly on citrus).

However, beginning in the 1920s, research toward use of entomopathogens for control also focused on invasives that were relatively new to North America at that time: the European corn borer, *Ostrinia nubilalis*, and the Japanese beetle,

Popillia japonica. Although research determined that both *B. thuringiensis* and *M. anisopliae* were effective for control of corn borer, products were not developed and put to use. In the US, research on bacterial pathogens of the invasive Japanese beetle (*Popillia japonica*) yielded a commercially available milky disease product based on *Paenibacillus popilliae*, a localized bacterial species from North America. This bacterium constituted the first insect pathogen that was approved for use by the US government, shortly after WWII (Federici 2005). To hasten spread of the pathogen, 109 tons of *P. popilliae* spore powder was distributed to 90 sites in 13 eastern states from 1939 to 1953 (Falcon 1971). A product based on *P. popilliae* is still available today in the US, since Japanese beetles continue to be a pest problem, however production is limited by *in vivo* methods. A nematode species *Steinernema glaseri* was also isolated from the Japanese beetle and appeared promising but, because of the success of milky disease, this nematode was not developed further. However, this initial work with *S. glaseri* is commonly credited as the beginning of microbial control with nematodes (Lord 2005).

Apparently during WWI, the Colorado potato beetle, *Leptinotarsa decemlineata*, was introduced into western Europe and it subsequently spread to the south and east. Because of much improved potato transportation, Colorado potato beetle was distributed across much of Europe by the 1940s. A *Beauveria bassiana*-based mycoinsecticide for control of the Colorado potato beetle (Boverin) was developed in 1965 in the former USSR (Kendrick 2000) and was used for many years (Ferron 1981) although it is not produced now (Faria & Wraight 2007). In the 1960s, *B. thuringiensis* products were initially commercialized and a few of the major pests targeted in the US (Federici 2005) were invasive species: gypsy moth and diamondback moth (*Plutella xylostella*), both native to Europe. Development of biopesticides has continued since these earlier projects and invasive arthropods continue to be among the principal targets for which microbial biopesticides are developed and used.

1.4 An Overview of Use of Microbes for Control and Eradication of Invasive Arthropods

This book is organized in sections, beginning in Part I with this introductory chapter, followed by two chapters in Part II discussing instances and implications of infection of invasives by endemic pathogens and then modeling dynamics of invasive arthropods with implications for their eradication and control with entomopathogens. The next 12 chapters all present case histories of eradication and control programs using arthropod specific pathogens; covering a diversity of approaches for a diversity of hosts and pathogens in a diversity of ecosystems. Part III includes two chapters describing and discussing eradication programs using *Bacillus thuringiensis*. The next ten chapters (Part IV) are case histories of control programs using entomopathogens. Chapters in Part IV are organized by the hosts, beginning with Hemiptera (1 chapter), Orthoptera (1 chapter), Coleoptera (3 chapters), Lepidoptera (1 chapter), and

Hymenoptera (2 chapters), and finishing with 2 chapters on use of pathogens for control of mites. One of the chapters on beetles (Coleoptera) is not specifically concerned with an invasive beetle but with an invasive nematode that is vectored by a native wood-boring beetle (Chapter 9). Among these chapters, all of the major groups of pathogens are included: viruses, bacteria, fungi, nematodes and protists. Some of the chapters on control cover one host and one pathogen (e.g. Chapter 14 on use of *Neozygites tanajoae* against cassava green mite) while others cover one host and numerous pathogens (e.g. Chapter 11 on gypsy moth includes *Bacillus thuringiensis, Lymantria dispar* nucleopolyhedrovirus, the fungus *Entomophaga maimaiga* and microsporidia). Control strategies that are covered in these chapters range from inundative augmentation to classical biological control.

In some instances, the public has been concerned about use of microbes for control of invasive arthropods, especially when control measures must take place where people live and work. Chapter 16 presents a discussion of the human health effects of *B. thuringiensis*, the entomopathogen that has been used most extensively for inundative applications against invasive arthropods in urban areas. Another concern that is raised is whether use of entomopathogens against invasive arthropods will affect our environment; Chapter 17 addresses non-target impacts of microbial control agents.

There have been some great successes in use of pathogens for control of invasive arthropods. The final chapter synthesizes the material presented in earlier chapters to discuss the constraints experienced in use of entomopathogens against invasive pests and improvements and new approaches for increasing success with entomopathogens against invasive arthropods in the future.

References

Alsop FJ, Laughlin TF (1991) Changes in the spruce-fir avifauna of Mt. Guyot, Tennessee, 1967–1985. J Tenn Acad Sci 66:207–209
Barlow ND, Goldson SL (2002) Alien invertebrates in New Zealand. In: Pimentel D (ed) Biological invasions: Economic and environmental costs of alien plant, animal and microbe species. CRC Press, Boca Raton, FL. pp 195–216
Beggs JR, Rees JS (1999) Restructuring of insect communities by introduced *Vespula* wasps in a New Zealand beech forest. Oecologia 119:565–571
Brewer GJ, Charlet LD (1999) Introduction and overview. In: Charlet LD, Brewer GJ (eds) Biological control of native or indigenous insect pests: Challenges, constraints, and potential. Entomol Soc Amer, Lanham, MD. pp 1–3
Cameron JWM (1973) Insect pathology. In: Smith RF, Mittler TE, Carroll NS (eds) History of entomology. Annual Reviews, Palo Alto, CA. pp 285–306
Cook A, Weinstein P, Woodward A (2002) The impact of exotic insects in New Zealand. In: Pimentel D (ed) Biological invasions: Economic and environmental costs of alien plant, animal and microbe species. CRC Press, Boca Raton, FL. pp 217–265
Cunningham JC. Entwistle PF (1981) Control of sawflies by baculovirus. In: Burges HD (ed) Microbial control of pests and plant diseases 1970–1980. Academic Press, London. pp 379–407
Eilenberg J, Hajek A, Lomer C (2001) Suggestions for unifying the terminology in biological control. BioControl 46:387–400

Falcon LA (1971) Use of bacteria for microbial control. In: Burges HD, Hussey NW (eds) Microbial control of insects and mites. Academic Press, London. pp 67–95

Faria M de, Wraight SP (2007) Mycoinsecticides and mycoacaricides: A comprehensive list with worldwide coverage and international classification of formulation types. Biol Control 43:237–256

Federici B (2005) Insecticidal bacteria: An overwhelming success for invertebrate pathology. J Invertebr Pathol 89:30–38

Ferron P (1981) Pest control by the fungi *Beauveria* and *Metarhizium*. In: Burges HD (ed) Microbial control of pests and plant diseases 1970–1980. Academic Press, London. pp 465–482

Follett PA, Neven LG (2006) Current trends in quarantine entomology. Annu Rev Entomol 51:359–385

Glare TR, O'Callaghan M (2000) *Bacillus thuringiensis*: Biology, ecology and safety. John Wiley & Sons, Chichester, UK

Hajek AE, McManus ML, Delalibera Júnior I (2005) Catalogue of introductions of pathogens and nematodes for classical biological control of insects and mites. USDA, For. Serv. FHTET-2005-05. 59 pp http://www.fs.fed.us/foresthealth/technology/pdfs/catalogue.pdf [accessed 28 February 2008]

Hajek AE, McManus ML, Delalibera Júnior I (2007) A review of introductions of pathogens and nematodes for classical biological control of insects and mites. Biol Control 41:1–13

Haugen DA, Underdown MG (1990) *Sirex noctilio* control program in response to the 1987 Green Triangle outbreak. Aust For 53:33–40

Hufbauer RA, Torchin ME (2007) Integrating ecological and evolutionary theory of biological invasions. In: Nentwig W (ed) Biological invasions. Springer, Berlin. pp 79–96

Liebhold AM, Tobin PC (2008) Population ecology of insect invasions and their management. Annu Rev Entomol 53:387–408

Lord JC (2005) From Metchnikoff to Monsanto and beyond: The path of microbial control. J Invertebr Pathol 89:19–29

Kendrick M (2000) The fifth kingdom, 3rd edn. Mycologue Publ., Sydney, Australia

Myers JH, Simberloff D, Kuris AM, Carey JR (2000) Eradication revisted: Dealing with exotic species. Trends Ecol Evol 15:316–320

National Invasive Species Council (2000) Definition of invasive species. http://www.invasivespecies.org/resources/DefineIS.html [accessed 27 February 2008]

Nentwig W (2007a) Biological invasions: Why it matters. In: Nentwig W (ed) Biological invasions. Springer, Berlin. pp 1–6

Nentwig W (2007b) Pathways in animal invasions. In: Nentwig W (ed) Biological invasions. Springer, Berlin. pp 11–27

OTA (Organization of Technology Assessment) (1993) Harmful non-indigenous species in the United States. Office of Technology Assessment, United States Congress, Washington, D.C.

Pimentel, D (1993) Habitat factors in new pest invasions. In: Kim KC, McPheron BA (eds) Evolution of insect pests – Patterns of variation. John Wiley & Sons, NY. pp 165–181

Pimentel D, Lach L, Zuniga R, Morrison D (2002) Environmental and economic costs associated with non-indigenous species in the United States. In: Pimentel D (ed) Biological invasions: Economic and environmental costs of alien plant, animal and microbe species. CRC Press, Boca Raton, FL. pp 307–329

Pimentel D, Zuniga R, Morrison D (2005) Update on the environmental and economic costs associated with alien-invasive species in the United States. Ecol Econ 52:273–288

Ragenovich IR, Mitchell RG (2006) Balsam woolly adelgid. USDA, Forest Service, Forest Insect & Disease Leaflet 118. 12 pp http://www.na.fs.fed.us/pubs/fidls/bwa.pdf [accessed 27 February 2008]

Shields E, Testa A, Neumann G (2004) Biological control of alfalfa snout beetle: a small scale field trial. Report to the Northern New York Agricultural Development Program. http://www.nnyagdev.org/reportarchives/NNYADPBioASB04Shields.pdf [Accessed 16 October 2007]

Simberloff D, Gibbons L (2004) Now you see them, now you don't!—population crashes of established introduced species. Biol Invasions 6:161–172

Torchin, ME, Mitchell CE (2004) Parasites, pathogen, and invasions by plants and animals. Front Ecol Environ 2:183–190

USDA Forest Service (2007) Emerald ash borer. http://www.emeraldashborer.info/http://www.emeraldashborer.info/[accessed 5 November 2007]

van der Putten WH, Yeates GW, Duyts H, Reis CS, Karssen G (2005) Invasive plants and their escape from root herbivory: A worldwide comparison of the root-feeding nematode communities of the dune grass *Ammophila arenaria* in natural and introduced ranges. Biol Invasions 7:733–746

Van Driesche J, Van Driesche R (2004) Nature out of place: Biological invasions in the global age. Island Press, Washington, D.C.

Williamson M (1996) Biological invasions. Chapman & Hall, London

Wojcik DP, Allen CR, Brenner RJ, Forys EA, Jouvenaz DP, Lutz RS (2001) Red imported fire ants: impact on biodiversity. Amer Entomol 47:16–23

Zimmermann G, Papierok B, Glare T (1995) Elias Metschnikoff, Elie Metchnikoff or Ilya Ilich Mechnikov (1845–1916): A pioneer in insect pathology, the first describer of the entomopathogenic fungus *Metarhizium anisopliae* and how to translate a Russian name. Biocontr Sci Technol 5:527–530

Part II
Ecological Considerations

Chapter 2
Naturally Occurring Pathogens and Invasive Arthropods

Ted E. Cottrell and David I. Shapiro-Ilan

Abstract Establishment of introduced pest arthropods has been attributed, in part, to the pest arthropods' separation from natural control agents in their native ranges. Here we focus on the role of endemic pathogens in establishment and population regulation of exotic pest and beneficial arthropods and explore factors affecting their regulation by endemic pathogens. We do not attempt an exhaustive list of examples but illustrate some instances showing diverse aspects of the host-pathogen relationships involved. As a case study, we discuss establishment of the multicolored Asian lady beetle and its rapid spread across North America as related to its resistance to an endemic fungal pathogen to which some native lady beetle species are susceptible. It is clear that advances in our knowledge about the epizootiology of endemic pathogens with exotic arthropods will enhance our understanding of invasion biology and assist in regulation of invasive pests.

2.1 Introduction

The successful establishment of introduced pest arthropods has been attributed, in part, to the pest arthropods' separation from natural control agents in their native ranges (Williamson 1996, Ehler 1998). This concept, i.e. the 'enemy release hypothesis', is commonly referenced in the literature as a mechanism that fosters invasive species (Keane & Crawley 2002, Torchin *et al.* 2003, Clay 2003, Prenter *et al.* 2004). For example, in a study of invasive plant species, Mitchell and Power (2003) found that each invasive species was infected by 77% fewer fungal and viral pathogen species in naturalized versus native ranges. Similarly, Torchin *et al.* (2003) reported that invasive species possess about half the number of parasites as compared with native species. Based on examples such as these, one might extend the argument to pathogen load in invasive arthropods, i.e. one would predict a low prevalence

---

T.E. Cottrell
USDA, Agricultural Research Service, SE Fruit and Tree Nut Research Lab, 21 Dunbar Rd., Byron, Georgia 31008 USA
e-mail: ted.cottrell@ars.usda.gov

of disease in invasive arthropods. And indeed this is the case in certain invasive arthropods as we illustrate later in the chapter. However, we would be remiss not to acknowledge that isolates of some endemic pathogens can be quite virulent to exotic insects (Lacey *et al.* 2001, Koppenhöfer & Fuzy 2003, Duncan *et al.* 2003). In this chapter we focus on the role of endemic pathogens in establishment and population regulation of exotic arthropods.

For this discussion the term endemic refers to an organism that naturally occurs in the area and has not been introduced. Estimating the impact of endemic pathogens on introduced arthropods is difficult. One difficulty is that it is not always clear whether a pathogen is endemic or whether it may have been introduced along with its host. Certainly some introduced pathogens have become established in particular regions and may have significant impact on their host populations (Hajek *et al.* 2007), e.g. the case of the fungus *Entomophaga maimaiga* Humber, Shimazu, and Soper and its effect on the gypsy moth, *Lymantria dispar* (L.) (Weseloh & Andreadis 1992, Hajek 1997). Yet for our purposes these established introductions are not considered endemic. In cases where the host-pathogen relationship is highly specific (e.g. many baculoviruses and microsporidia) it is likely the pathogen was introduced along with its host. Yet even in cases where the pathogen has a broad host range (e.g. many entomopathogenic fungi and nematodes), it may not be clear if the particular strain or species of pathogen was present prior to the arthropod's introduction. In this chapter, the pathogens we discuss are, to the best of our knowledge, endemic. Here we first address, in a general sense, factors that contribute to endemic pathogen impact on invasive arthropods. We then offer a case study, i.e. the establishment and spread of *Harmonia axyridis* (Pallas) (Coleoptera: Coccinellidae) across North America, to serve as a basis for discussion of natural enemy release and the role of entomopathogens in invasion biology.

2.2 Factors Affecting Endemic Entomopathogen Regulation of Introduced Arthropod Populations

This section explores factors that may affect the regulation of exotic arthropods by endemic pathogens. The intent is not to provide an exhaustive list, but rather to use several examples to illustrate some of the diverse aspects of the host-pathogen relationships involved. Various abiotic and biotic factors are known to affect the ability of entomopathogens to cause disease in host populations (Fuxa 1987, Fuxa & Tanada 1987, Lacey & Shapiro-Ilan 2008). Generally, many factors affecting regulation of introduced arthropods by endemic entomopathogens can be expected to be similar to other pathogen-host relationships that include endemic hosts or introduced pathogens. Yet some nuances may be anticipated given that, in the case of interest here, it is the host species that has the challenge of adapting to the new environment whereas the endemic pathogen is already established and has managed to exist and evolve in the native ecosystem. Indeed, it is arguable that environmental barriers influencing population regulation may not be as pronounced in endemic pathogens as compared with introduced pathogens (due to the former's inherent establishment in the environment). Examples of endemic pathogens exhibiting superior persistence and efficacy in controlling introduced arthropod pests include entomopathogenic nematode control of the Japanese beetle, *Popillia japonica* Newman (Coleoptera: Scarabaeidae) (Lacey *et al.* 2001, Koppenhöfer & Fuzy 2003), and control of a root weevil, *Diaprepes abbreviatus* (L.) (Coleoptera: Curculionidae), by *Steinernema diaprepesi* Nguyen & Duncan (Duncan *et al.* 2003).

The impact of endemic pathogens on introduced arthropods can vary. Some examples of endemic pathogens that have caused substantial mortality in introduced insect pests include: 84% of soybean aphid, *Aphis glycines* Matsumura (Hemiptera: Aphididae), infected with entomopathogenic fungi (Nielsen & Hajek 2005), up to approximately 53% mortality in *D. abbreviatus* infected by entomopathogenic nematodes (Duncan *et al.* 2003), and a 79% reduction in a whitefringed beetle population, *Graphognathus leucoloma* (Boheman) (Coleoptera: Curculionidae) associated with the presence of *Heterorhabditis* sp. (Sexton & Williams 1981). Additionally, the introduced millipede *Ommatoiulus moreletii* (Lucas) has declined in Australia as a result of natural biological control by the native nematode *Rhabditis necromena* Sudhaus and Schulte (McKillup *et al.* 1988, Schulte 1989, Sudhaus & Schulte 1989). In contrast, contributions by endemic entomopathogens can also be minimal; for example, in southern Michigan, entomopathogenic nematode infections in *P. japonica* were <1% (Cappaert & Smitley 2002), yet substantially higher impact by entomopathogenic nematodes has been observed in the eastern US (Campbell *et al.* 1998). The reasons for varying impacts of endemic pathogens are unclear. Some factors that have been documented to affect population regulation by endemic pathogens include the environment, host density and host range.

Although endemic pathogens are adapted to their native ecosystem, environmental factors still contribute to epizootics and population regulation. For example, Wraight *et al.* (1993) observed that prevalence of fungal infection in the Russian wheat aphid, *Diuraphis noxia* (Mordvilko) (Hemiptera: Aphididae), was substantially higher in irrigated fields relative to non-irrigated fields. Additionally, soil type was found to impact the prevalence and relative distribution of endemic entomopathogenic nematodes in citrus (Duncan *et al.* 2003).

Host density may also be a factor affecting population regulation in some systems. Generally, increases in host density are expected to increase the chances of an epizootic (Fuxa 1987). Nielsen and Hajek (2005) observed that mycosis induced by endemic fungi was proportional to the population density of the soybean aphid, *A. glycines*. Contrarily, mycosis was not correlated to host density of *D. noxia* (Feng *et al.* 1991).

Perhaps most importantly, the potential for population regulation is likely to be limited by the innate host range of the endemic pathogen in question. However, establishment of pathogenicity in the laboratory does not necessarily predict an endemic pathogen-exotic host relationship as physiological host range is not equivalent to ecological host range (Federici & Maddox 1996, Solter & Maddox 1998). In order for an endemic pathogen to impact an introduced host's population, the two must be biologically and environmentally compatible in terms of the niche they occupy in the ecosystem.

Based on known host ranges and the biology/ecology of endemic pathogens, we can predict or make some generalizations regarding which groups of endemic pathogens have potential to impact particular arthropods. For example, disease prevalence may be predicted in part by a pathogen's adaptation to infection in soil versus above ground. Aphid species tend to be susceptible to certain fungi that are capable of effectively causing infection aboveground, e.g. fungi in the order Entomophthorales. Indeed, a number of endemic Entomophthorales, such as *Pandora* (=*Erynia*) *neoaphidis* (Remaudiere & Hennebert) and *Conidiobolus* spp., can cause substantial mortality in established populations of exotic *D. noxia* (Wraight *et al.* 1993, Feng *et al.* 1991, Hatting *et al.* 1999, 2000) and *A. glycines* (Nielsen & Hajek 2005). In contrast, ground-dwelling larvae of Coleoptera tend to be susceptible to entomopathogenic nematodes and hypocrealean fungi (Ascomycota: Clavicipitaceae) (e.g. *Beauveria bassiana* (Balsamo) Vuillemin and *Metarhizium anisopliae* (Metchnikoff) Sorokin). Examples of introduced Coleoptera that support this generalization include larvae of *D. abbreviatus* (Beavers *et al.* 1983, McCoy *et al.* 2000, Duncan *et al.* 2003) and *P. japonica* (Campbell *et al.* 1998, Lacey *et al.* 2001, Koppenhöfer & Fuzy 2003). One interesting example where an endemic pathogen affects an introduced insect both in the soil and above ground is *B. bassiana* infection of the European corn borer, *Ostrinia nubilalis* (Hübner) (Lepidoptera: Crambidae). In corn fields, *B. bassiana* can infect a substantial proportion of overwintering larvae in the soil (up to 84%), but also infects *O. nubilalis* larvae within the corn plant, a process that is facilitated by the ability of the endemic fungus to function as an endophyte and colonize corn (Bing & Lewis 1991, 1993).

The ability of endemic pathogens to regulate introduced pests can be complex. Not only can the pathogens be differentially affected by environmental factors and other endemic soil organisms (Fuxa 1987, Koppenhöfer & Grewal 2005) but endemic pathogens may interact (indirectly) with introduced pathogen populations as well. In some cases introduced pathogens can out-compete endemic pathogen populations, potentially reducing the overall level of arthropod suppression (Duncan *et al.* 2003). Yet in other cases introduced pathogens may not cause displacement of endemic pathogens since they occupy different niches in the soil (Millar & Barbercheck 2001). Our knowledge and understanding of the complexities affecting population regulation of introduced arthropods by endemic pathogens is severely limited by the paucity of studies addressing the issue. Clearly additional research in this area is required.

2.3 A Case Study: Endemic Pathogens and the Multicolored Asian Lady Beetle

Here we will examine the case of an introduced species of lady beetle, *Harmonia axyridis* (Pallas) (Coleoptera: Coccinellidae) (Fig. 2.1), and the emerging evidence that this species exhibits resistance to some endemic entomopathogens in North America. It is likely that freedom from endemic predators, parasites and possibly pathogens benefited those early populations of *H. axyridis* that became established.

Fig. 2.1 An adult of the highly invasive *Harmonia axyridis*

Pest management through classical biological control has led to the introduction of numerous exotic insect predators and parasites into the U.S. Some of these introduced natural enemy species became established but many did not, even when apparently suited to the target pest species and the new habitat. In order for an introduced species to be considered established, Andow *et al.* (1997) proposed that more than one generation of that species must have been recovered far from introduction sites. When introduced natural enemy species fail to establish, specific reasons for failure generally are not known. Sometimes the assumption is made that numbers of released individuals were too low or attempts to establish natural enemies were hampered by weather, insecticide usage, competition with native natural enemies, diseases decreasing the population of the target pest, and the disruptive effects of crop harvest (van den Bosch *et al.* 1959, Simmonds 1966, DeBach & Rosen 1991). Other ambiguous reasons stated for failure of an introduced species to establish relate to inadequate biological information about the introduced natural enemy, indifference by researchers, lack of persistence by the introduced natural enemy, technical problems and climatic factors (van den Bosch 1968). The possible role of pathogens, specifically endemic pathogens, is rarely broached in the research literature regarding success or failure of an introduced natural enemy. Thus, examples of endemic pathogens impacting exotic natural enemies by hampering their establishment are lacking.

If we consider the process of intentionally introducing natural enemies through classical biological control, these natural enemies should be free from most of their own natural enemies. Before the release of natural enemies into a new environment, Simmonds (1966) states that the removal of parasites from predators and hyperparasites from parasites is necessary. Additionally, the culturing of insects before release can free them from pathogens that are associated with their native habitats (Simmonds 1966), especially pathogens transmitted horizontally. Accidentally released arthropods that establish may also be free from a percentage of their natural complement of pathogens due to the expected small size of the establishing population and due to complex life cycles of pathogens that may not be compatible with the transport and establishment process.

When data are presented concerning the effect of naturally-occurring pathogens upon beneficial arthropods, i.e. parasitoids and predators, it concerns rates of infection by nematodes or gregarine protozoa (see Ceryngier & Hodek 1996), mortality of overwintering populations when beetles aggregate (Iperti 1966a, b, Ceryngier 2000) or male-killing bacteria (Hurst & Jiggins 2000). But by far, most information regarding the impact of pathogens on beneficial arthropods concerns the non-target impacts of commercially-available, biorational products (e.g. fungi) applied in an insecticidal fashion (Magalhaes *et al.* 1988, Giroux *et al.* 1994, James & Lighthart 1994, Todorova *et al.* 1994, James *et al.* 1995, Pingel & Lewis 1996, Poprawski *et al.* 1998, James *et al.* 1998, Cagáň & Uhlík 1999, Todorova *et al.* 2000, Smith & Krischik 2000, Pell & Vandenberg 2002, Laird *et al.* 1990, Ludwig & Oetting 2001). Relatively little information exists with regard to how endemic pathogens affect populations of beneficial arthropods (native and exotic), especially with regard to the arthropod's viability in biological control programs (Vinson 1990). Kuznetsov (1997) reviewed research on pathogens of coccinellids in eastern Russia and found that *Coccinella septempunctata* L., *H. axyridis, Calvia quatuordecimguttata* (L.) and *Hippodamia tredecimpunctata* (L.) were infected by *B. bassiana* and *Beauveria brongniartii* (Saccardo) Petch (=*B. tenelle*). He considered their impact minimal on those predaceous coccinellids. Yinon (1969) found adult *Chilocorus bipustulatus* (L.) (Coleoptera: Coccinellidae) infected by the laboulbenialean fungus *Hesperomyces virescens* and larvae infected by 'white mycelium' on trunks in citrus groves but no supporting data were provided to indicate the prevalence of these diseases in field populations or the impact of these pathogens on predation by *C. bipustulatus*.

The introduction of lady beetles into North America for pest control was common following successful control of the exotic cottony-cushion scale, *Icerya purchasi* (Homoptera: Coccidae), on citrus by *Rodolia cardinalis* (Coleoptera: Coccinellidae) in California during the late 1880s (Hagen 1974, Gordon 1985, DeBach & Rosen 1991). During this pre-insecticide era, 179 foreign species of Coccinellidae were purposely released into the U.S. As an end result of these intentional introductions, 16 exotic species of Coccinellidae were established in the U.S. (Gordon 1985). *Harmonia axyridis* was first recorded being intentionally introduced into North America as early as 1916 and more releases were made by various federal and state researchers into the 1980s (Gordon 1985). Establishment of *H. axyridis* via accidental introductions at seaports remains a possibility (Krafsur *et al.* 1997). Nonetheless, established *H. axyridis* populations in North America were not detected until 1988 in Louisiana and thereafter *H. axyridis* quickly spread across much of the United States and southern Canada (Tedders & Schaefer 1994, Coderre *et al.* 1995, Lamana & Miller 1996, Brown & Miller 1998, Colunga-Garcia & Gage 1998). Although the intent of establishing *H. axyridis* was for control of plant pests, several negative impacts have developed. The overwhelming successful establishment of this insect negatively affects competing native lady beetles (Cottrell & Yeargan 1998), it can overwinter in buildings including homes where it can be a great nuisance (Nalepa *et al.* 2000) and it has been implicated as a pest of midwestern wine grapes when adults feed on damaged grapes and are harvested along with the crop, resulting in the wine flavor being tainted (Koch 2003).

It is likely that the initial population of *H. axyridis* that became established in North America faced many hurdles commonly encountered when attempting to establish any beneficial arthropod in a new habitat, and for those reasons, many of the numerous, early introductions failed to establish. But once a population did establish, inherent traits gave this species a competitive edge over native guild species. Traits that suggest why *H. axyridis* has apparently out-competed native coccinellid species in some habitats include its polyphagous diet, high fecundity (Michaud 2002), aggressive behavior (Cottrell & Yeargan 1999, Michaud 2002), high mobility (With *et al*. 2002) and large body size (Cottrell & Yeargan 1999, Michaud 2002). Another factor influencing population dynamics of native and exotic species may involve attack by parasites. Obrycki (1989) reported on differential susceptibility of certain native and exotic coccinellid species to the cosmopolitan parasitoid *Dinocampus coccinellae* Schrank (Hymenoptera: Braconidae). Following this trend, Hoogendorn and Heimpel (2002) showed that successful parasitism of *H. axyridis* by *D. coccinellae* was lower than parasitism of the native *Coleomegilla maculata* (DeGeer) (Coleoptera: Coccinellidae) even though the proportion of each species attacked was similar. However, it is questionable whether a dearth of endemic pathogens adapted to attack this exotic species could have contributed, in part, to its successful establishment and rapid dispersal across much of North America. Recently, we suggested that reduced susceptibility to endemic pathogens facilitated the establishment and quick dispersal of *H. axyridis* (Cottrell & Shapiro-Ilan 2003, Shapiro-Ilan & Cottrell 2005).

Initial support for our hypothesis was based on differential susceptibility of *H. axyridis* and the native coccinellid *Olla v-nigrum* (Mulsant) to *B. bassiana*. The suspected differences were first based on preliminary field observations of these two species' susceptibility to *B. bassiana*. It was discovered that some adults of the native *O. v-nigrum* overwintering under pecan bark (*Carya illinoinensis*) were infected with *B. bassiana* (Cottrell & Shapiro-Ilan 2003). These infected, dead adults had visible mycosis typical of *B. bassiana* (Fig. 2.2). Some *O. v-nigrum* adults collected

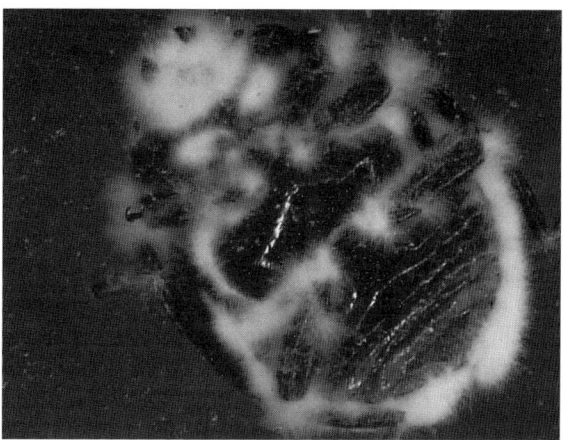

Fig. 2.2 An adult *Olla v-nigrum* infected by *Beauveria bassiana*

alive from under pecan bark and held in the laboratory with aphids and water died and also exhibited mycosis. However, mycosis of *H. axyridis* had never been observed whether insects came from overwintering aggregations in buildings or from field collections in various habitats including pecan orchards. These observations prompted Cottrell and Shapiro-Ilan (2003) to test the pathogenicity of *B. bassiana* collected from infected *O. v-nigrum* against the exotic *H. axyridis* and against the native *O. v-nigrum*. We discovered that *B. bassiana* isolated from *O. v-nigrum* was substantially more virulent to *O. v-nigrum* than *H. axyridis*. A concentration of 2.5×10^5 conidia/ml was found to be the LC_{50} for *O. v-nigrum* but this concentration did not cause any mortality in *H. axyridis*. In fact, even a concentration of 3.9×10^8 conidia/ml was not pathogenic to *H. axyridis*. The commercially-available GHA strain of *B. bassiana* was not virulent against either the native *O. v-nigrum* or the exotic *H. axyridis* when assayed at the LC_{50} for *O. v-nigrum*. This is interesting given the incredibly broad host range of *B. bassiana* (Goettel *et al.* 1990). Nonetheless, the results when using *B. bassiana* isolated from the native *O. v-nigrum* indicate lower susceptibility of the exotic *H. axyridis* to an endemic pathogen and support our hypothesis. James & Lighthart (1994) comment that information from the field is needed to determine how direct effects seen in the laboratory play out in habitats where multiple species interact. Again, field data from Cottrell & Shapiro-Ilan (2003), although geographically limited, was suggestive that isolates of *B. bassiana* exist that are pathogenic to native coccinellids but not to *H. axyridis*.

Harmonia axyridis occupies many habitats and overlaps both temporally and spatially with many native species of North American Coccinellidae. If our hypothesis regarding *H. axyridis* is correct, we should expect that some of these other native competitors will be negatively affected by endemic pathogens. To partially address this issue, Cottrell and Shapiro-Ilan (2008) used an isolate of *B. bassiana* collected from a naturally-infected *O. v-nigrum* adult (as used by Cottrell & Shapiro-Ilan 2003) and tested it against four species of Coccinellidae native to North America, i.e. *O. v-nigrum, Cycloneda munda* (Say), *Hippodamia convergens* (Guérin-Méneville) and *C. maculata*. The hypothesis was supported in that *C. munda, O. v-nigrum* and *H. convergens* were susceptible but *H. axyridis* was not susceptible. Further support comes from the fact that another exotic lady beetle established in North America, *C. septempunctata*, was not susceptible to this same isolate of *B. bassiana*. *Coleomegilla maculata*, however, was the only native species tested that was not susceptible. Additionally, only the native species *C. munda, O. v-nigrum* and *H. convergens* exhibited visible signs of mycosis when treated with *B. bassiana* isolated from *O. v-nigrum*. Results such as these show that specific isolates could play a key role in the population regulation of native natural enemies while allowing resistant, newly-established natural enemies to gain a competitive edge over native species.

In addition to the observed differential susceptibility to an isolate of *B. bassiana*, more evidence of *H. axyridis* being less susceptible to disease than native species is provided by Shapiro-Ilan and Cottrell (2005). In this study, two species of native coccinellids (i.e. *C. maculata* and *O. v-nigrum*) and two species of exotic coccinellids (i.e. *H. axyridis* and *C. septempunctata*) were assessed for their relative susceptibility to entomopathogenic nematodes. The nematode strains used, i.e. *Steinernema carpocapsae* (Cxrd) and *Heterorhabditis bacteriophora* (VS), were both originally isolated in North America. Given the widespread geographic distribution of both the nematode species and the native coccinellids *C. maculata* and *O. v-nigrum*, it is likely that natural interaction between these organisms would occur. Although most coccinellid species would be expected to spend more time on plants, coccinellid-nematode interaction is most likely to occur with larval stages that can be found on the soil. Interactions between the nematodes and adults would be limited but possible, especially for *C. maculata* adults that spend time on low-growing plants. Larval coccinellids often fall to the soil from host plants while searching for prey and will even disperse from plants via the soil (Cottrell & Yeargan 1999). Thus, the native species would be expected to have had more contact with these endemic pathogens compared with the exotic coccinellids. Indeed, Shapiro-Ilan and Cottrell (2005) showed that mortality of native coccinellids was generally higher than mortality of exotic coccinellids when treated with *S. carpocapsae* (Cxrd), whereas *H. bacteriophora* (VS) had less effect overall on coccinellids with the only difference being higher mortality of the native *O. v-nigrum* versus *H. axyridis*. Again, our hypothesis that the establishment of some exotic species may be fostered by their resistance to endemic pathogens is supported.

2.4 Conclusions

The effect of endemic pathogens upon native versus exotic species can play out in different ways. For example, the endemic pathogens could exhibit high virulence to the exotic arthropod, regardless of virulence toward native arthropods, and thus negatively affect establishment of the exotic arthropod, or, the endemic pathogens could exhibit low virulence to the exotic arthropod and high virulence to native competitors thus facilitating establishment of the exotic arthropod. The former could provide a partial explanation why only 16 of the 179 species of coccinellids intentionally introduced into North America have become established (Gordon 1985). Although there is no direct evidence to support this supposition regarding all of those introduced coccinellids, the impact of endemic pathogens on introduced arthropods can be highly significant as demonstrated with entomopathogenic fungi against *A. glycines* (Nielsen & Hajek 2005) and entomopathogenic nematodes infecting *D. abbreviatus* (Duncan *et al*. 2003) and *G. leucoloma* (Sexton & Williams 1981). The latter, i.e. low virulence to the exotic arthropod and high virulence to native competitors, provides an explanation for the success of *H. axyridis* in North America and favors the enemy release hypothesis. This hypothesis has been used to explain the success of some invading plants and animals but has not (previous to our work) been suggested specifically as a mechanism fostering establishment of exotic natural enemies such as *H. axyridis* where lower susceptibility to specific strains or isolates of endemic pathogens has been demonstrated compared with some native competitors. It is likely that the successful establishment and widespread distribution of another exotic coccinellid, *C. septempunctata*, across North America was also fostered by its reduced susceptibility to endemic pathogens. Thus the role of endemic pathogens upon establishment of exotic arthropod species should be examined further. This can be done by examining the natural prevalence of pathogens in native and exotic species where both exist at different geographic locations and comparing their relative competitiveness as correlated with disease prevalence. It should also be determined if differential susceptibility between native and exotic species occurs with more isolates of endemic pathogens. Additionally, the virulence of pathogens from the native range of the exotic species could be tested to further determine the relevance of the enemy release hypothesis in this situation.

An understanding of the physiological and ecological host range of endemic pathogens can have positive impacts on the success of biological control. With respect to potential biological control agents that are being considered for introduction, it would be useful to know their physiological susceptibility to a suite of endemic pathogens known to attack similar native species at the proposed introduction site. Although the physiological susceptibility will likely exceed the ecological susceptibility, knowing which pathogens will not infect the host under optimum conditions is useful and would allow for further refining of ecological susceptibility assays and could aid in decision-making regarding releases. Additionally, little information is available on the prevalence of endemic pathogens naturally infecting beneficial arthropods in the field and what impact they may have upon beneficial arthropod populations.

Advances in our knowledge about the epizootiology of endemic pathogens with exotic arthropods will enhance our understanding of invasion biology, and assist in regulation of invasive pests. The threat of exotic arthropod species becoming established may be related to the pathogenicity and virulence of endemic pathogens to those exotic species. Thus, we may be able to predict the outcome of endemic pathogens on potential exotic arthropod threats in order to categorize risk levels relating to the potential for those exotic arthropods to establish and have a positive (e.g. predators and parasitoids) or negative (e.g. pests) ecological, economical or social impact. For example, if we know that certain endemic pathogens tend to control certain orders, families or genera of pests (and specific virulence can be tested under quarantine) then conceivably the risk level for exotic arthropod species, within that order, family or genera, would be deemed low. With continued globalization, there has been a parallel increase in the number of invasive pests (Perrings *et al.* 2005); therefore the importance of elucidating the role of endemic pathogens in preventing establishment or controlling exotic pests post-establishment carries an increasingly greater weight.

References

Andow DA, Ragsdale DW, Nyvall RF (1997) Biological control in cool temperate regions. In: Andow DA, Ragsdale DW, Nyvall RF (eds) Ecological interactions and biological control. Westview Press, Boulder, CO. pp 1–28

Beavers JB, McCoy CW, Kaplan DT (1983) Natural enemies of subterranean *Diaprepes abbreviatus* (Coleoptera: Curculionidae) larvae in Florida. Environ Entomol 12:840–843

Bing LA, Lewis LC (1991) Suppression of *Ostrinia nubilalis* (Hübner) (Lepidoptera: Pyralidae) by endophytic *Beauveria bassiana* (Balsamo) Vuillemin. Environ Entomol 20:1207–1211

Bing LA, Lewis LC (1993) Occurrence of the entomopathogen *Beauveria bassiana* (Balsamo) Vuillemin in different tillage regimes and in *Zea mays* L. and virulence towards *Ostrinia nubilalis* (Hübner). Agric Ecosys Environ 45:147–156

Brown MW, Miller SS (1998) Coccinellidae (Coleoptera) in apple orchards of eastern West Virginia and the impact of invasion by *Harmonia axyridis*. Entomol News 109:136–142

Cagáň L, Uhlík V (1999) Pathogenicity of *Beauveria bassiana* strains isolated from *Ostrinia nubilalis* Hbn. (Lepidoptera: Pyralidae) to original host larvae and to ladybirds (Coleoptera: Coccinellidae). Plant Prot Sci 35:108–112

Campbell JF, Orza G, Yoder F, Lewis E, Gaugler R (1998) Spatial and temporal distribution of endemic and released entomopathogenic nematode populations in turfgrass. Entomol Exp Appl 86:1–11

Cappaert DL, Smitley DR (2002) Parasitoids and pathogens of Japanese beetle (Coleoptera: Scarabaeidae) in southern Michigan. Environ Entomol 31:573–580

Ceryngier P, Hodek I (1996) Enemies of Coccinellidae. In: Hodek I, Honěk A (eds) Ecology of Coccinellidae. Kluwer Academic Publishers, Boston. pp 319–350

Ceryngier P (2000) Overwintering of *Coccinella septempunctata* (Coleoptera: Coccinellidae) at different altitudes in the Karkonosze Mts, SW Poland. Eur J Entomol 97:323–328

Clay K (2003) Parasites lost. Nature 421:585–586

Coderre D, Lucas E, Gagne I (1995) The occurrence of *Harmonia axyridis* (Coleoptera: Coccinellidae) in Canada. Can Entomol 127:609–611

Colunga-Garcia M, Gage SH (1998) Arrival, establishment, and habitat use of the multicolored Asian lady beetle (Coleoptera: Coccinellidae) in a Michigan landscape. Environ Entomol 27:1574–1580

Cottrell TE, Yeargan KV (1998) Intraguild predation between an introduced lady beetle, *Harmonia axyridis* (Coleoptera: Coccinellidae), and a native lady beetle, *Coleomegilla maculata* (Coleoptera: Coccinellidae). J Kans Entomol Soc 71:159–163

Cottrell TE, Yeargan KV (1999) Factors influencing the dispersal of larval *Coleomegilla maculata* from the weed *Acalypha ostryaefolia* to sweet corn. Entomol Exp Appl 90:313–322

Cottrell TE, Shapiro-Ilan DI (2003) Susceptibility of a native and an exotic lady beetle (Coleoptera: Coccinellidae) to *Beauveria bassiana*. J Invertebr Pathol 84:137–144

Cottrell TE, Shapiro-Ilan DI (2008) Susceptibility of endemic and exotic North American ladybirds to endemic fungal entomopathogens. Eur J Entomol 105:455–460

DeBach P, Rosen D (1991) Biological control by natural enemies, 2nd edition. Cambridge University Press, New York

Duncan LW, Graham JH, Dunn DC, Zellers J, McCoy CW, Nguyen K (2003) Incidence of endemic entomopathogenic nematodes following application of *Steinernema riobrave* for control of *Diaprepes abbreviatus*. J Nematol 35:178–186

Ehler LE (1998) Invasion biology and biological control. Biol Control 13:127–133

Federici BA, Maddox J (1996) Host specificity in microbe-insect interactions: Insect control by bacterial, fungal, and viral pathogens. Bioscience 46:410–421

Feng M-G, Johnson JB, Halbert SE (1991) Natural control of cereal aphids (Homoptera: Aphididae) by entomopathogenic fungi (Zygomycetes: Entomophthorales) and parasitoids (Hymenoptera: Braconidae and Encyrtidae) on irrigated spring wheat in southwestern Idaho. Environ Entomol 20:1699–1710

Fuxa JR (1987) Ecological considerations for the use of entomopathogens in IPM. Annu Rev Entomol 32:225–251

Fuxa JR, Tanada Y (1987) Epizootiology of insect diseases. John Wiley & Sons, New York

Giroux SJ, Côté C, Vincent C, Martel P, Coderre D (1994) Bacteriological insecticide M-One effects on predation efficiency and mortality of adult *Coleomegilla maculata lengi* (Coleoptera: Coccinellidae). J Econ Entomol 87:39–43

Goettel MS, Poprawski TJ, Vandenberg JD, Li Z, Roberts DW (1990) Safety to nontarget invertebrates of fungal biocontrol agents. In: Laird M, Lacey LA, Davidson EW (eds), Safety of microbial insecticides. CRC Press, Boca Raton, FL. pp. 209–231

Gordon RD (1985) The Coccinellidae (Coleoptera) of America north of Mexico. J New York Entomol Soc 93:1–912

Hagen KS (1974) The significance of predaceous Coccinellidae in biological and integrated control of insects. Entomophaga Mem H S 7:25–44

Hajek AE (1997) Fungal and viral epizootics in gypsy moth (Lepidoptera: Lymantriidae) populations in central New York. Biol Control 10:58–68

Hajek AE, McManus ML, Delalibera I Jr (2007) A review of introductions of pathogens and nematodes for classical biological control of insects and mites. Biol Control 41:1–13

Hatting JL, Humber RA, Poprawski TJ, Miller RM (1999) A survey of fungal pathogens of aphids from South Africa with special reference to cereal aphids. Biol Control 16:1–12

Hatting JL, Poprawski TJ, Miller RM (2000) Prevalences of fungal pathogens and other natural enemies of cereal aphids (Homoptera: Aphididae) in wheat under dryland and irrigated conditions in South Africa. BioControl 45:179–199

Hoogendorn M, Heimpel GE (2002) Indirect interactions between an introduced and a native ladybird beetle species mediated by a shared parasitoid. Biol Control 25:224–230

Hurst GDD, Jiggins FM (2000) Male-killing bacteria in insects: Mechanisms, incidence, and implications. Emerg Infect Dis 6:329–336

Iperti G (1966a) Natural enemies of aphidophagous coccinellids. In: Hodek I (ed) Ecology of aphidophagous insects. Academia, Prague and Dr. W. Junk, The Hague. pp 185–187

Iperti G (1966b) Protection of coccinellids against mycosis. In: Hodek I (ed) Ecology of Aphidophagous Insects. Academia, Prague and Dr. W. Junk, The Hague. pp 189–190

James RR, Lighthart B (1994) Susceptibility of the convergent lady beetle (Coleoptera: Coccinellidae) to four entomogenous fungi. Environ Entomol 23:190–192

James RR, Shaffer BT, Croft B, Lighthart B (1995) Field evaluation of *Beauveria bassiana*: Its persistence and effects on the pea aphid and a non-target coccinellid in alfalfa. Biocontrol Sci Tech 5:425–437

James RR, Croft BA, Shaffer BT, Lighthart B (1998) Impact of temperature and humidity on host-pathogen interactions between *Beauveria bassiana* and a coccinellid. Environ Entomol 27:1506–1513

Keane RM, Crawley MJ (2002) Exotic plant invasions and the enemy release hypothesis. Trends Ecol Evol 17:164–169

Koch RL (2003) The multicolored Asian lady beetle, *Harmonia axyridis*: A review of its biology, uses in biological control, and non-target impacts. J Insect Sci 3:1–16

Koppenhöfer AM, Fuzy EM (2003) *Steinernema scarabaei* for the control of white grubs. Biol Control 28:47–59

Koppenhöfer AM, Grewal PS (2005) Compatibility and interactions with agrochemicals and other biocontrol agents. In: Grewal P, Ehlers RU, Shapiro-Ilan D (eds) Nematodes as biological control agents. CABI, New York. pp 363–381

Krafsur ES, Kring TJ, Miller JC, Nariboli P, Obrycki JJ, Ruberson JR, Schaefer PW (1997) Gene flow in the exotic colonizing ladybeetle *Harmonia axyridis* in North America. Biol Control 8:207–214

Kuznetsov VN (1997) Lady beetles of the Russian Far East, memoir no. 1, Center for Systematic Entomology. The Sandhill Crane Press, Inc, Gainesville, FL. 248p

Lacey LA, Rosa JS, Simoes NO, Amaral JJ, Kaya HK (2001) Comparative dispersal and larvicidal activity of exotic and Azorean isolates of entomopathogenic nematodes against *Popillia japonica* (Coleoptera: Scarabaeidae). Eur J Entomol 98:439–444

Lacey LA, Shapiro-Ilan DI (2008) Microbial control of insect pests in temperate orchard systems: Potential for incorporation into IPM. Annu Rev Entomol 53:7.1–7.24

Laird M, Lacey LA, Davidson EW (1990) Safety of microbial insecticides. CRC Press Inc., Boca Raton, FL, USA

Lamana ML, Miller JC (1996) Field observations on *Harmonia axyridis* Pallas (Coleoptera: Coccinellidae) in Oregon. Biol Control 6:232–237

Ludwig SW, Oetting RE (2001) Susceptibility of natural enemies to infection by *Beauveria bassiana* and impact of insecticides on *Iphesuius degenerans* (Acari: Phytoseiidae). J Agric Urban Entomol 18:169–178

Magalhaes BP, Lord JC, Wraight SP, Daoust RA, Roberts DW (1988) Pathogenicity of *Beauveria bassiana* and *Zoophthora radicans* to the coccinellid predators *Coleomegilla maculata* and *Eriopis connexa*. J Invertebr Pathol 52:471–473

McCoy CW, Shapiro DI, Duncan LW, Nguyen K (2000) Entomopathogenic nematodes and other natural enemies as mortality factors for larvae of *Diaprepes abbreviatus* (Coleoptera: Curculionidae). Biol Control 19:182–190

McKillup SC, Allen PG, Skewes MA (1988) The natural decline of an introduced species following its initial increase in abundance; an explanation for *Ommatoiulus moreletii* in Australia. Oecologia 77:339–342

Michaud JP (2002) Invasion of the Florida citrus ecosystem by *Harmonia axyridis* (Coleoptera: Coccinellidae) and asymmetric competition with a native species, *Cycloneda sanguinea*. Environ Entomol 35:827–835

Millar LC, Barbercheck ME (2001) Interaction between endemic and introduced entomopathogenic nematodes in conventional-till and no-till corn. Biol Control 22: 235–245

Mitchell CE, Power AG (2003) Release of invasive plants from fungal and viral pathogens. Nature 421:625–627

Nalepa CA, Kidd KA, Hopkins DI (2000) The multicolored Asian lady beetle (Coleoptera: Coccinellidae): Orientation to aggregation sites. J Entomol Sci 35:150–157

Nielsen C, Hajek AE (2005) Control of invasive soybean aphid, *Aphis glycines* (Hemiptera: Aphididae), populations by existing natural enemies in New York State, with emphasis on Entomophthorales fungi. Environ Entomol 34:1036–1047

Obrycki JJ (1989) Parasitization of native and exotic coccinellids by *Dinocampus coccinellae* (Schrank) (Hymenoptera: Braconidae). J Kans Entomol Soc 62:211–218

Pell JK, Vandenberg JD (2002) Interactions among the aphid *Diuraphis noxia*, the entomopathogenic fungus *Paecilomyces fumosoroseus* and the coccinellid *Hippodamia convergens*. Biocontr Sci Technol 12:217–224

Perrings C, Dehnen-Schmutz K, Touza J, Williamson M (2005) How to manage biological invasions under globalization. Trend Ecol Evol 20:212–215

Pingel RL, Lewis LC (1996) The fungus *Beauveria bassiana* (Balsamo) Vuillemin in a corn ecosystem: Its effect on the insect predator *Coleomegilla maculata* DeGeer. Biol Control 6:137–141

Poprawski TJ, Legaspi JC, Parker PE (1998) Influence of entomopathogenic fungi on *Serangium parcesetosum* (Coleoptera: Coccinellidae), an important predator of whiteflies (Homoptera: Aleyrodidae). Environ Entomol 27:785–795

Prenter J, MacNeil C, Dick JTA, Dunn AM (2004) Roles of parasites in animal invasions. Trends Ecol Evol 19:385–390

Sexton SB, Williams P (1981) A natural occurrence of parasitism of *Graphognathus leucoloma* (Boheman) by the nematode *Heterorhabditis* sp. J Aust Entomol Soc 20:253–255

Schulte F (1989) The association between *Rhabditis necromena* Sudhaus & Schulte, 1989 (Nematoda: Rhabditidae) and native and introduced millipedes in South Australia. Nematologica 35:82–89

Shapiro-Ilan DI, Cottrell TE (2005) Susceptibility of lady beetles (Coleoptera: Coccinellidae) to entomopathogenic nematodes. J Invertebr Pathol 89:150–156

Simmonds FJ (1966) Insect parasites and predators. In: Smith CN (ed) Insect colonization and mass production. Academic Press, New York, NY, USA. pp 489–499

Smith SF, Krischik VA (2000) Effects of biorational pesticides on four coccinellid species (Coleoptera: Coccinellidae) having potential as biological control agents in interiorscapes. J Econ Entomol 93:732–736

Solter LF, Maddox JV (1998) Physiological host specificity of Microsporidia as an indicator of ecological host specificity. J Invertebr Pathol 71:207–216

Sudhaus W, Schulte F (1989) *Rhabditis (Rhabditis) necromena* sp. n. (Nematoda: Rhabditidae) from South Australian Diplopoda with notes on its siblings *R. myriophila* Poinar, 1986 and *R. caulleryi* Maupas, 1919. Nematologica 35:15–24

Tedders WL, Schaefer PW (1994) Release and establishment of *Harmonia axyridis* (Coleoptera: Coccinellidae) in the southeastern United States. Entomol News 105:228–243

Todorova SI, Côté JC, Martel P, Coderre D (1994) Heterogeneity of two *Beauveria* strains revealed by biochemical tests, protein profiles and bio-assays on *Leptinotarsa decemlineata* (Col.: Chrysomelidae) and *Coleomegilla maculata lengi* (Col.: Coccinellidae) larvae. Entomophaga 39:159–169

Todorova SI, Coderre D, Côté J (2000) Pathogenicity of *Beauveria bassiana* isolates toward *Leptinotarsa decemlineata* [Coleoptera: Chrysomelidae], *Myzus persicae* [Homoptera: Aphididae] and their predator *Coleomegilla maculata lengi* [Coleoptera: Coccinellidae]. Phytoprotection 81:15–22

Torchin ME, Lafferty KD, Dobson AP, McKenzie VJ, Kuris AM (2003) Introduced species and their missing parasites. Nature 421:628–630

van den Bosch R, Schlinger EI, Dietrick EJ, Hagen KS, Halloway JK (1959) The colonization and establishment of imported parasites of the spotted alfalfa aphid in California. J Econ Entomol 52:136–141

van den Bosch R (1968) Comments on population dynamics of exotic insects. Bull Entomol Soc Amer 14:112–115

Vinson SB (1990) Potential impact of microbial insecticides on beneficial arthropods in the terrestrial environment. In: Laird M, Lacey LA, Davidson EW (eds) Safety of microbial insecticides. CRC Press Inc., Boca Raton, FL, USA. pp 43–64

Weseloh RM, Andreadis TG (1992) Epizootiology of the fungus *Entomophaga maimaiga*, and its impact on gypsy moth populations. J Invertebr Pathol 59:133–141

Williamson M (1996) Biological invasions. Chapman and Hall, Cornwall, UK

With KA, Pavuk DM, Worchuck JL, Oates RK, Fisher JL (2002) Threshold effects of landscape structure on biological control in agroecosystems. Ecol Appl 12:52–65

Wraight SP, Poprawski TJ, Meyer WL, Peairs FB (1993) Natural enemies of Russian wheat aphid (Homoptera: Aphididae) and associated cereal aphid species in spring-planted wheat and barley in Colorado. Environ Entomol 22:1383–1391

Yinon U (1969) The natural enemies of the armored scale lady-beetle *Chilocorus bipustulatus* (Col. Coccinellidae). Entomophaga 14:321–328

Chapter 3
Population Ecology of Managing Insect Invasions

Andrew M. Liebhold and Patrick C. Tobin

Abstract Invasions are characterized by three phases: arrival, establishment and spread. In this chapter we focus on the establishment and spread phases with consideration of how population processes operating during each stage influence the selection of management strategies. Typically, the establishment phase is dominated by the Allee effect in which population growth rates decrease with decreasing abundance. Allee effects can arise from several different mechanisms and are capable of driving low-density populations to extinction. Strategies to eradicate newly established populations should focus on either enhancing Allee effects or suppressing populations below Allee thresholds, such that extinction proceeds without further intervention. Spread of invading populations results from the coupling of population growth with dispersal. The spread of most non-indigenous insects is characterized by "stratified dispersal" in which occasional long-distance dispersal results in the formation of isolated colonies ahead of the continuously infested range boundary. These colonies grow, coalesce and greatly increase spread rates. Allee effects also affect spread, in part, by contributing to the extinction of colonies formed through stratified dispersal. One approach to containing the spread of an invading species focuses on eradicating these isolated colonies. Microbial control is one management tactic that is very appropriate for suppressing populations of invasive species below Allee thresholds and consequently preventing their establishment or limiting their rate of spread.

3.1 Introduction

Given ever-increasing trends in world trade and travel, invasions by non-indigenous insect species are likely to continue to increase (Levine & D'Antonio 2003). While efforts continue to close various pathways by which alien forest pests arrive (e.g. quarantines on solid wood packing material), it is unlikely that new arrivals

---

A.M. Liebhold
USDA, Forest Service, Northern Research Station, 180 Canfield St., Morgantown, West Virginia 26505, USA
e-mail: aliebhold@fs.fed.us

Table 3.1 Three phases of any invasion with corresponding management activities

Invasion Phase	Management activities
Arrival	International quarantines
	Inspection
Establishment	Detection
	Eradication
Spread	Domestic quarantines
	Barrier zones

can be completely stopped. Therefore it is clear that new pest species are likely to continue to arrive and many of these species are likely to have catastrophic ecological and economic consequences (Strong & Pemberton 2000, Pimentel *et al.* 2000).

Three distinct population processes occur during the course of virtually every invasion: arrival (the process by which individuals are transported from their native to non-native habitat), establishment (the process by which populations grow to sufficient levels such that extinction is no longer feasible), and spread (the expansion of a population's range) (Dobson & May 1986, Shigesada & Kawasaki 1997, Liebhold & Tobin 2008) (Table 3.1). In this chapter we focus on the population ecology principles that are operating during the establishment and spread phases and implications for management through eradication (i.e. forced extinction of a population) and containment (i.e. quarantine and barrier zones implemented to minimize the spread of the population to new areas). We explore these processes in the context of managing invading populations using microbials. The arrival phase is also an important invasion process but is less relevant to management using microbials; readers may wish to consult other sources about invasion pathways and arrival processes (McCullough *et al.* 2006, Everett 2000, Jerde & Lewis 2007).

3.2 Establishment

Considerable evidence indicates that arrival of alien species occurs at a much higher rate than the rate of establishment; most introduced species fail to establish (Williamson & Fitter 1996, Simberloff & Gibbons 2004). The establishment phase thus represents a critical period during which populations grow and expand their distribution such that extinction is highly unlikely. Founder populations typically are small and consequently are at great risk of extinction. All populations (both sparse and abundant) are affected by random abiotic influences (e.g. weather), but low-density populations are particularly influenced by such effects. Mathematically, change in population density from one time period to the next can be represented as

$$N_{t+1} = f(N_t) + \varepsilon_t, \qquad (3.1)$$

where N is population density in year t or $t+1$, $f(N_t)$ is a function that encompasses birth and death processes, and ε_t random variation in the environment. In addition to environmental stochasticity, all populations are affected by demographic stochasticity, which refers to random variation in birth and death processes (Engen *et al.* 1998). Because sparse populations, by definition, exist at densities near zero, both demographic and environmental stochasticity can cause these populations to go extinct. In addition, Allee effects can often have a significant impact on low-density populations.

Warder Allee (1931) was fascinated by cooperative behavior in animals and is credited as the first individual to suggest that cooperation is significant at the population level. In particular, Allee noticed that at low densities, survival and reproduction were often limited by the lack of conspecifics and this could lead to population decline. Certain processes may lead to decreasing net population growth with decreasing density, and thus, there may exist a threshold below which low-density populations are driven toward extinction (Fig. 3.1A). This phenomenon, known as the Allee effect (Courchamp *et al.* 1999, Dennis 1989), has been critically important in our understanding of extinction in low-density populations of rare species and therefore is highly relevant in conservation efforts (Stephens & Sutherland 1999). However, because of their role in low-density populations, Allee effects have been recognized recently to be important during the establishment phase of biological invasions since founder populations of introduced species are generally at low densities (Drake 2004, Leung *et al.* 2004, Taylor & Hastings 2005).

The many causes of Allee effects include failure to locate mates (Berec *et al.* 2001, Hopper & Roush 1993), inbreeding depression (Lande 1998), the failure to satiate predators (Gascoigne & Lipcius 2004), and the lack of cooperative feeding (Clark & Faeth 1997). The term "component Allee effect" refers to the decrease in population growth with decreasing density caused by a single mechanism. The term "demographic Allee effect" refers to decreasing growth with decreasing density at the population level. In many cases, demographic Allee effects arise from several components, although multiple sources of Allee effects do not necessarily act additively and instead may interact in complex ways (Berec *et al.* 2007).

Because most insects reproduce sexually, failure to locate mates when populations are at low densities can limit population growth and contribute to a demographic Allee effect (Hopper & Roush 1993, Berec *et al.* 2001, Wells *et al.* 1998). Even though many insect species have evolved highly efficient mate-location systems (e.g. sex pheromones), males and females may be simply too far apart in space to find each other and mate, which may ultimately prevent a low-density population of invaders from establishing. Mate-finding can also be influenced by seasonal asynchrony of adult male and female sexual maturation, and species with pronounced protandry (i.e. male emergence precedes that of females) and variation in developmental times are likely to exhibit stronger Allee effects (Robinet *et al.* 2007, Calabrese & Fagan 2004).

Many herbivorous insect species also utilize cooperative feeding behaviors, in which an aggregation of individuals increases the ability of populations to overcome host defenses and successfully establish themselves, as is the case in tree-killing bark beetles (Raffa & Berryman 1983). Larval aggregations, especially neonates or younger larvae for several species, are better able to exploit food resources protected

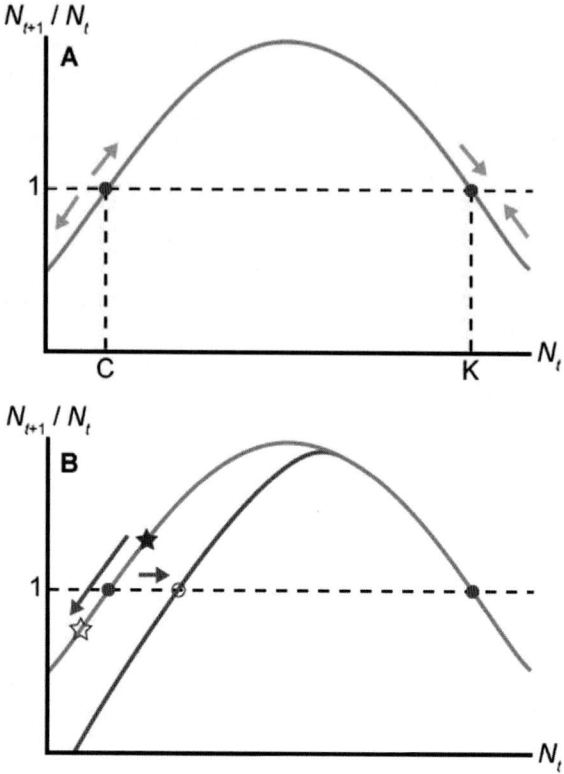

Fig. 3.1 Schematic representation of the Allee effect modified from Liebhold & Tobin (2008). Change in population density, N_{t+1}/N_t is plotted as a function of density at the beginning of the generation, N_t. **A**. Illustration of equilibria. When density is less than C, the Allee threshold (*black circle*), it will decrease toward extinction. When density exceeds C it will increase toward K, the carrying capacity. When populations exceed the carrying capacity, they will decrease. **B**. Illustration of eradication strategies. The first strategy is to reduce the population density (*solid star*) to a new density (*open star*) that is below the Allee threshold. In the second strategy, the Allee threshold is increased to a new level (*open circle*) that exceeds the population density. Both strategies result in population extinction

by the host (either chemically or physically) because large groups are more efficient at initializing successful feeding sites (Ghent 1960, Clark & Faeth 1997). Thus, Allee effects may occur when founder populations arrive in numbers insufficient to capitalize on cooperative feeding advantages, e.g. if numbers of bark beetles are insufficient to overcome host tree defenses. Similarly, Allee effects may arise at low densities because the presence of conspecifics may promote population growth via saturation of predators. In general, predation can be expected to contribute a component Allee effect in any low-density invading population (Gascoigne & Lipcius 2004).

3.3 Spread

Spread of a non-indigenous species is a process by which a species expands its range from a habitat that it currently occupies to one that it does not. In its simplest form, spread is continuous and simply results from the coupling of dispersal with population growth. This was first described in the ecological literature by Skellam (1951), whose model specifically combined exponential population growth with diffusive (random) movement (Fig. 3.2A). The magnitude of dispersal is characterized by the diffusion coefficient, D, which can be estimated as the standard deviation of dispersal distances, typically compiled from mark-recapture experiments (Kareiva 1983). In Skellam's model, population growth was characterized by r, which is the intrinsic rate of population increase under ideal conditions. Skellam (1951) applied his model to show that the radial rate of range expansion, V, is asymptotically constant and can be estimated according to

$$V = 2\sqrt{rD}. \qquad (3.2)$$

Although there are several examples in which Eq. 3.2 has provided spread estimates that were similar to observed rates of spread (Shigesada & Kawasaki 1997, Andow et al. 1990), there are many cases where it fails (Shigesada & Kawasaki 1997). This failure can be attributed to the simplicity of Skellam's biological assumptions, and in particular, the failure to account for long-distance dispersal.

Insect spread may not always be represented by simple diffusion; instead, long-range movement of insects, such as through anthropogenic or other mechanisms, may occur with considerable frequency. Such leptokurtic or "fat-tailed" dispersal-distance distributions can lead to much faster rates of spread than those predicted by simple diffusion models (Clark 1998, Kot et al. 1996). Moreover, long-distance dispersal events are more often than not caused by a completely different mecha-

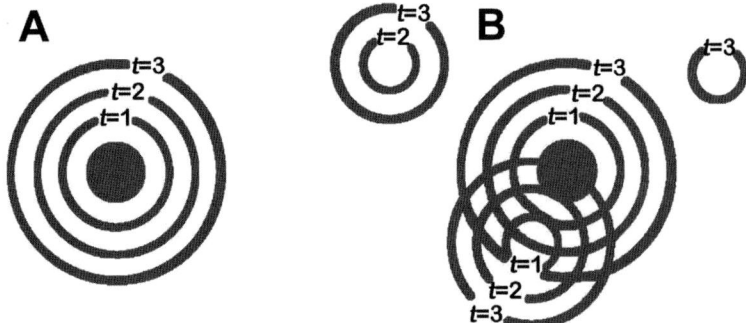

Fig. 3.2 Schematic of spread between successive generations (t_1 to t_3). The solid innermost circle represents the initial range at time 0: **A** shows spread according to Skellam's (1951) model of successive time steps; **B** illustrates spread predicted using a stratified dispersal model (Hengeveld 1989, Shigesada et al. 1995, Shigesada & Kawasaki 1997)

nism, such as through the movement of goods by humans or through atmospheric transport mechanisms, than short-distance movement events, such as those that occur through active insect flight. The combined processes of short-range and long-range dispersal is referred to as "stratified dispersal," and is a major driver of spread dynamics (Shigesada & Kawasaki 1997, Shigesada et al. 1995, Hengeveld 1989). The occurrence of long-distance dispersal, even at a relatively low rate, has been shown to cause much greater rates of spread than would occur with only short-range dispersal (Shigesada et al. 1995). This is because long-range dispersal causes populations to "jump" well ahead of the continuously infested boundary of the expanding species range. These jumps result in isolated populations that grow (via population growth coupled with short-range dispersal) and ultimately coalesce. This phenomenon is well illustrated by the spread of the gypsy moth (*Lymantria dispar*) in North America: short-distance dispersal of airborne 1st instars on silken threads carries many individuals relatively short distances but humans sometimes accidentally transport life stages long distances, well ahead of the advancing population front (Liebhold et al. 1992, Whitmire & Tobin 2006). Spread is thus enhanced by the growth of isolated colonies ahead of the expanding population front and their eventual coalescence with the population front. It is the rate of formation and growth of these colonies that ultimately determines the rate of gypsy moth spread (Sharov & Liebhold 1998).

Low-density, isolated colonies founded ahead of the expanding population, through long-distance dispersal mechanisms, are typically influenced by Allee effects analogous to newly arrived populations that are in the early stages of establishment (see Section 3.2). Theoretical studies have shown that Allee dynamics can negatively affect the growth and persistence of isolated colonies, driving a large proportion to extinction, thereby reducing spread (Taylor et al. 2004, Hui & Li 2004). Johnson et al. (2006) noted a curious phenomenon of periodically-pulsed spread every 3–4 years in historical records of gypsy moth range expansion, and contended that this behavior was caused by an interaction between stratified dispersal and Allee effects. Specifically, they observed that most isolated colonies founded from long-distance dispersal went extinct because of Allee effects unless populations at the advancing population front were at sufficient levels such that they could found isolated populations that exceeded the Allee threshold. At this point, spread would pulse forward but then stop again until populations once again grew to sufficient levels. For more details of how Allee effects influence spread, the reader should consult Taylor and Hastings (2005).

3.4 Management of Establishment

The failure of invading populations to establish following arrival results in extinction and this should be the goal of any effort to prevent establishment. We define eradication as the total elimination of a species from a geographical area (Simberloff 2003, Knipling 1979). Despite the importance of eradication in the management of biological invasions, the entire concept has been controversial for many years (Myers *et al.* 1998, 2000, Simberloff 2003, DeBach 1964). One source of this controversy is a belief held by many that it is either impossible or impractical to eliminate every individual in a population (Dahlsten *et al.* 1989). Another problem that is often raised is that eradication is impossible if there is a steady flow of new invaders ready to recolonize an area where this species was previously eradicated (Myers *et al.* 2000, Carey 1991). But perhaps the greatest reason for skepticism is the occurrence of several failed eradication programs, such as efforts to eliminate the imported fire ant, *Solenopsis invicta* (Perkins 1989), and the gypsy moth (Forbush & Fernald 1896, Spear 2005) from the United States. However, there are also many examples of highly successful eradication efforts (Simberloff 2003, 2001, Myers *et al.* 1998). Thus, it critical to clarify when and where eradication is practical and how it might be most efficiently accomplished. The basic population processes associated with the establishment process are critical to these questions.

A frequent criticism of eradication as a strategy is the impracticality of killing every individual in a population. The critical basis for this argument is that eradication can only occur by eliminating all individuals within the population (Dahlsten *et al.* 1989, Knipling 1979). But our current knowledge of the population biology of sparse, invading populations indicates that most insect species exhibit both stochastic dynamics and Allee effects at low densities, and these processes are likely to drive extremely small populations toward extinction without intervention. Thus, the success of eradication does not depend on achieving 100% reduction of a population but rather on reducing a population below some threshold, below which they will proceed toward extinction without further effort (Fig. 3.2B). This concept was illustrated by Liebhold and Bascompte (2003) using a simple model of stochastic population growth coupled with Allee dynamics. They applied this model to invading gypsy moth populations as a case study to parameterize their model and explored the relationship between population size and the "killing power" (i.e. percent mortality) necessary to achieve eradication. Their model indicated that if invading populations were reduced below a threshold, then most populations went extinct without further intervention. These results explain the consistent success that has been achieved in eradicating sparse gypsy moth populations using only one or two aerial applications of *Bacillus thuringiensis*, a treatment that is considered to achieve only 95% mortality in a best-case scenario (Miller 1990, Liebhold & McManus 1999).

Successful eradication can thus be an attainable goal if the application of any treatment (e.g. a microbial pesticide) kills a sufficient proportion of a population such that the residual population falls below the Allee threshold and extinction proceeds. In addition to the approach of suppressing populations below the Allee threshold, an alternative strategy is increasing the Allee threshold (via manipulation of population processes) to a point that it exceeds the current population level and therefore the population decreases toward extinction (Fig. 3.2B). Some examples of this approach are provided by various tactics designed to increase the strength of Allee effects, such as interfering with mating. This can be accomplished via mating disruption (Yamanaka 2007), mass-trapping (El-Sayed *et al.* 2006) and the release of sterile insects (Klassen 2005). There are many examples of how these methods have been utilized in successful eradication programs even though no individuals were actually killed (El-Sayed *et al.* 2006, Klassen 2005). In addition to altering mating success, there are probably several other approaches that could be used to manipulate the Allee threshold and thereby achieve eradication. Technically, any agent that increases mortality or reduces reproductive output can be expected to reduce population growth (i.e. decrease along the y-axis in Fig. 3.1B) and potentially, these agents could reduce the population size below the Allee threshold (C in Fig. 3.1A, defined here as the density below which population growth is <1) and consequently drive the population to extinction. Augmentation of microbial organisms that increase host mortality (e.g. *Beauveria bassiana*) or reduce reproductive output (e.g. microsporidia) are approaches that have not received attention as eradication approaches but warrant further consideration.

The concept of eradication also is criticized because in some areas where a species is removed, re-invasion by the same species is likely to occur, thus negating the eradication effort (Myers *et al.* 1998). This issue is particularly important in areas where the propagule pressure is high (e.g. ports of entry), which obviously makes the task of eradication more difficult (Drake & Lodge 2005). But Simberloff (2003) argued that even though re-invasion detracts from the value of eradication, it does not necessarily make it a futile endeavor. Alternatively, the economic and social benefits of eradication can be addressed in cost-benefit analyses that account for the probability of re-invasion. There are several examples of insect species (e.g. the gypsy moth in the western United States, Mediterranean fruit fly, *Ceratitis capitata* in California) that are repeatedly eradicated from areas following re-invasion, but the benefits of eradication in these cases vastly outweigh the cost of establishment (Simberloff 2003). Rates of reinvasion may be particularly high or low in certain areas; geographically isolated areas, such as islands, typically have lower rates of re-invasion, making eradication particularly economically feasible, especially when weighed against the typically large impacts of invasive species establishment on island biodiversity.

Another important factor in determining eradication pragmatism is the availability of sensitive survey methods. Some type of survey method is necessary, both to delimit the spatial extent of the nascent population (and target suppression) as well as to evaluate the success of the effort. Chemical attractants are used by many insect species and these are often used in inexpensive yet highly effective traps for detecting their presence. Attractants may not be available for some species and surveys would have to rely upon more expensive techniques (e.g. visual examination) to detect and delimit a population. In one such example, pheromone-baited traps were not available during eradication efforts in New Zealand against the painted apple moth (*Teia anartoides*); nevertheless, eradication was successfully achieved (at great expense) using traps baited with live females (produced in a rearing facility) as the primary survey tool (Kean & Suckling 2005) (see Chapter 4). When sensitive survey tools are not available for pest species of potentially high economic importance, it may be beneficial to regularly apply control measures in areas of known high propagule pressure, even in the absence of survey data indicating the presence of a population. For example, sterile male releases are made regularly as prophylactic treatments against establishment of invasive fruit flies in selected urban locations in the USA, Australia and Japan (Hendrichs *et al*. 2002). Methods using prophylactic applications of microbials could also be used in a similar way, though we are not aware of any systems where this has been implemented.

3.5 Management of Spread

Various approaches may be available for slowing, stopping or reversing the spread of an invading species. Reversal or stopping spread can be expected to require greater resources than slowing spread so the latter may be the most realistic objective in many cases. Because there is considerable variation among species in critical parameters, such as dispersal, growth rates and Allee effects, optimal spread management strategies are likely to vary considerably among species.

Because the occurrence of spread is the result of coupling dispersal with population growth, its management should logically focus on reducing dispersal, growth, or some combination of both. As mentioned earlier, long-distance dispersal events, though proportionally less common, can have a large influence on spread. As a result, expending a little effort to minimize long-distance dispersal may greatly reduce spread. Long-distance dispersal is frequently caused by humans moving goods on which invaders hitchhike, so domestic quarantines and similar approaches that minimize the risks associated with anthropogenic movement may be useful in reducing spread rates. In some systems, there may be other approaches available for reducing spread by mitigating long-distance dispersal. For example, Krushelnycky *et al*. (2004) found that budding dispersal of Argentine ant colonies at the expanding population front could be stopped through pesticide applications and in some cases, this tactic has completely eliminated spread into adjoining uninfested areas.

Another approach to limiting dispersal is reducing population growth. There are two distinct strategies for limiting population growth that may be used to reduce spread: (1) suppression of populations at or behind the advancing population front that serve as source populations from which propagules originate and invade new areas through long-distance dispersal and (2) suppression of isolated colonies that are formed through long-distance dispersal. In some cases, the former strategy may not be practical because of the sheer magnitude of the area that must be suppressed. An exception to this is the effort in the United States to reverse the spread of the boll weevil (*Anthonomus grandis*), in which populations have been suppressed over large areas to systematically reduce the boll weevil range (Hardee & Harris 2003). In the second strategy, spread may be reduced by suppressing the growth of outlying colonies. Sharov and Liebhold (1998) proposed a generalized model for slowing the spread of an invading species by eradicating isolated colonies (formed through long-distance dispersal) located ahead of the expanding population front. One benefit of this strategy is that the treatment of relatively small areas (isolated colonies) has a large, negative impact on spread. A good example of this strategy is provided by the gypsy moth in the USA where isolated populations are located (using grids of pheromone-baited traps in an approximately 100-km wide band along the advancing gypsy moth front) and then suppressed. Roughly $8–10 million is spent on the program annually and gypsy moth spread has been reduced by over 60% (Tobin *et al.* 2004, Sharov *et al.* 2002, Tobin & Blackburn 2007).

Again, an important prerequisite to the success of spread management efforts is the availability of practical methods for detecting low-density populations. Sensitive survey methods are particularly crucial for strategies that focus on finding and eradicating isolated populations formed through long-distance dispersal because densities of such populations are often low. The success of both the current gypsy moth and boll weevil containment efforts may be largely due to the availability of pheromone-baited traps, that are inexpensive yet highly sensitive for detecting low-density populations (Tobin & Blackburn 2007, Hardee & Harris 2003). Containment may not be practical for other insect species where such survey methods are not available.

Another prerequisite for successful management of spread is the availability of control tactics that are effective at low densities. Preferably, such control tactics should have minimal adverse non-target and environmental impacts so that they may be applied in all potential habitats. The ultimate effectiveness of any eradication or barrier zone approach may be compromised by allowing some populations to go untreated. Another crucial consideration, once again, is the Allee effect; isolated populations forming from long-distance dispersal along the expanding population front may be eradicated by manipulating Allee thresholds (e.g. reducing population growth or reproductive output) and this would greatly reduce spread (Tobin *et al.* 2007). As we suggested earlier for eradication, augmentation of microbial organisms that reduce population growth, either through elevated mortality or reduced reproductive output, merit some consideration as alternative approaches to managing isolated populations that contribute to spread.

3.6 Conclusions

Several past investigations and reviews have focused on the population ecology of biological invasions, providing many examples of how mathematical ecology can make useful predictions that have substantial applied value (Shigesada & Kawasaki 1997, Sakai *et al.* 2001, Puth & Post 2005, Lockwood *et al.* 2007). Unfortunately, there still is a disconnect between the current state of knowledge of invasion population ecology and the management of invasions. Successful management of invasions depends on more than just developing new treatment methods (e.g. new strains of microbial pesticides) because it is also crucial to understand when and where to apply treatments. In the future, more work is needed to bridge this gap between population theory and management by utilizing ecological concepts in the formulation of invasion management strategies. In particular, Allee dynamics present a potential "weak link" during the establishment phase that can be greatly exploited to achieve successful eradication. Similarly, the fact that many exotic species spread through stratified dispersal provides excellent opportunities for limiting the spread of these organisms through management programs, including those that likewise manipulate Allee dynamics as a means to reduce spread.

As the field of invasion biology continues to grow, it is likely there will be important new advances in understanding the population ecology of invading species. Advances are also likely to be made in the development and optimization of microbial pesticides, which hold great potential not only for suppressing invading pest insect populations, but also minimizing undesirable environmental effects. However, for these advances to translate into progress in minimizing the current explosion of damaging non-indigenous pests, it will be necessary for pest management specialists to work with insect population ecologists to identify, develop and implement effective, science-based strategies that utilize our understanding of population processes operating during biological invasions.

Acknowledgments We are grateful to Leah Bauer and R. Talbot Trotter III for reviewing this manuscript.

References

Allee WC (1931) Animal aggregations: a study in general sociology. University of Chicago Press, Chicago, Illinois

Andow DA, Kareiva PM, Levin SA, Okubo A (1990) Spread of invading organisms Landscape Ecol 4:177–188

Berec L, Boukal DS, Berec M (2001) Linking the Allee effect, sexual reproduction, and temperature-dependent sex determination via spatial dynamics. Am Nat 157:217–230

Berec L, Angulo E, Courchamp F (2007) Multiple Allee effects and population management. Tr Ecol Evol 22:185–191

Calabrese JM, Fagan WF (2004) Lost in time, lonely, and single: reproductive asynchrony and the Allee effect. Am Nat 164:25–37

Carey JR (1991) Establishment of the Mediterranean fruit fly in California. Science 253:1369–1373

Clark BR, Faeth SH (1997) The consequences of larval aggregation in the butterfly *Chlosyne lacinia*. Ecol Entomol 22:408–415

Clark JS (1998) Why trees migrate so fast: confronting theory with dispersal biology and the paleorecord. Am Nat 152:204–224

Courchamp F, Clutton-Brock, T, Grenfell B (1999) Inverse density dependence and the Allee effect. Tr Ecol Evol 14:405–410

Dahlsten DL, Garcia R, Lorraine H (1989) Eradication as a pest management tool: concepts and contexts. In Dahlsten DL, Garcia R (eds) Eradication of exotic pests. Yale University Press, New Haven, Connecticut. pp 3–15

DeBach P (1964) Some ecological aspects of insect eradication. Bull Entomol Soc Am 10:221–224

Dennis B (1989) Allee effects: population growth, critical density, and the chance of extinction. Nat Res Modeling 3:481–538

Dobson AP, May RM (1986) Patterns of invasions by pathogens and parasites. In Mooney HA, Drake JA (eds) Ecology of biological invasions of North America and Hawaii. Springer-Verlag, New York. pp 58–76

Drake JA, Lodge DM (2005) Allee effects, propagule pressure and the probability of establishment: risk analysis for biological invasions. Biol Invasions 8:365–375

Drake JM (2004) Allee effects and the risk of biological invasion. Risk Anal 24:795–802

El-Sayed AM, Suckling DM, Wearing CH, Byers JA (2006) Potential of mass trapping for long-term pest management and eradication of invasive species. J Econ Entomol 99:1550–1564

Engen S, Blakke O, Islam A (1998) Demographic and environmental stochasticity – Concepts and definitions. Biometrics 54:840–846

Everett RA (2000) Patterns and pathways of biological invasions. Tr Ecol Evol 15:177–178

Forbush EH, Fernald CH (1896) The gypsy moth, *Porthetria dispar* (Linn.). Wright and Potter Printing Co., Boston, MA

Gascoigne JC, Lipcius RN (2004) Allee effects driven by predation. J Appl Ecol 41:801–810

Ghent AA (1960) A study of the group-feeding behavior of the jack pine sawfly *Neodiprion pratti banksianae*. Behaviour 16:110–148

Hardee DD, Harris FA (2003) Eradicating the boll weevil (Coleoptera: Curculionidae): a clash between a highly successful insect, good scientific achievement, and differing agricultural philosophies. Amer Entomol 49:82–97

Hendrichs J, Robinson AS, Cayol JP, Enkerlin W (2002) Medfly areawide sterile insect technique programmes for prevention, suppression or eradication: the importance of mating behavior studies. Fla Entomol 85:1–13

Hengeveld R (1989) Dynamics of biological invasions. Chapman and Hall, London

Hopper KR, Roush RT (1993) Mate finding, dispersal, number released, and the success of biological control introductions. Ecol Entomol 18:321–331

Hui C, Li Z (2004) Distribution patterns of metapopulation determined by Allee effects. Popn Ecol 46:55–63

Jerde CL, Lewis MA (2007) Waiting for invasions: a framework for the arrival of non-indigenous species. Am Nat 170:1–9

Johnson DE, Liebhold AM, Tobin PC, Bjørnstad ON (2006) Pulsed invasions of the gypsy moth. Nature 444:361–363

Kareiva PM (1983) Local movement in herbivorous insects: applying a passive diffusion model to mark-recapture field experiments. Oecologia 57:322–327

Kean JM, Suckling DM (2005) Estimating the probability of eradication of painted apple moth from Auckland. NZ Plant Prot 58:7–11

Klassen W (2005) History of the sterile insect technique. In Dyck VA, Hendrichs J, Robinson AS (eds) Sterile insect technique: principles and practice in area-wide integrated pest management. Springer, Netherlands. pp 3–36

Knipling EF (1979) The basic principles of insect population suppression and management. US Department of Agriculture, Washington, DC

Kot M, Lewis MA, van den Driessche P (1996) Dispersal data and the spread of invading organisms. Ecology 77:2027–2042

Krushelnycky PD, Loope LL, Joe SM (2004) Limiting spread of a unicolonial invasive insect and characterization of seasonal patterns of range expansion Biol Invasions 6:47–57

Lande R (1998) Anthropogenic, ecological and genetic factors in extinction and conservation. Res Popn Ecol 40:259–269

Leung B, Drake JM, Lodge DM (2004) Predicting invasions: propagule pressure and the gravity of Allee effects. Ecology 85:1651–1660

Levine JM, D'Antonio CM (2003) Forecasting biological invasions with increasing international trade. Conserv Biol 17:322–326

Liebhold A, Bascompte J (2003) The Allee effect, stochastic dynamics and the eradication of alien species. Ecology Lett 6:133–140

Liebhold A, McManus M (1999) The evolving use of insecticides in gypsy moth management. J For 97(3):20–23

Liebhold AM, Halverson JA, Elmes GA (1992) Gypsy moth invasion in North America: a quantitative analysis. J Biogeog 19:513–520

Liebhold AM, Tobin PC (2008) Population ecology of insect invasions and their management. Annu Rev Entomol 53:387–408

Lockwood J, Hoopes M, Marchetti M (2007) Invasion ecology. Blackwell Publ. Ltd., Malden, Massachusetts

McCullough DG, Work TT, Cavey JF, Liebhold AM, Marshall D (2006) Interceptions of nonindigenous plant pests at U.S. ports of entry and border crossings over a 17 year period. Biol Invasions 8:611–630

Miller JC (1990) Field assessment of the effects of a microbial pest control agent on nontarget Lepidoptera. Amer Entomol 36:135–139

Myers JH, Savoie A, Van Randen E (1998) Eradication and pest management. Annu Rev Entomol 43:471–91

Myers JH, Simberloff D, Kuris AM, Carey JR (2000) Eradication revisited: dealing with exotic species. Tr Ecol Evol 15:316–320

Perkins JH (1989) Eradication: scientific and social questions. In Dahlsten DL, Garcia R (eds) Eradication of exotic pests. Yale University Press, New Haven, Connecticut. pp 16–40

Pimentel D, Lach L, Zuniga R, Morrison D (2000) Environmental and economic costs of nonindigenous species in the United States. BioScience 50:53–65

Puth LM, Post DM (2005) Studying invasion: have we missed the boat? Ecology Lett 8:715–721

Raffa KF, Berryman AA (1983) The role of host plant resistance in the colonization behaviour and ecology of bark beetles. Ecol Monogr 53:27–49

Robinet C, Liebhold A, Grey D (2007) Variation in developmental time affects mating success and Allee effects. Oikos 116:1227–1237

Sakai AK, Allendorf FW, Holt JS, Lodge DM, Molofsky J, With KA, Baughman S, Cabin RJ, Cohen JE, Ellstrand NC, McCauley DE, O'Neil P, Parker IM, Thompson JN, Weller SG (2001) The population biology of invasive species. Annu Rev Ecol Syst 32:305–332

Sharov AA, Leonard D, Liebhold AM, Roberts EA, Dickerson W (2002) "Slow the spread": a national program to contain the gypsy moth. J For 100:30–35

Sharov AA, Liebhold AM (1998) Model of slowing the spread of gypsy moth (Lepidoptera: Lymantriidae) with a barrier zone. Ecol Appl 8:1170–1179

Shigesada N, Kawasaki K (1997) Biological invasions: theory and practice. Oxford University Press, New York, NY

Shigesada N, Kawasaki K, Takeda Y (1995) Modeling stratified diffusion in biological invasions. Am Nat 146:229–251

Simberloff D (2001) Eradication of island invasives: practical actions and results achieved. Tr Ecol Evol 16:273–274

Simberloff D (2003) Eradication: preventing invasions at the outset. Weed Sci 51:247–253

Simberloff D, Gibbons L (2004) Now you see them, now you don't! – Population crashes of established introduced species. Biol Invasions 6:161–172

Skellam JG (1951) Random dispersal in theoretical populations. Biometrika 38:196–218

Spear RJ (2005) The great gypsy moth war: a history of the first campaign in Massachusetts to eradicate the gypsy moth, 1890–1901. University of Massachusetts Press, Amherst, MA

Stephens PA, Sutherland WJ (1999) Consequences of the Allee effect for behaviour, ecology and conservation. Tr Ecol Evol 14:401–405

Strong DR, Pemberton RW (2000) Biological control of invading species: risk and reform. Science 288:1969–1970

Taylor CM, Hastings A (2005) Allee effects in biological invasions. Ecol Lett 8:895–908

Taylor CM, Davis HG, Civille JC, Grevstad FS, Hastings A (2004) Consequences of an Allee effect on the invasion of a Pacific estuary by *Spartina alterniflora*. Ecology 85:3254–3266

Tobin PC, Blackburn LM (2007) Slow the Spread: a national program to manage the gypsy moth. U.S. Department of Agriculture, Forest Service, Northern Research Station, Newtown Square, PA

Tobin PC, Sharov AA, Liebhold AM, Leonard DS, Roberts EA, Learn MR (2004) Management of the gypsy moth through a decision algorithm under the STS Project. Amer Entomol 50:200–209

Tobin PC, Whitmire SL, Johnson DM, Bjørnstad ON, Liebhold AM (2007) Invasion speed is affected by geographic variation in the strength of Allee effects. Ecology Lett 10:36–43

Wells H, Strauss EG, Rutter MA, Wells PH (1998) Mate location, population growth and species extinction. Biol Conserv 86:317–324

Whitmire SL, Tobin PC (2006) Persistence of invading gypsy moth populations in the United States. Oecologia 147:230–237

Williamson M, Fitter A (1996) The varying success of invaders. Ecology 77:1661–1666

Yamanaka T (2007) Mating disruption or mass trapping? Numerical simulation analysis of a control strategy for lepidopteran pests. Popn Ecol 49:75–86

Part III
Eradication

Chapter 4
Use of Pathogens for Eradication of Exotic Lepidopteran Pests in New Zealand

Travis R. Glare

Abstract New Zealand has a history of invasive pests causing significant damage both economically and environmentally. As an island nation with an agriculturally-based economy, preventing invasion by arthropods and other pests has a high priority. Discovery of three lymantriid moth species in the North Island of New Zealand resulted in eradication programmes, based on the use of the *Bacillus thuringiensis kurstaki*-based product, Foray 48B. New Zealand has no lymantriids and early evaluations of risk determined that each introduced species had the potential to cause economic loss and environmental damage. Populations of two of these pests, *Orgyia thyellina* (white spotted tussock moth) and *Teia anartoides* (painted apple moth), were first found in Auckland in 1996 and 1999, respectively, while a single *Lymantria* sp. male was trapped in Hamilton, 70 km to the south of Auckland, in 2003. This single male was originally identified as the Asian strain of *L. dispar* and a tentative morphological identification suggested it conformed to *L. dispar praetera*, a subspecies subsequently synonymised with *L. umbrosa*. However, molecular comparisons need to be completed to confirm specific identity and most publications in New Zealand still refer to the Asian gypsy moth in Hamilton. *L. dispar* had been frequently found on cargo and ships in New Zealand ports prior to the single male trap catch. Extensive spraying, both aerial and ground-based, led to the eradication of *O. thyellina* and *T. anartoides* from Auckland. The single gypsy moth find in Hamilton led to a limited aerial spraying of Foray 48B. No populations of gypsy moth were ever found. The successful eradication of the three lymantriid populations from New Zealand demonstrates that with sufficient treatment, eradication using a microbial-based insecticide is possible, even in urban environments.

T.R. Glare
AgResearch Ltd., Biocontrol, Biosecurity & Bioprocessing, Lincoln Science Centre, Christchurch 8140, New Zealand
e-mail: travis.glare@agresearch.co.nz

4.1 Introduction

New Zealand is an island nation in the South Pacific. It has a land area of approximately $268,680 \text{ km}^2$, which primarily consists of two islands, known unimaginatively as North Island and South Island. Having separated from Gondwanaland millions of years ago, many indigenous and novel species have evolved. Since European colonization around 200 years ago, agriculture has become the mainstay of the country, with animal production (meat, wool and dairy) the largest export earner. This combination of indigenous and valued fauna and flora (tourism remains an important earner of export dollars) and a large agricultural sector mean the country is at risk from introductions of pests and diseases. Hence, biosecurity is considered particularly important and is a focus for government, industry and science. There are many very obvious examples of noxious introductions of vertebrates and invertebrates in the past two hundred years (e.g. Australian possums, European wasps, rats, rabbits and stoats), which serve to remind the country of the importance of excluding new pests.

Being an island nation, exclusion of new exotic organisms is potentially easier than in countries that share land borders with other countries. However, despite rigorous quarantine procedures, some pests arrive and, in some cases, have become established. Recent examples include the clover root weevil (*Sitona lepidus*) and *Varroa* bee mite. These two examples were discovered when populations were well established and it was decided that eradication would not be possible. However, New Zealand authorities have been successful in eradicating several exotic introductions of tussock moths (Lepidoptera: Lymantriidae), which is quite remarkable, given that in each case the population was discovered in a city of some size. Of the three invasive moth species found in New Zealand in recent years, two (*Orgyia thyellina* and *Teia anartoides*) were in Auckland, the largest city and main entry point for visitors and goods, and one (*Lymantria* sp., possibly *umbrosa*; see below) was in Hamilton (about 70 km south of Auckland).

White spotted tussock moth (*O. thyellina*; WSTM), a species known in North Asia, and painted apple moth (*T. anartoides*; PAM), from Australia, are sometimes pests in their home ranges but little literature exists on their control or natural enemies. WSTM was discovered in New Zealand in 1996 and PAM was first discovered in 1999. Egg masses and other life stages of the third species, the Asian gypsy moth (*L.* sp., possibly *umbrosa*; AGM), had been found on ships in New Zealand ports and on imported automobiles and other cargo for a number of years (Cowley *et al.* 1993). However, the finding of a single male AGM (although there is now some doubt about the exact species designation – see Section 4.4.1.) in Hamilton in 2003 was of more concern because this suggested the existence of a reproducing population on land. These were not the only discoveries of new invertebrates in New Zealand during this period but, given the perceived threat to the country's economy from any forest defoliator, and the potential for successful eradication due to the nature of the insect (in two cases restricted ability to spread due to flightless females) and localised infestations at the time of discovery, eradication campaigns were conducted. Herein, the three programmes are briefly reviewed.

New Zealand has a long history of use of biological control, especially the introduction of exotic insects (often parasites) for control of exotic pests (e.g. Cameron *et al*. 1989). There have also been many importations of insect pathogens, with the first records of the importation of the fungus *Beauveria brongniartii* from France dating back to the 1890s (Anonymous 1893). Given this history of biological options for pest control, a public view of New Zealand as "clean and green" and a government focused on sustainable agricultural solutions, it was not surprising that the bacterium *Bacillus thuringiensis kurstaki* (*Btk*) was chosen as the agent for use in eradication programmes. *Btk* has a long history of environmental and mammalian safety when used for insect control around the world (Glare & O'Callaghan 2000). The fact that the caterpillars of all three species were susceptible to *Btk* was probably the most compelling factor in even attempting eradication.

4.2 White Spotted Tussock Moth

4.2.1 Detection

Orgyia thyellina (Lepidoptera: Lymantriidae) was found to be established in Auckland's eastern suburbs on the 17 April 1996 (Hosking *et al*. 2003). WSTM is a native of eastern Asia (Japan, Korea and Taiwan), where it is occasionally a pest. However, it had potential to become a pest of plantation forestry, horticulture and native forests in New Zealand, and its discovery raised serious concern. This was exacerbated by awareness of another Lymantriidae, the Asian gypsy moth, which had been the subject of much public and government discussion due to incursions elsewhere in the world. New Zealand has no indigenous Lymantriidae.

There is little published literature on the WSTM eradication effort, but the programme was reviewed by some members of the eradication team (Hosking *et al*. 2003) and recently by the New Zealand Ombudsman (2007) in a report investigating public complaints of the spray programmes against exotic Lepidoptera. Much of the detail below is derived from these two publications.

In its home range, *O. thyellina* hosts include oak, elm, willow, larch, apple, cherry, pear, plum, chestnut, walnut and mulberry. It has three generations a year, the third generation females being flightless and laying diapausing eggs. Because the NZ infestation was detected in April as the final generation was emerging, there was time to develop strategies and consider response while the pest was overwintering (May–October). Diapausing eggs were not expected to hatch until spring (October–November). During the winter, researchers and the government debated options for control and/or eradication and held discussions with the public and health agencies (Hosking *et al*. 2003).

4.2.2 Response

The discovery of *O. thyellina* led the Ministry for Forestry (now incorporated into the Ministry of Agriculture and Forestry) to form an Initial Response Group consisting of representatives from science, government and the forestry sector to recommend a course of action. Government and scientific impact assessments determined that this species could be a serious problem in forestry and horticulture industries in New Zealand, if allowed to spread. After consideration, the Ministry recommended to the government that eradication be attempted using *Btk*, and the programme was named Operation Evergreen.

One problem inherent in assessing the severity of the threat posed by infestations of exotic insects is the lack of experts within the country on new invasive species. In the case of WSTM, the Ministry and scientists involved in the programme quickly established international links with experts in lymantriids around the world. These links were crucial in decision making, as the risk to New Zealand was difficult to determine due to the paucity of data in the literature (Hosking *et al.* 2003). The identification of the insect as *O. thyellina* was confirmed by Dr. Paul Schaefer of the USDA (Hosking *et al.* 2003).

Visual delimitation surveys, to establish the size of the population and its distribution, were undertaken early in the programme. Survey results established that the infestation covered an area of approximately $7\,km^2$. Early surveys were conducted by a search of easily accessible parks and reserves, as well as suburban gardens, with intensive searches at the outer areas of the infestation. According to Hosking *et al.* (2003), *O. thyellina* was thought to be mainly established in a $3\,km^2$ core zone in the suburbs of Auckland. Later surveys used caged live females to trap male moths. At this time, a pheromone development programme was conducted, which eventually led to the discovery and use of a synthetic pheromone in monitoring (see below).

The decision to eradicate was assisted by the low natural dispersal rate of the insect. The final generation females are flightless, which restricted spread. The major risk immediately after discovery of the pest was spread by human activities. As the infestation was in an urban environment, movements of gardening rubbish out of the area had a high likelihood of moving eggs. Therefore, a zone around the infestation was established, restrictions on moving green waste were imposed and a disposal site was established within this zone. In addition, *O. thyellina* lays eggs on inanimate objects, so field staff inspected household goods (e.g. furniture) that had been moved from the infested areas and carried out pre-movement inspections (Hosking *et al.* 2003). Of course, the risk of undetected movements still existed and was recognised.

After considering the costs and benefits of responses and risk of *O. thyellina* to New Zealand, the Initial Response group recommended attempting eradication to the Government, an approach that was formally accepted by the government on the 22 July 1996 (Hosking *et al.* 2003).

4.2.2.1 Eradication Programme

A project management team was established to oversee the eradication programme, Operation Evergreen, which included the operational aspects of eradication, research and public relations needs. Separately, the government set up a Science Panel to provide independent advice and a Community Advisory Group to represent community concerns.

The eradication strategy was based on the heavy use of Foray 48B (Nufarm, NZ), which contains *Btk* as the active agent. *Btk* has a long track record of use against other members of the Lymantriidae, including aerial use against gypsy moth, *L. dispar*, in the Northern Hemisphere (Reardon *et al.* 1994, Chapters 5 and 11). *Btk* is considered specific to Lepidoptera in the field, although toxicity to a few other insects has been found in the laboratory (Glare & O'Callaghan 2000). The aerial application against AGM invasions in the city of Vancouver was well publicized (Chapter 5; Ministry of Forests and Range 2007), which made *Btk* an obvious choice for urban spraying for WSTM. *Btk* was also available in the large quantities required for an eradication campaign; over 150,000 L were used in 6 months. In fact, without a non-chemical agent such as *Btk* with a long history of safe use, it is difficult to see how the eradication could have been attempted. *Btk* was already registered in New Zealand for use as ground spray on some food crops, but obviously not for control of WSTM.

Efficacy of *Btk* against *O. thyellina* was determined in New Zealand, early in the programme. The team also established that susceptibility dropped off after the 3rd instar, with exposure to a 5 L/ha equivalent of Foray 48B. This finding established that spray periods should not exceed 10 days between sprays in order to cover larvae from egg hatch to 3rd instar twice (Hosking *et al.* 2003). However, subsequent research indicated that persistence of *Btk* on foliage in the spray zone was longer than anticipated, which indicted that the retreatment period of less than 10 days was quite conservative (Hosking *et al.* 2003).

During Operation Evergreen, an area of up to 4,000 ha was treated in the eastern suburbs of Auckland. Over 30,000 households and an estimated 86,000 people were within the treatment zone (Ombudsman 2007). In total, 158,000 L of Foray 48B were used, in both aerial and ground sprays. A four engine DC6 was used for most of the aerial application, which was the mainstay of the eradication programme. The plane applied nine sprays, flying at a height of 60 m at a speed of approximately 200 knots, with a swath width of 120 m (Hosking *et al.* 2003). *Btk* was applied at an undiluted rate of 5 L of Foray 48B/ha (60 BIU/ha), and aiming for a droplet size of 50 microns. Spray applications were carried out at approximately weekly intervals between October and December 1996, although the schedule was weather dependent. In addition to the DC6, a Hughes 500C aircraft and a BK 117 helicopter were used, mainly to treat non-residential gullies, cliffs and bush area. The helicopter was also used to treat a 300 ha area that still had a heavy infestation of WSTM at the end of the programme. A total of 23 applications were made by the second aircraft and helicopter (Ombudsman 2007) at the same application rate. Application by helicopter continued until April 1997 (Hosking *et al.* 2003).

Over 500 properties with identified infestations of WSTM were also treated from the ground using mistblowers, at rates of a 1:9 dilution or undiluted Foray 48B. These treatments were carried out between October 1996 and April 1997, within the area that had been sprayed from the air (Hosking *et al.* 2003).

4.2.3 Monitoring

Monitoring of the WSTM population was essential to measure efficacy of the treatments. Colonies of WSTM were maintained in the laboratory at Rotorua (Forest Research Institute, now known as Scion), to supply larvae for bioassays and females for use in attraction traps. Live females were used extensively for most of the programme as no pheromone was available. An array of 350 traps, each containing a single live female, was maintained to monitor the population. The females were renewed every 3–5 days. In this programme and the PAM eradication programme (described in Section 4.3.), rearing of colonies was a major undertaking and was crucial to the success of the programme.

Monitoring of the spray deposition after aerial application was also conducted. Potted roses and pine seedlings were placed in the spray zone in various situations (shaded, open, etc.), then removed to quarantine facilities and seeded with 1st – 3rd instar larvae. Spray depositions were modelled using computer software and checked in the field (Hosking *et al.* 2003). Deposition data were collected after each spray run. Persistence of *Btk* was also monitored from environmental samples for 2 years after the spray programme (Gribben *et al.* 2002). There was only a slight reduction in numbers of *Btk* spores in the environment over the 2 years.

There was a formal requirement from the Government that the health of those exposed to the spray be monitored. Self reporting, reports from medical practitioners in the area, and a review of collected health data including birth outcomes, were all included. These results have been reviewed elsewhere (e.g. Aeraqua Medical Services 2001, Hosking *et al.* 2003, Ombudsman 2007). In brief, the formal monitoring indicated no significant outcomes attributable to the spraying (Aeraqua Medical Services 2001), but the Ombudsman's report on investigations into health impacts was less conclusive, quoting health studies finding that 375 people (out of 80,000 thought to be exposed to sprays) reported some medical symptoms including respiratory complaints (Ombudsman 2007). The Health Surveillance following Operation Evergreen (Aeraqua Medical Services 2001) did not find "adverse patterns" in a data comparison between medical practises inside and outside the spray zone. The extensive use of *Btk*-based products against gypsy moth in the Northern Hemisphere was often cited regarding the safety of these sprays.

4.2.4 Regulatory Issues

As has been mentioned above, there were a number of regulatory issues to be considered before and during the WSTM eradication programme. At the time, regulations were inadequate to cover the rapid response requirements of an eradication campaign (see Chapter 18). Use of *Btk* for eradication was covered by the Pesticide Act of 1979. The regulatory body under that Act, the Pesticide Board, granted an "experimental use permit (Limited sale)" for the WSTM eradication programme, a provision that was really designed for efficacy trials with new pesticides prior to full sales. As pointed out by the Ombudsman (2007), the application of 85,000 L under a limited sales permit, and several extensions to that permit, was probably not the original intention of the Act. Laws in New Zealand were changed as a result of the WSTM infestation; new laws to specifically cover biosecurity breaches were passed in 1997 (the BioSecurity Act 1996).

Other regulatory issues of note were the limiting of movement of vegetation and other risk items out of the infestation area and the operation of aircraft over urban areas. This latter issue involved the Civil Aviation Authority and also dictated the choice of aircraft. Single engine planes would no longer be allowed to fly low over urban and residential areas. Access to private land was also an issue for ground survey and spray teams (Hosking *et al.* 2003).

4.2.5 Other Research

A number of research efforts accompanied the discovery and eradication of WSTM, including feeding studies on the insect, susceptibility to other pathogens and development of a synthetic pheromone. Some were directly applicable to Operation Evergreen, while others were unrelated and were aimed at control measures if eradication failed.

At the commencement of the eradication programme, it was unclear from the literature which plants in New Zealand could act as hosts for WSTM. Larvae were thought to have a preference for Rosaceae. However, feeding trials in New Zealand have shown that the caterpillar will eat the native red beech (*Nothofagus fusca*), garden roses, fruit trees and many other hardwood trees, and larvae can also complete development on *Pinus radiata* and *Eucalyptus nitens* (M Kay & W Faulds personal communication). According to Hosking *et al.* (2003), over 40 species of plants could act as hosts, especially including species in the families Aceraceae, Fagaceae, Salicaceae, Betulaceae, Rosaceae and Fabaceae.

Trials were also conducted to test the susceptibility of WSTM to nucleopolyhedroviruses from prominent tussock moth pests in North America: *Orgyia pseudotsugata* and *Orgyia leucostigma*, as no pathogens were available from *O. thyellina* (Walsh *et al.* 1999). The *O. pseudotsugata* nucleopolyhedrovirus *Op*MNPV ("Virtuss^{TM}") and the *O. leucostigma* nucleopolyhedrovirus *Orle*SNPV were imported from Canada and tested for virulence against *O. thyellina*. Complete mortality by 15 days was achieved at doses of $\geq 10^5$ polyhedral inclusion bodies/cm^2 of artificial diet against 1–2nd instar larvae for both viruses (Walsh *et al.* 1999); however the viruses were never used. The rearing of WSTM for the programme also uncovered another pathogen, almost disastrously, as it caused a crash in the rearing colonies. This was a cypovirus (cytoplasmic polyhedrovirus), likely introduced through field-collected foliage. It is likely to be the same virus that affected the painted apple moth colony reared years later and was characterised by Vernon Ward (Markwick *et al.* 2005).

At the outset, the isolation of a pheromone was considered important to the success of the programme, although in the absence of a pheromone using live females became the mainstay of the monitoring and trapping efforts for the majority of the operation. Gries et al. (1999) were successful in determining the two compounds, (Z)-6-heneicosen-11-one (Z6-11-one) and (Z)-6-heneicosen-9-one, which, when combined in a 100:5 ratio, attracted male WSTM. Field tests were conducted in the field in Japan and in cages in New Zealand. Over the summer of 1997–1998, 45,000 commercially produced lures containing these compounds were used to monitor for remaining WSTM, although it is not reported if any WSTM were caught (Gries et al. 1999).

4.2.6 Public and Government Response

As stated by Hosking et al. (2003) "The ability to undertake the eradication operation was dependant on its acceptance by the affected community". During Operation Evergreen and subsequent eradication attempts, much effort and funding was used on communication with the wider community. On reading the review of Hosking et al. (2003), written by those directly involved in the programme, and the Ombudsman report (2007) based on complaints against the spray programmes against WSTM, PAM and AGM (detailed below), there are slightly different views on the success of community consultation, although both agree on a generally high level of public support for the programme. While the majority of residents were in favour of WSTM eradication, opposition to spraying increased in New Zealand following WSTM Operation Evergreen. The communication effort during Operation Evergreen was extensive and included free phone hotlines, public meetings and availability of environment (Ministry of Forests 1996) and health (Jenner Consultants Ltd 1996) impact assessments on the use of Foray 48B. Technical bulletins on the progress of the eradication effort, as well as on specific issues such pheromone trapping, choice of *Btk* as the agent and potential health issues were produced and distributed. However, the eradication programme was extensive, highly visible and prolonged, which led to an increase in opposition. Groups in New Zealand aligned to the Canadian *Btk* opposition groups emerged, such as the Society Targeting Overuse of Pesticides (STOP). Public health was the main concern in the community and remains the most controversial part of the three spray campaigns covered in this chapter. While some concern was expressed regarding the use of low flying aircraft, most concern was focused on the Foray 48B product. The lack of public disclosure of the ingredients in Foray 48B was a source of public concern for this and subsequent eradication campaigns.

The health impacts of the urticating hairs of the tussock moth received little attention, although workers in the rearing facility were affected to the point of withdrawing from the programme (Hosking et al. 2003).

At one point, the Operation Evergreen team released 1,000 monarch butterflies, a public event recognising that widespread application of *Btk* to the area reduced Lepidoptera numbers. In reality, the application mostly in a relatively small residential area did not threaten any well-established species and emigration into the zone would quickly re-establish butterfly populations (e.g. Butler et al. 1995).

4.2.7 Outcome

While the spray programme ceased in 1996, it was 2 years before it could be judged successful. An intensive pheromone trapping programme based on the newly developed pheromone failed to find any residual populations. In June 1998, with no individuals trapped for around 2 years, the WSTM was declared eradicated. The moth has not been seen in New Zealand since the end of Operation Evergreen.

There were many aspects of the programme that lead to a successful outcome. According to the leaders of the response team, the six key elements were: definition of infestation, prevention of spread, application of pesticide, intensive monitoring, strong science input and communication initiatives (Hosking *et al*. 2003).

4.3 Painted Apple Moth

After the successful eradication of the WSTM from Auckland, a second exotic lymantriid was found in 1999. The painted apple moth (PAM), *Teia anartoides* (formerly *Orgyia anartoides*), is a native of south-eastern Australia, from as far north as southern Queensland to Tasmania in the south. It is usually a minor problem as a pest of urban garden plants (Elliott *et al*. 1998) and occasionally orchards. It is, however, polyphagous and can feed on key horticultural and forestry species. In April 1999, PAM was found in the suburb of Glendene in Auckland. Insects of Australian origin are routinely found established in New Zealand, often blown over by prevailing winds. However, after the WSTM programme, the discovery of another lymantriid in Auckland, one which was considered polyphagous (Common 1990, Elliott *et al*. 1998), triggered serious concerns. As with third generation WSTM females, female PAM do not fly, which generally limits the spread. PAM can have five generations a year, increasing the risk of colonization compared to WSTM, which had three or less.

Much of the information on the response to PAM can be found at the Ministry of Agriculture and Forestry (MAF) website (MAF 2007). It is possible to track the progress of the response to PAM through the MAF media releases, albeit with a very positive view of the programme. The Ombudsman report (2007) also provides a summary of the programme, in terms of timing, decisions and statistics. Other than those documents, no review of PAM eradication campaign has been compiled.

4.3.1 Detection

PAM was discovered in three properties in an industrial area in the western Auckland suburb of Glendene on the 5 May 1999. All life cycle stages of the moth were found (eggs, caterpillars, pupae and adults), suggesting the insect had been established locally for some time. The MAF had the properties and surrounding areas treated with the organophosphate insecticide, chlorpyrifos and/or Decis (a synthetic pyrethroid insecticide) (MAF 2002a). A delimiting survey (visual ground search) was conducted within 2 weeks and resulted in several more properties in Glendene being treated. Another delimitation survey found no new populations and the threat was thought to be contained. However, by October 1999, PAM had been found both near the original infestation and a small population was also found in the suburb of Mt. Wellington, 15 km away. Because females don't fly, it was thought this was a separate infestation.

While ground spraying with chlorpyrifos and Decis was ongoing, the officials and advisors were considering whether PAM was a serious threat to New Zealand. A preliminary economic impact assessment suggested costs of more than $NZ 48M (present value in 2000) could be incurred in plantation forestry over 20 years if PAM became widely established (MAF 2000). A second estimate using more detailed information put the potential economic cost of PAM at between $NZ 58–356 M over 20 years (MAF 2002b), mainly as a result of production losses and treatment costs in forestry. The conservation estate (native areas) was not included in the impact assessment, but the Department of Conservation considered potential impacts in these areas to also be significant (MAF 2001a).

As with WSTM, no pheromone was available and for detection live females reared in laboratory colonies were used exclusively throughout the years that the PAM operation continued, a truly mammoth rearing task. Use of live females in traps started in December 2000, and eventually, over 600 traps covered the region from Glendene to Mt. Wellington. These live females were replaced approximately weekly. Between May and October 2001, trap catches reached almost 1,000 male moths, indicating spread of the populations, especially towards a large area of native bush on the edge of the city (Waitakere Ranges) (Ombudsman 2007). At the height of the programme, up to 1,880 traps were baited and serviced weekly, covering 62,000 ha (Anonymous 2007).

4.3.2 Response

The decision to attempt eradication was made very soon after the original discovery, but had to be revisited several times as new populations were found outside the treatment zone. In August 2000, Cabinet (made up of senior government Ministers) approved an eradication programme, using mainly ground sprays of the chemical insecticides and host plant removal, targeting preferred species. However, PAM remained and by October 2001, Cabinet had formed an advisory committee to make a recommendation on the preferred response.

The decision to use Foray 48B for eradication was enabled by evidence of high potency against PAM (M Kay *et al.* unpublished data quoted in the Environmental Impact Assessment, MAF 2003a). The success of the WSTM eradication programme and control of gypsy moth overseas using Foray 48B (both lymantriids like PAM), made the decision easier. It was likely that no agent other than *Btk* would have been acceptable to community groups. In fact, some members of the community advisory group, which was formed during the programme, requested that *Btk* be used for ground spraying instead of chlorpyrifos. Later there was opposition from some members of the community advisory group to aerial applications of *Btk*, partly on the basis of the more broad-scale nature of aerial application.

All PAM populations could not be reached by early ground spraying efforts, especially those in riparian strips and parks where high trees and terrain made ground application difficult. Therefore, a limited aerial approach was taken. Because of previous opposition to aerial spraying during the WSTM programme, the spray area was concentrated on 600 ha around the Whau River riparian strips and a large park-like cemetery (Ombudsman 2007). Targeted spraying commenced in January 2002, aiming for seven to eight applications at 3- to 4-week intervals, affecting about 800 properties. One application took around 7 hours to cover the entire aerial spray zone of around 60 ha (MAF 2002a). Spraying continued until early June 2002. By May 2002, MAF had reported that the spray programme was having an impact, with total traps of male moths down from a high of around 900/week before spraying commenced to only 73/week in April 2002 (MAF 2002c). However, a significant population still remained as winter reduced the activity of the moths. The treated area was gradually increased to 900 ha by September 2002. In total nine aerial sprays were completed in this period, mainly by helicopter and a fixed wing Air Tractors.

It was clear that targeted spraying was not containing PAM, so more extensive aerial applications began in October, 2002. The area was increased to 8,000 ha and eventually covered 12,000 ha, with ten aerial applications at approximately 21-day intervals using a Fokker Friendship, an Air Tractor aircraft and a helicopter. Ground spraying continued wherever new populations were located, including spraying a new discrete population in Hobsonville, Auckland, in December 2002.

By the completion of spraying in May 2003, up to 43,000 homes and 193,000 people had been exposed to Foray 48B in the treated areas (Ombudsman 2007). Spraying had occurred over a 104-week period, with some 40 "operations" completed in that time (Ombudsman 2007). The spraying was concluded officially in Auckland on 13 May 2004 (MAF 2004), when no moths had been captured for 4 months.

During the spray programme, restrictions were placed on the removal of garden waste from the infestation zone, to limit the spread. MAF offered free disposal of material. These restrictions were not lifted until March 2006 (MAF 2006).

The environmental impact assessment prepared for WSTM eradication using *Btk* was revised so that it was appropriate for the PAM eradication (MAF 2003a), including analysis of particular non-target impacts in the region (e.g. Glare & Hoare 2003). A Health Advisory Steering group was established to oversee a health monitoring programme. A Health Risk Assessment, which examined the potential impacts (Auckland District Health Board 2002) of the PAM programme, was completed.

4.3.3 Other Research

While eradication with *Btk* was the primary goal of the PAM operation, MAF continued to fund aligned research, both to provide tools and data to assist the eradication, and to examine other control and eradication options. Through the ground surveys, data were collected on the host plants (and inanimate surfaces) preferred by PAM (Stephens *et al.* 2007). PAM was found mostly on wattle species (*Acacia* spp.), which are native to Australia, but also on several New Zealand native species.

Another study looked for other potential pathogens and parasites of PAM. The overall goal was to determine if any parasites or microbial pathogens present in New Zealand or identified in Australia had potential as control agents to slow the spread of PAM. A team collected PAM in Australia and attempted to rear and identify all parasitoids. They managed to rear 12 species of egg, larval, larval/pupal and possibly pupal parasites, mostly Hymenoptera from the families Braconidae, Ichneumonidae and super-family Chalcidoidea (P Gerard personal communication). Another team examined the potential to use other insect pathogens for PAM control: a nucleopolyhedrovirus had been previously identified from PAM in Australia (Teakle 1973) and a sample was obtained and shown to be virulent against PAM, using the New Zealand rearing colonies (Markwick *et al.* 2008). Similarly, the tussock moth nucleopolyhedroviruses (*Op*MNPV "Virtuss^{TM}" and *Orle*SNPV) from Canada were effective in the laboratory against PAM larvae (Markwick *et al.* 2005). A cypovirus (CPV) was also isolated from the rearing colonies of PAM, as occurred with WSTM. This time the virus was characterized using polyhedron sequences, and a new cypovirus was identified (Markwick *et al.* 2005).

Only one active pathogen was identified from wild populations of PAM in New Zealand, the fungus *Beauveria bassiana*. Isolates of *B. bassiana* from PAM in New Zealand were isolated, genotyped and bioassayed, showing potential as a control agent (Markwick *et al.* 2005).

A new approach was taken in the final stages of the eradication programme, the release of sterile male moths. The sterile male programme was initiated to determine the flight potential of male moths, as trap data using live females would often turn up single males from outlier traps, making it difficult to determine the population spread. Other useful data such as temperature limits for flight were determined (Suckling *et al.* 2005). Irradiation of females for trapping (to reduce risk of using reproductively active females) was also investigated (Suckling *et al.* 2002, 2006). The Sterile Insect Technique (SIT) was aimed at releasing sterile males to disrupt the breeding cycle of PAM. In 2003, SIT was used in conjunction with spraying to complement the final stages of eradication (Suckling *et al.* 2005, Wee *et al.* 2005).

The search for a synthetic pheromone was conducted by two competing teams. The two teams published preliminary results in the same issue of the Journal of Chemical Ecology (Gries *et al.* 2005, El-Sayed *et al.* 2005), but no synthetic pheromone was ever available for use in the field.

4.3.4 Public and Government Response and Regulation

There were several steps in the decision to eradicate PAM, such as scientific advice on eradication feasibility and estimates of potential impact of this exotic insect. However, overall, eradications that require millions of dollars are essentially political decisions based on the best advice available. The Government made the decision to support aerial spraying operations after over a year of ground operations failed to contain the PAM population.

The eradication operation was the subject of a number of legislations. Subsequent to the WSTM Operation Evergreen, the Biosecurity Act 1996 was passed into law. The Act provides MAF with the power to carry out aerial spraying and meant MAF could declare "controlled areas", for example to restrict garden waste movement. Other legislation, such as the Resource Management Act that governed discharge of substances into the environment (including application of *Btk*), was bypassed by government decree, which provided exemption. The Civil Aviation Act covered approvals needed for low flying aircraft involved in the spray programme. One district council found that the proposed use of helicopters for spraying would breach the bylaws about noise control and would require applications for resource consent (i.e. legislation would need to be overridden), which could have delayed the control process by many months. There were emergency override clauses in the Biosecurity Act (essentially allowing the government to bypass the law), but instead MAF had advice from government lawyers that the council plan was unlawful and spraying continued. Foray 48B was registered under the Pesticide Act for the WSTM programme and the registration needed to be modified to include PAM. Several other acts (e.g. the Reserves Act 1977 and the Conservation Act 1987) required Ministerial approval for use of biological agents.

Health issues had been raised during the WSTM operation and there were further concerns raised during the PAM eradication. A Health Risk Assessment of potential impacts was made available during the programme (Auckland District Health Board 2002). According to the Government media release, MAF "was aware the medical conditions of some people could be exacerbated by some of the components in Foray 48B" so independent health support services were set up (MAF 2004). Health studies were also conducted to monitor effects of the spray programme (Aeraqua Medical Services 2005a, b). Much debate continued about health effects from the *Btk* sprays and the Ombudsman study (2007) was created in response to continued public complaints about the spray programme and perceived health effects.

A PAM Community Advisory Group was set up to provide feedback about community concerns to the Ministry. It first met in September 2001, but became a focus for opposition to the spray programme. This caused a disconnection between the technical advice and advice from the community advisory group, with the technical group backing the spray programme and the community group largely opposing spraying. Interestingly, MAF-funded surveys of the affected homes found 89% of people were neutral, agreed or strongly agreed with the eradication programme (MAF 2001b). MAF made efforts to reduce the impact of such a massive and highly visible eradication campaign, such as spraying on weekends when possible to minimize disruption to school children (MAF 2006).

The effect of the uricating hairs from PAM had a similar effect on laboratory workers as WSTM. One technician's reaction was so severe she had to be removed from the programme. Some ground spraying workers also suffered from the effects of the hairs (MAF 2003b).

4.3.5 Outcome

After the final spray application in May 2003, it was not until March 2006 that the government officially declared the eradication successful (MAF 2006, Stephens *et al.* 2007). The criterion used for declaring successful eradication was no moths found for 2 years. MAF reported that Auckland suburbs had been aerially treated 69 times between 1999 and 2003 (MAF 2006). The eradication campaign cost $NZ 62.4 M.

However, the declaration of eradication became complicated by the capture of single moths in other parts of Auckland in 2005 (MAF 2005). Investigation using molecular typing and stable isotope approaches determined these moths were not part of the original infestation, but were new arrivals. Isotope analysis indicated the moth trapped at Otahuhu in 2005 had pupated in a climate significantly more arid that Auckland. This indicated the moth was a recent immigrant, rather than having originated in Auckland.

Kean & Suckling (2005) used a modelling approach to determine what interval was necessary to declare that PAM had been successfully eradicated. Based on trap data, they estimated that after around 9 months of negative trap data the probability that the core area had a resident population was very slight. Less concentrated trapping at the outer areas increased the time before eradication could confidently be declared to around 2 years. The authors also included the effect of finding a single moth in Mt. Eden in 2005. This resulted in extra traps being deployed, which in turn decreased the time required before eradication could confidently be declared.

Much had been learnt from the PAM eradication campaign and associated research. The use of specific regions of mtDNA (Ball & Armstrong 2006) and stable isotope testing (Iso-trace New Zealand Limited, Dunedin) to analyse the origin of moths was particularly useful. The SIT programme was also a success in assisting with the final phase of eradication and in supplying vital data on the flight range of male moths. In retrospect, wider areas of application were needed than initially used, but the public pressure against the use of aerial spraying meant that a limited application was considered the best first option.

4.4 Asian Gypsy Moth

Much of the response to WSTM and PAM as lymantriid pests in New Zealand had stemmed from concern about the Asian strain of gypsy moth, *Lymantria dispar* (AGM), reaching New Zealand. Therefore, it was not unexpected but somewhat ironic that an adult gypsy moth should be found in New Zealand during the response to PAM. The focus on the gypsy moth as a threat dated back to the early 1990s, which meant that resources were in place to respond to these other moth incursions more rapidly than if they had been other types of insects.

The threat of AGM to New Zealand had been reviewed in the 1990s (Cowley *et al.* 1993, Horgan 1994) and assessments concluded that almost inevitably, gypsy moth would reach New Zealand and that it would be a devastating pest. As a major outbreak pest of native and plantation forests and horticultural tree crops throughout the Northern Hemisphere, gypsy moth was not considered a threat (i.e. would not survive the transportation to New Zealand) until the "Asian" strain emerged, probably from North Asia/Far East with characteristics making survival across the equator more likely. The AGM strain was also considered more devastating than the North America/European gypsy moth strain, having a wider host range, females that fly and was generally larger with a higher premature eclosion. From 1993, viable egg masses and other life stages of AGM were routinely found on ships and cargo in New Zealand ports (Kay *et al.* 2002).

A pheromone-based nationwide trapping programme was initiated in 1993, concentrating on ports and surroundings, with over 1,000 traps at one time. Somewhat surprisingly, the first detection of a male moth was not from one of the international points of entry, but from one of only seven traps maintained in Hamilton, a medium-sized inland city in the North Island.

4.4.1 Detection

In March 2003, a single male gypsy moth was discovered in a pheromone trap in Hamilton. Detection of a single male moth led to much debate among the technical and scientific panel convened by MAF to determine a recommended response. There was no evidence from ground surveys of an infestation, although Hamilton has one of the largest populations of oak trees (*Quercus* spp.; a favoured host) in New Zealand. Many of these trees are tall, making location of egg masses difficult. The single specimen was sent to Dr Robert Hoare (Landcare Research, New Zealand), who tentatively identified it as *Lymantria dispar* subspecies *praeterea* based on morphology (Bain & Ross 2003). *L. d. praeterea* has recently been reclassified as *L. umbrosa* (Pogue & Schaefer 2007), a species found naturally in Siberia and northern Japan. This species forms part of a *L. dispar* complex. Molecular analysis using a variable region of the 16S was completed, but lack of specific species representatives within the complex for comparison limited the study. The molecular analysis did indicate that the lone gypsy moth in Hamilton was similar to samples obtained from Honshu, Japan (outside of the known range for *L. umbrosa*), and East Russia (within the range), but dissimilar to those from southwest Hokkaido (at the limit of the range). However, these markers may not have been discriminatory enough to distinguish *L. umbrosa* from other sister species (K Armstrong personal communication). Given the lack of specific identification, we continue the usage of the name AGM for this find.

The advisory group debated the likelihood of a single moth representing a local population. Previous economic impact assessments had placed the cost to New Zealand of an AGM colonization at between $NZ 5M and 400 M (Ombudsman 2007), a large range. Given the risk, the technical advisory group recommended eradication using aerial spraying, as no population was found to target on the ground, but there was great likelihood that a local population existed. Given flight times in Japan/Russia for male gypsy moth, the chances of an adult AGM being transported to New Zealand was very low.

4.4.2 Response

The response to the Hamilton find of an AGM male was similar to responses to WSTM and PAM, except with more rapid decision-making and confined to an aerial and trapping programme as no ground infestation was ever located. The difficulty in planning a response based on only a single trapped male with no ground population, was apparent as defining the area to target was dependent on information on flight patterns and behaviour of AGM overseas. Balancing the risk of allowing AGM to establish against spraying another urban population based on little tangible threat was problematic. The technical advisory group recommended an eradication programme, partly because the risk was high there was a population undetected and that every generation the response was delayed, eradication would become that much more difficult.

The actual decision on spraying was taken by the Government, after weighing the potential risks against the public attitude to another spraying. In September 2003, Cabinet approved an aerial spray programme for Hamilton. Between October and November 2003, eight aerial applications were conducted over specific parts of Hamilton, using a total of 65,300 L of Foray 48B (5–7 L/ha) (Ross 2005). The area of application was around 1,253 ha, and some 30,600 people were estimated to be present in the application zone (Ombudsman 2007). The spraying occurred over 52 days at around 7-day intervals.

Because of the previous WSTM and concurrent PAM eradication campaigns using the same agent, Foray 48B, much of the background information on the spray programme was already available, such as a general Environmental Impact Assessment and Health Assessments on Foray 48B. Additions to the Environmental Impact Assessment were completed with specific focus on the Hamilton spray zone (e.g. Glare & Hoare 2003) and a newly revised Health Risk Assessment was available in 2003 (Auckland Regional Public Health Service 2003). In previous eradications, maintaining colonies of the pest were crucial to success, for use in lures and for bioassays. It would have been a difficult decision to maintain a laboratory colony of AGM, given that no ground population was ever found and there is always the risk of escape. Approval to import AGM to start a colony was unlikely. As Foray 48B was extensively used overseas for AGM control based on abundant research, bioassays were unnecessary. As synthetic lures for the gypsy moth pheromone, Disparlure, were available, female AGM were not required.

The spray applications were conducted under the same regulations as for PAM, including use of the Biosecurity Act, which eliminated the need for any consent process under the Resource Management Act.

4.4.3 Other Research

Because AGM had been considered a likely invader of NZ since the early 1990s (e.g. Cowley *et al*. 1993, Horgan 1994), considerable research and thought had gone into potential responses. As stated above, the discovery of a lone AGM male was due to the extensive trapping programme based on the synthetic lure, Disparlure. Extensive studies examining New Zealand plants as potential hosts for AGM were carried out (Kay *et al*. 2002). Barlow *et al*. (2000) modeled the likely establishment and spread of AGM in New Zealand and the effects of control agents such as viral pathogens. Molecular identification systems for AGM were developed and tested (Armstrong 2000, Armstrong & Cameron 2000, Armstrong *et al*. 2003).

Several control agents were considered and in some cases, preliminary work was conducted. *Btk* was always considered the most likely agent for control or eradication, but insect growth regulators, mating disruption using pheromones, chemical pesticides and microbial pathogens other than *Btk* were all considered (Glare *et al*. 1998, 2003). Of the known pathogens of gypsy moth, two were considered most likely potential agents for use in control or even eradication. The nucleopolyhedrovirus of *L. dispar*, *Ld*MNPV, is produced in several countries and used as a spray-applied pesticide against GM. The virus was imported into containment in New Zealand and tested against a range of non-target insects to provide environmental safety data for any future use (Glare *et al*. 1995). Efforts were made to determine the availability and potential sources of *Ld*MNPV from the USA and the Czech Republic (Glare *et al*. 2003). Similarly, the fungus *Entomophaga maimaiga*, which had proved so effective in the USA against gypsy moth (Chapter 11), was imported into containment but was never used (TR Glare unpublished data).

4.4.4 Outcome

The eradication operation was completed by November 2003 and cost approximately $NZ 7M. No subsequent AGM were trapped and a ground population was never found. Some Hamilton residents applied to the High Court of New Zealand to stop the spraying before it commenced, but were unsuccessful. According to the Ombudsman (2007) analysis, Hamilton residents within the spray zone received 37% more Foray 48B per ha than those exposed to the PAM eradication programme. MAF declared eradication of the gypsy moth from Hamilton on 26 May 2005, after enough time for two generations to have passed with no further trap catches or other stages found (Ross 2005).

4.5 Conclusions

With the eradication of two established and one possibly established lymantriids in two cities, New Zealand has shown that urban eradication using microbial pathogens is possible. It was remarkable that aerial application (also with ground-based efforts) was able to be conducted over these cities. There were many lessons learnt from the three eradication programmes. The logistics involved in organization of all aspects of the programmes were substantial. From acquiring the hundreds of thousands of litres of Foray 48B, organizing planes, liaising with various relevant authorities, to community communications and answering the thousands of queries, the teams involved were effective. There was a learning process for the relevant government agencies about how to best integrate opinions from technical, community and government perspectives, which often have very different inputs.

In all cases the most contentious area of these eradication programmes was human health aspects of spraying large populations with a pesticide, even if biological, from the air. The debate and investigation into possible health impacts is ongoing, as demonstrated by the public complaints that led to the Ombudsman's (2007) investigation (published numerous years after the actual spraying occurred). The Ombudsman's report concluded that "insufficient attention was paid to the impacts of [the spray] operations" (summary, page 8). As the Ombudsman points out, continued support for such programmes from the public will be critical for any future programmes to be conducted. The recognition that sprays will cause some medical issues, such as respiratory complaints or allergic reactions, seems to be understood, but was not, in the Ombudsman view, sufficiently publicised. While most of the community recognized the need for aerial spraying, there appeared to be a decline in support over time. There was also a lack of trust of the government agencies and science teams, much of it around the lack of disclosure of all of the ingredients of Foray 48B (which is a trade secret). Disclosure to officials was not deemed enough for some of the New Zealand public. There is an obvious dilemma around the lack of disclosure of formulations to be used in large scale eradication programmes.

Overall, the Ministry of Health reported "no adverse health patterns found, when considered at the population level" (Auckland District Health Board 2002). There remain a limited number of people implacably opposed to spray programmes, especially aerial spraying over high density urban environments.

References

Aeraqua Medical Services (2001) Health surveillance following Operation Ever Green: a programme to eradicate the white-spotted tussock moth from the eastern suburbs of Auckland. Report to the Ministry of Agriculture and Forestry. 60pp + Appendices

Aeraqua Medical Services (2005a) A study of presentations of householder concerns to the painted apple moth (PAM) health service and Auckland summer symptom survey. Report to AgriQuality Ltd. 131pp

Aeraqua Medical Services (2005b) A comparison of presentations of householder concerns to the painted apple moth (PAM) and Asian gypsy moth (AGM) health services. Report to AgriQuality Ltd. 68pp

Anonymous (1893) White grub culture. NZ Farmer, Bee and Poultry Journal, February, 1893:45

Anonymous (2007) Painted Apple Moth – Auckland New Zealand May 1999. Submissions received to date from Parties and organizations for the Convention on Biological Diversity in-depth review on IAS for consideration at COP 9. 19pp http://www.cbd.int/doc/submissions/ias/ias-nz-moth-2007-en.pdf [accessed March 2008]

Armstrong KF (2000) DNA diagnostic procedures for the identification of selected species of *Lymantria* and *Orgyia* moths from intercepted egg masses. MAF Operational Research Final Report MBS301. 23pp

Armstrong KF, Cameron CM (2000) Molecular kit for species identification: Lymantriidae. MAF Operational Research MBS301. 32pp

Armstrong KF, McHugh P, Chinn W, Frampton ER, Walsh PJ (2003) Tussock moth species arriving on imported used vehicles determined by DNA analysis. NZ Plant Protection 56: 16–20

Auckland District Health Board (2002) Health risk assessment of the 2002 aerial spray eradication programme for the painted apple moth in some western suburbs of Auckland. Report to the Ministry of Agriculture and Forestry, 66pp + Appendices

Auckland Regional Public Health Service (2003) Human health considerations in the use of *Btk*-based insecticide Foray 48B for Asian gypsy moth in Hamilton. Summary report prepared for the Ministry of Health, Ministry of Agriculture and Forestry, and Waikato DHB Public Health Unit. October 2003. 45pp

Bain J, Ross M (2003) Asian gypsy moth in Hamilton. Forest Health News 128, April 2003. ISSN 1175–9755

Ball SL, Armstrong KF (2006) DNA barcodes for insect pest identification: a test case with tussock moths (Lepidoptera: Lymantriidae) Can J For Res 36:337–350

Barlow ND, Caldwell NP, Kean JM, Barron MC (2000) Modelling the use of NPV for the biological control of Asian gypsy moth *Lymantria dispar* invading New Zealand. Agric For Entomol 2:173–184

Butler L, Zivkovich C, Sample BE (1995) Richness and abundance of arthropods in the oak canopy of West Virginia's Eastern Ridge and Valley Section during a study of impact of *Bacillus thuringiensis* with emphasis on macrolepidoptera larvae. Bull Agric & Forestry Experiment Station, West Virginia University. No. 711. 19pp

Cameron PJ, Hill RL, Bain J, Thomas WP (eds) (1989) A review of biological control of invertebrate pests and weeds in New Zealand 1874 to 1987 Tech. Commun Commonw Inst Biol Control 10. CAB International, Wallingford, UK

Common IFB (1990) Moths of Australia. Melbourne University Press, Melbourne, Australia

Cowley J, Bain J, Walsh P, Harete DS, Barker RT, Hill CF, Whyte CF, Barber CJ (1993) Pest risk assessment for Asian gypsy moth, *Lymantria dispar* L. (Lepidoptera: Lymantriidae). MAF internal document. 29pp

Elliott HJ, Ohmart CP, Wylie FR (1998) Insect pests of Australian forests: ecology and management, Reed International, Australia

El-Sayed AM, Gibb AR, Suckling DM, Bunn B, Fielder S, Comeskey D, Manning LA, Foster SP, Morris BD, Ando T, Mori K (2005) Identification of sex pheromone components of the painted apple moth: a tussock moth with a thermally labile pheromone component. J Chem Ecol 31(3): 621–646

Glare TR, Hoare RJB (2003) Appendix to the painted apple moth environmental impact assessment. Eradication of gypsy moth from Hamilton. Report for MAF. 11pp

Glare TR, O'Callaghan M (2000) *Bacillus thuringiensis*, biology, ecology and safety. John Wiley and Sons, Chichester, UK. 350pp

Glare TR, Barlow ND, Walsh PJ (1998) Possible agents for use in New Zealand for the eradication or control of gypsy moth, *Lymantria dispar*. Proc 51th NZ Plant Prot Conf. pp 224–229

Glare TR, Newby EM, Nelson TL (1995) Safety testing of a nuclear polyhedrosis virus for use against gypsy moth, *Lymantria dispar*, in New Zealand. Proc 48th NZ Plant Prot Conf. pp 264–269

Glare TR, Walsh PJ, Kay M, Barlow ND (2003) Strategies for the eradication or control of gypsy moth in New Zealand. Report for the Forest Health Research Collaborative of New Zealand. 178pp

Gribben JR, Lewis GD, Wigley PJ, Broadwell AH (2002) Environmental persistence and growth dynamics of the *Bacillus thuringiensis* Foray 48B biopesticide. In: Akhurst RJ, Beard CE, Hughes PA (eds) Biotechnology of *Bacillus thuringiensis* and its environmental impact. Proc 4th Pacific Rim Conf, CSIRO, Canberra. pp 200–205

Gries R, Khaskin G, Clearwater J, Hasman D, Schaefer PW, Khaskin E, Miroshnychenko O, Hosking G, Gries F (2005) (Z, Z)-6,9-Heneicosadien-11-One: major sex pheromone component of painted apple moth, *Teia anartoides*. J Chem Ecol 31: 603–620

Gries G, Clearwater J, Gries R, Khaskin G, King S, Schaefer P (1999) Synergistic sex pheromone components of whitespotted tussock moth, *Orgyia thyellina*. J Chem Ecol 25: 1091–1104

Horgan G (1994) Economic impact of Asian gypsy moth. Report prepared for NZ Min of Forestry, August 1994. 20pp

Hosking GJ, Clearwater J, Handiside J, Kay M, Ray J, Simmons, N (2003) Tussock moth eradication: a success story from New Zealand. Int J Pest Mgt 49:17–24

Jenner Consultants Ltd (1996) Health risk assessment of *Btk* (*Bacillus thuringiensis* var. *kurstaki*) spraying in Auckland's eastern suburbs to eradicate white spotted tussock moth. (*Orgyia thyellina*). Report to the Ministry of Health and the Ministry of Forestry, commissioned by the Northern Regional Health Authority, North Health, Jenner Consultants, 4 September 1996

Kay M, Matsuki M, Serin J, Scott JK (2002) A risk assessment of the Asian Gypsy moth to key elements of the New Zealand flora. Forest Health Collaborative Report, NZ. 32pp

Kean JM, Suckling DM (2005) Estimating the probability of eradication of painted apple moth from Auckland. NZ Plant Prot 58:7–11

MAF (2000) Potential economic impact on New Zealand of the painted apple moth, Ministry of Agriculture and Forestry, New Zealand, July 2000. 8pp

MAF (2001a) Targeted aerial spraying to go ahead in West Auckland. Ministry of Agriculture and Forestry media release New Zealand, 23 October 2001

MAF (2001b) Majority of Residents Support Targeted Aerial Spraying. Ministry of Agriculture and Forestry New Zealand media release, 18 December 2001

MAF (2002a) Ministry of Agriculture and Forestry Response to Meriel Watts article "MAF bungles the biosecurity in West Auckland". Ministry of Agriculture and Forestry, New Zealand media release, 14 January 2002

MAF (2002b) Painted apple moth: reassessment of potential economic impacts. Report for the Ministry of Agriculture and Forestry New Zealand, 7 May 2002, 18pp

MAF (2002c) Faster aerial spray operation planned. Ministry of Agriculture and Forestry New Zealand media release, 6 May 2002

MAF (2003a) Environmental Impact Assessment of Aerial Spraying *Btk* in NZ for painted apple moth. February 2003. http://www.biosecurity.govt.nz/pests-diseases/forests/painted-apple-moth/environmental-impact.html

MAF (2003b) Technician Removed from Painted Apple Moth Rearing Project after Reaction to Moth Hairs. Ministry of Agriculture and Forestry New Zealand media release, 3 February 2003

MAF (2004) Painted Apple Moth spray programme ends, Ministry of Agriculture and Forestry New Zealand media statement, 13 May 2004

MAF (2005) Testing confirms Painted Apple Moth find a new arrival, Ministry of Agriculture and Forestry New Zealand media statement, 2 June 2005

MAF (2006) Two Auckland Pest Moth Populations Declared Eradicated: Ministry of Agriculture and Forestry New Zealand media release. http://www.maf.govt.nz/mafnet/press/200306pam.htm [Cited May 2006]

MAF (2007) Ministry of Agriculture and Forestry New Zealand website. http://www.biosecurity. govt.nz/pest-and-disease-response/pests-and-diseases-watchlist/painted-apple-moth [accessed January 2007]

Markwick N, Ward V, Kay N, Glare T (2005) Microbial control of painted apple moth: the virulence and safety of *Oran*NPV: final Report for Ministry of Agriculture and Forestry New Zealand. PAM/602/2002. 35pp

Markwick NP, Glare TR, Hauxwell C, Li Z, Poulton J, Ward JM, Young VL, Ward VK (2008) The infectivity and host-range of *Orgyia anartoides* nucleopolyhedrovirus. Biol Control (submitted)

Ministry of Forests and Range (2007) Gypsy moth in British Columbia, Government of British Columbia. http://www.for.gov.bc.ca/hfp/gypsymoth/index.htm [accessed January 2007]

Ombudsman (2007) Report of the opinion of Ombudsman Mel Smith on complaints arising from aerial spraying of the biological insecticide Foray 48B on populations of parts of Auckland and Hamilton to destroy incursions of painted apple moths and Asian gypsy moths respectively during 2002–2004. Office of the Ombudsman, Wellington, New Zealand. 108pp

Pogue MG, Schaefer PW (2007) A review of selected species of *Lymantria* (Hübner) [1819] (Lepidoptera: Noctuidae: Lymantriinae) from subtropical and temperate regions of Asia, including three new species, some potentially invasive to North America. USDA Forest Service FHTET-2006-07

Reardon RC, Dubois N, McLane W (1994) *Bacillus thuringiensis* for managing gypsy moth: a review. USDA, Tech. Trans. FHM-NC-01-94. 32pp

Ross, MG (2005). Response to a gypsy moth incursion within New Zealand. A paper presented at the IUFRO conference, Hanmer, 2004, and updated July 2005. 10pp http://www.biosecurity.govt.nz/files/pests/gypsy-moth/residents/response-gm-incursion.pdf

Stephens AEA, Suckling DM, Burnip GM, Richmond J, Flynn A (2007) Field records of painted apple moth (*Teia anartoides* Walker: Lepidoptera: Lymantriidae) on plants and inanimate objects in Auckland, New Zealand. Aust J Entomol 46: 152–159

Suckling DM, Charles J, Allan D, Chaggan A, Barrington A, Burnip GM, El-Sayed AM (2005) Performance of irradiated *Teia anartoides* (Lepidoptera: Lymantriidae) in urban Auckland, New Zealand. J Econ Entomol 98:1531–1538

Suckling DM, Hackett JK, Barrington AM, Daly JM (2002) Sterilisation of painted apple moth *Teia anartoides* (Lepidoptera: Lymantriidae) by irradiation. NZ Plant Prot 55:7–11

Suckling DM, Hackett JK, Chhagan A, Barrington A, El-Sayed AM (2006) Effect of irradiation on female painted apple moth *Teia anartoides* (Lepidoptera: Lymantriidae) sterility and attractiveness to males. J Appl Entomol 130:167–170

Teakle RE (1973) A nuclear-polyhedrosis virus of the painted apple moth (*Orgyia anartoides* (Walker)). Queensland J Agric Anim Sci 30 :179–190

Walsh PJ, Glare TR, Nelson TL, Sadler TJ, Ward VK (1999) Virulence of nucleopolyhedroviruses from *Orgyia pseudotsugata* and *O. leucostigma* (Lep., Lymantriidae) for early instars of the white-spotted tussock moth, *Orgyia thyellina*. J Appl Entomol 123:375–379

Wee SL, Suckling DM, Burnip GM, Hackett J, Barrington A, Pedley R (2005) Effects of substerilizing doses of gamma radiation on adult longevity and level of inherited sterility in painted apple moth *Teia anartoides* (Lepidoptera: Lymantriidae). J Econ Entomol 98:732–738

Chapter 5
North American Eradications of Asian and European Gypsy Moth

Ann E. Hajek and Patrick C. Tobin

Abstract Although European gypsy moth (*Lymantria dispar dispar*) is established in the northeastern and northern midwestern parts of North America, members of the three subspecies of gypsy moth are constantly being introduced into new locations. Between 1980 and 2007, multiple eradication efforts targeting gypsy moth populations were conducted in 24 states in the US. In more recent years, eradication efforts have also slowed the westward and southern spread of gypsy moth. Of particular concern are introductions of the Asian and Japanese gypsy moths. The former has a broader host range than *L. dispar dispar*, and in both strains females are capable of flight. Consequently, the threshold for initiating an eradication effort against Asian and Japanese gypsy moth populations is much lower than it is for European gypsy moth. In some cases, the detection of as few as one Asian gypsy moth could trigger an eradication effort. A critical component that enables eradication projects against gypsy moth to be successful is the availability of a sensitive monitoring tool that can detect very low density, newly-establishing gypsy moth populations. The primary method for eradication of gypsy moth today is use of the HD-1 strain of *Bacillus thuringiensis kurstaki* (*Btk*). For eradication programs, aerial applications of *Btk* are generally applied 2–3 times when early instars are present.

5.1 Introduction

Lymantria dispar, commonly known as the gypsy moth, is native from Europe and North Africa across Asia to Japan (Pogue & Schaefer 2007). The taxonomy of gypsy moth has been very contentious in the past but it was always generally agreed that the gypsy moth in Europe (*Lymantria dispar dispar*; EGM) is a separate entity from gypsy moths in Asia. EGM is the gypsy moth subspecies that was introduced (from France) to northeastern North America in 1869 (Riley & Vasey 1870, Forbush & Fernald 1896). This polyphagous tree defoliator is capable of outbreaks that have

---

A.E. Hajek
Department of Entomology, Comstock Hall, Cornell University, Ithaca, New York 14853-2601, USA
e-mail: aeh4@cornell.edu

periodically decimated the hardwood forests of the northeast (Elkinton & Liebhold 1990), while also spreading to the south and west (6–18 km/year; Tobin *et al.* 2007). It is generally assumed that one of the reasons that EGM has spread so slowly is because females are flightless (Sharov 2004). Analyses of the historical spread of EGM have suggested that in the absence of anthropogenic movement of life stages, natural spread (via ballooning early instars) could be as slow as 2.5 km/year (Liebhold *et al.* 1992).

Two subspecies of *Lymantria dispar* occur in Asia: *L. dispar asiatica* is known as Asian gypsy moth (AGM) and is distributed mostly east of the Ural Mountains and in China and Korea, while *L. dispar japonica* (Japanese gypsy moth) is distributed on the main islands of Japan, although the distribution on Hokkaido is limited (Pogue & Schaefer 2007). It is presumed that both of the Asian subspecies, if introduced to North America, would be more problematic than EGM. Although all of the *L. dispar* subspecies are polyphagous, studies have shown that *L. dispar asiatica* has a wider host plant range, which includes more conifers than EGM (Baranchikov 1989, Baranchikov & Montgomery 1994). All three subspecies are known for their abilities to increase to outbreak densities. Eggs of Russian gypsy moths (= AGM) required less exposure to cold before hatch compared with EGM (Keena 1996). However, in contrast with EGM, the females of both subspecies of Asian gypsy moth are also capable of either sustained level or ascending flight (Pogue & Schaefer 2007), which would likely facilitate their spread. There is evidence suggesting that nocturnal mass migrations of variable lengths occur in *L. dispar asiatica*. As an additional form of dispersal, newly hatched AGM caterpillars balloon on the wind, as do EGM neonates (Zlotina *et al.* 1999).

European and Asian gypsy moths have been flagged worldwide as species that should be prevented from entering new areas, largely due to the long history of negative ecological and socioeconomic impacts caused by EGM in North America and the history of *L. dispar* outbreaks in areas where these subspecies are native (Campbell & Sloan 1977, Doane & McManus 1981, Leuschner *et al.* 1996, Redman & Scriber 2000). Gypsy moth life stages, predominantly egg masses, are transported relatively frequently. When standard pathways can be identified, analyses have shown that gypsy moth introductions are more frequent in years following outbreaks in source areas (Myers & Rothman 1995). Programs to detect *L. dispar* in areas where it does not occur are constantly under way worldwide, most notably in areas that serve as primary ports-of-entry for international commerce and shipping routes. Also, several ports-of-egress, especially in Asia where AGM is native, are monitored through international collaborative programs to pre-emptively minimize the arrival of gypsy moth life stages into North America.

When populations of EGM are detected through monitoring programs in areas of North America where they are not established, one management response is to increase pheromone-baited trap density in the following year to better determine the extent of the infestation (= delimitation trapping). However, if male trap catch is high enough, or if other life stages such as egg masses are detected in subsequent surveys, eradication efforts are initiated. In the case of AGM and hybrids between AGM and EGM, the threshold for male moth trap catch that triggers eradication efforts is lower than the threshold for EGM. Thus, efforts to eradicate incipient populations of AGM are much more aggressive than those against EGM. AGM and AGM hybrids have been detected in western North America, such as in British Columbia, Washington, Oregon, California, Idaho and Texas, and also on the East Coast, at ports in North and South Carolina (Zlotina *et al.* 1999, Johnson *et al.* 2007). In addition, eradications against highly localized EGM populations are constantly being conducted in areas of North America along the leading edge of the population front where gypsy moth is not yet established. Today, regardless of whether AGM or EGM is the target for an eradication program, bacterial and viral pathogens of gypsy moth are the mortality agents used most frequently. Here, we will provide background information on the use of eradication as a management strategy against the gypsy moth, the biopesticides that are used in such endeavors, and then we describe a diversity of eradication programs that have been undertaken in North America.

5.1.1 General Requirements for Successful Eradication Campaigns

Programs to eradicate an unwanted species can be very challenging and costly to implement, and are not always successful (Dahlsten *et al.* 1989, Myers *et al.* 1998, 2000). Although some native species have been targeted for eradication, it is a tactic that is more commonly used against non-native (invasive) species. Myers *et al.* (2000) identified conditions necessary for a successful eradication campaign. First, eradication programs generally require more resources than standard pest management programs; thus, for success, available resources must be sufficient to see the eradication program through. Second, invasive species may be present on private land as well as public, so it is critically important that government agencies responsible for the eradication program have the authority to apply treatments where and when they are necessary. This can be especially problematic with invasive species that often occur in urban and suburban areas where they have been transported by humans. Efficient control measures and strategies must also be available to prevent reinvasion and/or reduce propagule pressure.

A critical element in successful eradication programs is the availability of a monitoring tool that is effective at detecting low density populations. Most founding populations of non-native species arrive at such low densities that a majority of (but not all) new invasions fail to establish (Williamson & Fitter 1996, Ludsin & Wolfe 2001, Simberloff & Gibbons 2004). This is also the case for gypsy moth, as many, but once again not all, very low density populations (as determined through male moth trap catch) go extinct in the following year without any management intervention (Liebhold & Bascompte 2003, Whitmire & Tobin 2006). Yet, for eradication of a new invader to be a feasible management goal, a rapid response to the invasion is paramount since the feasibility and costs of eradication are directly related to the degree of establishment (Rejmánek & Pitcairn 2002).

Gypsy moth has been successfully eradicated many times from many areas because most of these conditions are fulfilled, except for the condition regarding reintroduction. Gypsy moth females are very fecund, egg masses can be difficult to detect, and the eggs go into diapause and require a prolonged winter chill before hatching. Thus, gypsy moth egg masses can be deposited on ships or goods in summer, the eggs go into winter diapause and hatch many months later, during which time the ships or goods can easily have been transported to another place. Thus, through globalization, gypsy moth has been and repeatedly continues to be introduced to new areas.

5.2 Detection and Biopesticide Use

In the 1970s a sex pheromone attracting gypsy moth males was developed for detection of *L. dispar* (Bierl *et al.* 1970, Beroza & Knipling 1972). The same pheromone (disparlure) attracts all of the different subspecies of *L. dispar* (Pogue & Schaefer 2007) and this pheromone has been used extensively for detection (USDA APHIS 2001, Tobin & Blackburn 2007). Detection of low density populations is a major requirement for eradication campaigns. Pheromone-baited traps are the most sensitive method for detecting gypsy moth and are very effective at both low and high densities. As populations increase, land managers may also be able to find egg masses (which are present for ca 10 months of the year), larvae and pupae (which are present for only a limited time each year) and even remains of all life stages. The presence of egg masses and other immature life stages, for example, usually denotes a reproducing, and hence established, population for which an eradication effort would then be implemented. In contrast, the presence of only males in traps does not necessarily denote an established population, and depending on the numbers of males trapped and the subspecies, the appropriate management response could be to initiate delimitation trapping (more intensive trapping to understand the distribution in detail).

Although EGM and AGM are morphologically quite similar, on close examination, there are enough morphological differences that subspecies can sometimes be differentiated. However, molecular methods are now principally relied upon to confirm gypsy moth subspecies, especially when hybridization among subspecies might have occurred. Early techniques that were developed to distinguish between AGM and EGM used mitochondrial DNA (e.g. Bogdanowicz *et al.* 1993). However, it became clear that these subspecies can hybridize. Since mitochondria are passed only from females to offspring, using mitochondrial DNA identifies only the genotype of the maternal parent, thus providing incomplete information about hybrids. Therefore, methods for differentiation based on gypsy moth nuclear DNA were developed (e.g. Garner & Slavicek 1996, Schreiber *et al.* 1997, Reineke & Zebitz 1999). If the subspecies of gypsy moth males in pheromone-baited traps are in question, specimens are sent to labs operated by the US and Canadian governments and identities are confirmed.

5.2.1 Biopesticides Used for Eradication

For many years, scientists in North America compared strains of *Bacillus thuringiensis kurstaki* (*Btk*) to select the strains most active against gypsy moth and then worked on optimizing formulation and application methods (Reardon *et al.* 1994, see Chapter 11). A strain was identified (HD-1) that is highly active against gypsy moth larvae and this strain was developed for control. In the late 1970s and early 1980s, when eradication became more feasible due to availability of monitoring tools (i.e. pheromone-baited traps) that were effective at low densities, *Btk* was not used very widely in eradication. Instead, chemical insecticides such as carbaryl were more commonly used. However, as application delivery methods were refined and *Btk* products were improved, trends changed and now *Btk* is often the only product used in eradication programs (Fig. 5.1). Today, the principle *Btk* product that is used is Foray®, which contains HD-1 (Valent Biosciences Corporation 1998). It is used as a flowable formulation at either 12.7 BIU/L (48 BIU/gal) or 20 BIU/L (76 BIU/gal) in Foray 48B and Foray 76B, respectively, and is manufactured by Valent BioSciences. Present day usage is usually 4.67 L/ha (0.5 gal/acre) for Foray 48B or 3.1 L/ha (0.33 gal/acre) for Foray 76B. The label for Foray 48B products restricts its use to 20–99 BIU/ha (8–40 BIU/acre) per application. These products are applied for suppression programs (i.e. programs to mitigate gypsy moth outbreaks) as well as eradication but the amounts applied differ. Eradication programs usually use multiple aerial applications with doses ranging from 59 to 94 BIU/ha (24–38 BIU/acre) or higher, with 5–10 days between the multiple applications. It seems that there is a trend toward two applications in the southern US and three applications in the western US, with the aggressiveness of the treatments often increasing with increasing distance from established gypsy moth populations (i.e. gypsy moth infestations on the west coast, such as in Washington, Oregon

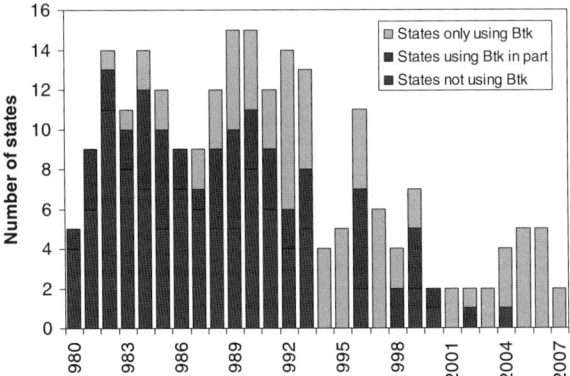

Fig. 5.1 Changes in materials used for eradication programs in the US, 1980–2007 (USDA Forest Service 2008). The numbers of states conducting eradication programs decreased markedly in the mid-1990s when the Slow the Spread program began taking over on eradications along the leading edge of gypsy moth spread

and California, are much further from the constantly spreading established gypsy moth population than infestations in the south, such as in North Carolina). In contrast, suppression programs often use single or double aerial applications of 59–89 BIU/ha.

Btk was selected for use in eradication campaigns because it is generally recognized as the safest insecticide to humans that is currently available in quantities for gypsy moth control. This is especially important in urban areas where, due to anthropogenic movement of life stages, most infestations are detected. Applicators must take care because this material must be applied when 2nd – 3rd instars are feeding to obtain optimal results. In addition, it should not be applied directly before rain. Second and 3rd instars are generally active for only approximately 2 weeks in early spring, and in periods of rain it can be difficult to apply *Btk* on the appropriate schedule.

The gypsy moth nucleopolyhedrovirus (*Ld*MNPV) is mass produced in the US by the USDA Forest Service and is formulated as the registered product Gypchek® (Chapter 11). Availability of Gypchek® is limited because it can only be produced *in vivo* and thus is very labor-intensive and costly to manufacture (Reardon *et al*. 1996). Because *Ld*MNPV is more host specific than *Btk*, this material is principally used for gypsy moth control in environmentally sensitive areas, when there is concern for non-target effects of *Btk* on native Lepidoptera, such as the endangered Karner Blue butterfly, *Lycaeides melissa samuelis* (USDA Forest Service 1995). Gypchek is also used for research. As for *Btk*, *Ld*MNPV must be ingested to be effective and early larval instars are more susceptible than later larval instars.

5.3 Case Studies of Eradication Programs

5.3.1 British Columbia

British Columbia has been concerned about introductions of gypsy moth since egg masses from Japan were first intercepted in 1911 (British Columbia Ministry of Forest & Range 2007). After pheromone-baited traps became available, the first male gypsy moth was trapped in 1978 (Bell 1994). Ground searches in the area found egg masses on a canoe and garden furniture that had originated from Quebec, a province hosting established populations of gypsy moth at that time. This became a regular pattern seen throughout North American areas that are outside of the gypsy moth established area: introductions of gypsy moth in uninfested areas are generally due to the accidental transport of life stages by humans, such as through commercial trade, household moves, and travel (Doane and McManus 1981, McFadden & McManus 1991). Since 1978, both AGM and EGM have repeatedly been detected in British Columbia. Between 1979 and 1999, male gypsy moths were captured at 120 sites and, at 20 of these sites, eradication programs were undertaken (Myers 2003). Two of these eradication programs are described below.

5.3.1.1 Asian Gypsy Moth Infestation in the Vancouver Area: 1991–1995

The interception of AGM egg masses on ships in Vancouver and Victoria in 1982 and 1989 presaged the infestation to come. Due to shipping patterns at the time, grain ships from the Russian Far East spent warmer months in the Soviet Arctic waters and then came to Canada the following spring to load grain, approximately when gypsy moth egg masses that had been laid on ships in Russia the previous summer would hatch. In 1991, Canadian inspectors found AGM egg masses on empty ships from the Russian Far East that were in the port of Vancouver waiting to load grain. It is assumed that some of the neonates hatching from egg masses blew to trees on the shore because adult male gypsy moths were subsequently trapped on shore. The specific pathway for this introduction was clarified based on the fact that adult gypsy moths are attracted to light (Wallner *et al.* 1995) and AGM females can fly. During the 1980s and 1990s, gypsy moth outbreaks were occurring in the Russian Far East. AGM adults were attracted to the well-lit grain terminals at the edges of forested areas where ships were loaded and unloaded and egg masses were laid on ships. After these ships made their way to Pacific Northwest ports, e.g. Vancouver and Seattle, hatching egg masses were visually detected on the ships.

In 1989, 80 AGM egg masses were found on a ship at Victoria, BC (British Columbia Ministry of Forest & Range 2007). By 1990, hundreds of AGM egg masses, along with some hatching larvae were found on another vessel and, by 1991, thousands of AGM egg masses with ballooning larvae were found on 29 infested grain ships in Vancouver harbor, at which time a ban was put into effect. By 1992, Canadian authorities had banned 16 ships from entering the port of Vancouver when AGM egg masses and some hatching larvae were found on them.

After 23 AGM males were detected in pheromone-baited traps in and around the port of Vancouver in 1991 (Fig. 5.2), in 1992 an area of 20,000 ha in the city of Vancouver was sprayed from the air three times with *Btk* at 50 BIU/ha at a cost of approximately $Canadian 6 million (Myers 2003). Subsequently, from 1993 through 1995, 8 male AGM were trapped in the Vancouver area but no further spraying was conducted (Fig. 5.2). After 1995, no more AGM were trapped and officials declared AGM eradicated from the Vancouver area.

5.3.1.2 European Gypsy Moth Eradication in the Victoria Area: 1999

Male EGM were found quite regularly in pheromone-baited traps in British Columbia from 1978 to 2007, but generally in low numbers (Fig. 5.2). Between 1995 and 1998, EGM populations were increasing on southern Vancouver Island. Numbers of males collected in traps increased from 5 to 503 at 14+ locations over this period, and egg masses were also found from 1996 to 1998 (British Columbia Ministry of Forest & Range 2007). In 1998, an aerial spray program was proposed but the action was blocked by citizen protest. However, in 1999, a $Canadian 3 million spray program was undertaken to eradicate the EGM infestation. Foray 48B was applied aerially to $> 13,400$ ha on Vancouver Island, of which 12,204 ha were in the greater Victoria region (de Amorim *et al.* 2001). The spray zone included a

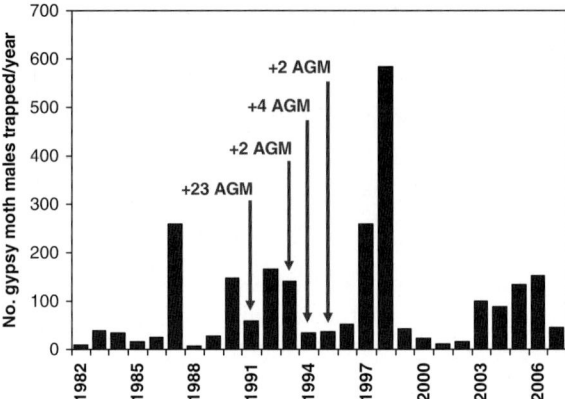

Fig. 5.2 Numbers of male European gypsy moths trapped/year in southern British Columbia, with the numbers of Asian gypsy moths trapped/year noted above (British Columbia Ministry of Forest & Range 2007)

mix of residential and rural areas, where approximately 75,420 people lived. The area considered to be infested was also regulated so that Christmas trees, nursery stock and outdoor household articles could not be moved to unregulated areas in the US without prior inspection. This EGM eradication program was considered a complete success (see de Amorim *et al.* 2001). One year after the *Btk* treatments, two adult males were trapped but no egg masses were detected in the treated areas; consequently, the threat of an embargo against lumber products from British Columbia was removed (USDA APHIS 2000).

5.3.2 Washington: Asian Gypsy Moth

The northwestern US state of Washington has undertaken aggressive EGM survey programs since 1974 (Alexander 1992). AGM was detected on the shore of North America late in 1991 near the port of Vancouver, British Columbia, Canada (see above), and male AGM were also found in traps in Washington and Oregon that year (USDA APHIS 2003). The exact source of the Washington and Oregon infestations is not known but it is thought that ships coming from the Russian Far East that were infested with egg masses probably introduced AGM to North America while visiting ports on the West Coast. Through a grid of 11,439 pheromone-baited traps deployed in 1991, nine AGM were found at eight locations in two counties near the port of Tacoma (Washington State Dept. Agriculture 2007). An eradication program was implemented between 21 April and 24 May1992, during which 47,146 ha were treated with *Btk* (Foray 48B) through aerial applications. The applications required 24 helicopters and the coordinated efforts of a team of > 100 people from eight different federal and state agencies (Alexander 1992). The gypsy moth trapping program became more aggressive in 1992, with 129,299 pheromone-baited traps deployed in Washington State (Washington State Dept. Agriculture 2007). No AGM males were trapped in the spray zone after the 1992 treatments and the eradication was considered a success (Alexander 1992). The number of traps decreased in 1993 to nearly 30,000 and from 1994 to 2006, the number of traps varied from 18,698 to 36,166 yearly. However, between 1992 and 1994, eight AGM/EGM hybrids were detected. These moths were genetically similar to hybrids found in Germany and it was suspected that they had been introduced during transport of military materials to Fort Lewis in Washington from Germany (see Section 5.3.3. below). One of the hybrid detection sites was treated in spring of 1995 to eradicate this infestation and, after 2 years of zero trap catch, eradication at this site was considered successful. Interestingly, between 1992 and 1994 the catches of EGM per year ranged from 173 to 458 but EGM was not a priority and no actions were taken to eradicate EGM populations (Washington State Dept. Agriculture 2007).

From 1995 to 1999 six AGM and three AGM/EGM hybrids were detected in Washington State (Washington State Dept. Agriculture 2007). Small-scale eradication programs were conducted at seven different sites during this period, each of which had trapped only one moth (all AGM or one of the hybrids). For example, in response to one AGM trapped in 1999, Foray 48B was applied aerially three times at 7 day intervals in Seattle in May 2000 (Washington State Dept. Health 2001). The eradication effort targeted approximately 259 ha of the Ballard and Magnolia areas of Seattle. This is an urban area with an estimated 2,200 businesses and properties and an estimated residential population of 6,600. For each eradication program between 1995 and 1999, eradication was declared to be successful following 2 years of zero trap catch from treated areas.

5.3.3 North Carolina: Hybrids of Asian and European Gypsy Moth

In addition to invasion pathways that arise through global trade and travel, the transport of military cargo can also introduce gypsy moth populations to North America. Such an introduction occurred on 4 July 1993 when a container ship hauling US military cargo arrived from Nordenham, Germany, and docked at the Military Ocean Terminal in Sunny Point, North Carolina (Garcia 1993). At the time of the ship's departure from Germany (21 June), many regions of Europe, including Germany, were experiencing gypsy moth outbreaks. Upon docking, moths were seen emerging from infested cargo containers, and subsequent inspections revealed that the ship, the USCS Advantage, was infested with a hybrid strain of adult gypsy moths containing both Asian and European DNA. At the time, AGM was not known to occur in Europe. However, tracebacks of this introduction and also two others (military freight from Germany that arrived in Baltimore, Maryland and Charleston, South Carolina, in 1994) subsequently confirmed the presence of AGM in Germany (Prasher & Mastro 1994). The ship was ordered back to sea on 6 July and remained approximately 5 miles from port until adult moth activity ceased. The USCS Advantage returned to port on 15 July, and fumigation of all cargo with methyl bromide began on 16 July.

In addition to fumigation, pheromone-baited and light traps were deployed at ≈ 350 m intervals from the mouth of the Cape Fear River (approximately 6.5 km south of the Sunny Point Military Ocean Terminal), to roughly 16 km north of Sunny Point. These traps were placed in addition to the North Carolina state-wide trapping grid, which, in 1993, included the deployment of pheromone-baited traps across the entire state (Roberts 2008). Traps from the vicinity of the port recorded > 400 moths, including > 300 from traps deployed on Sunny Point property. Several hundred egg masses were also removed from the ship and its cargo (Garcia 1993).

Based upon the trapping data and in consultation with a management and science advisory panel, the port and surrounding areas were divided into five priority areas. The highest priority area was in New Hanover and Brunswick counties, which included the port of Sunny Point. In 1994, 55,838 ha of this area were treated with two applications of Foray 48B. A third application was made over 2,637 ha at the port (South 1994). Two applications of the commercial formulation of the gypsy moth virus, Gypchek®, were used on an additional 2,388 ha deemed environmentally-sensitive (South 1994). In 1994 and 1995, delimiting grids were placed around treatment blocks, and post-treatment moth counts in the high priority area totalled < 10 in both 1994 and 1995 (Roberts 2008). Although the 1994 eradication program was mostly successful, supplemental applications of Foray 48B were considered necessary to completely eliminate Asian gypsy moth from the area; 2,566 hectares were treated in 1995 and 603 hectares were treated in 1996. No gypsy moths with Asian DNA were detected in 1996 and 1997 and this eradication was then considered successful (USDA Forest Service 1996, 1997).

5.3.4 The Slow the Spread Program: Eradication at the Edge of the Gypsy Moth Distribution

Gypsy moth currently occupies roughly one-third of the susceptible habitat in North America, and continues to invade new habitats at variable rates (Tobin et al. 2007). New colonies often arise along the leading edge of the population front (Sharov & Liebhold 1998) through a process known as stratified dispersal that combines short-range (usually through ballooning neonates) and long-range (usually through the anthropogenic movement of life stages or meteorological transport mechanisms) dispersal (Hengeveld 1989, Shigesada et al. 1995). Colonies that arise in a transition zone between the infested and uninfested areas of the gypsy moth distribution then grow and coalesce with the population front, which greatly contributes to and enhances the rate of gypsy moth spread. Under the gypsy moth Slow the Spread (STS) Program (Tobin & Blackburn 2007), these newly-established, spatially disjunct colonies are detected and delimited through the use of pheromone-baited traps, and then aggressively targeted for eradication to limit their influence on the rate of spread (Fig. 5.3).

One of the principle management tactics used in the STS Program is aerial applications of *Btk* (Reardon et al. 1994). During the pilot STS Program (1996–1999),

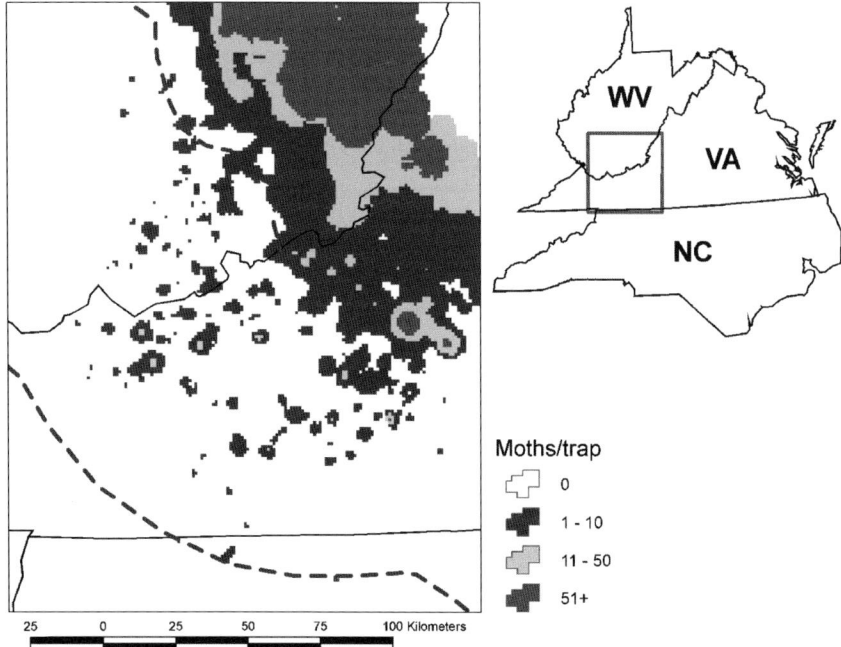

Fig. 5.3 Interpolated male moth density from pheromone-baited traps deployed in West Virginia (WV), Virginia (VA), and North Carolina (NC) under the Gypsy Moth Slow the Spread Program (Roberts 2008). Within a transition zone along the population front (outlined by *dashed lines*), incipient colonies are detected and eradicated to minimize their contribution to gypsy moth rates of spread

the use of *Btk* was the dominant treatment tactic (> 85, 389 ha treated with *Btk*, which represented 69.0% of all treated hectares, Tobin & Blackburn 2007). Today, *Btk* is often the tactic of choice, as opposed to other control tactics such as mating disruption (Thorpe *et al.* 2006), when infestations are abundant (i.e. generally maximum pheromone-baited trap catch > 30) or when life stages (i.e. egg masses) have been documented. In some cases, if a high infestation occurs within a larger area requiring treatment, *Btk* will be used in concert with mating disruption by using *Btk* to target the "hot spot" within a larger mating disruption block. Since 2000, approximately 258,594 ha have been treated with *Btk* under the STS Program.

In STS, *Btk* doses generally range from 59 to 94 BIU/ha using either one or two applications during the period of early instar activity. A summary of the use of *Btk* in STS, including doses, area treated, and pre- and post-treatment male moth density is presented in Table 5.1. There are several factors that are considered when determining the appropriate dose and number of applications, including the presence of egg masses (which generally increases the dose and/or number of applications) and budgetary constraints (which generally decreases the dose and/or number of applications). However, the primary factors are the initial male moth density, as recorded through pheromone-baited traps, and the distance of the treatment block

Table 5.1 Summary of the use of *Btk* in eradicating isolated colonies under the Gypsy Moth Slow the Spread Program, 2000–2007

Year	Dose[1]	ha Treated	Prior to treatment	Post-treatment	
			Mean maximum moth catch	Mean maximum moth catch	Mean index of treatment success[2]
2000	59BIU × 1	31,590	18.8	5.8	0.6
2001	59BIU × 2	22,984	65.9	19.8	0.7
2002	59BIU × 1	546	140.5	0.5	1.0
	59BIU × 2	11,163	188.3	7.6	0.8
	74BIU × 2	30	148.0	0.0	1.0
2003	59BIU × 1	1,950	60.2	29.3	0.9
	59BIU × 2	19,885	88.4	14.8	0.7
	74BIU × 2	166	6.8	4.2	0.5
	94BIU × 1	6,262	215.6	83.8	0.8
	94BIU × 2	329	65.0	1.5	1.0
2004	59BIU × 2	48,783	154.3	46.8	0.6
	74BIU × 2	923	135.8	3.8	0.9
2005	59BIU × 1	18,628	244.0	15.0	0.8
	59BIU × 2	49,883	126.1	8.5	0.9
	74BIU × 1	374	25.5	1.5	1.0
	74BIU × 2	2,257	106.6	41.1	0.7
	94BIU × 1	116	202.5	134.8	0.9
2006	59BIU × 1	3,001	80.1	24.3	0.7
	59BIU × 2	24,418	106.3	22.8	0.8
	94BIU × 1	14,753	52.8	16.1	0.8
2007	59BIU × 1	947	190.0	0.0	1.0
	59BIU × 2	23,201	57.9	25.8	0.7
	94BIU × 1	4,052	34.2	0.0	1.0

[1] Dose per hectare in BIU × the number of applications.
[2] The Index of Treatment Success considers the change in moth density in the treated area before and after treatment while adjusting for corresponding changes in density in nearby, untreated areas that serve as a "control." Values > 0.7 indicate a treatment success (Sharov *et al.* 2002, Tobin & Blackburn 2007).

from the infested area (Fig. 5.4). Higher density infestations are generally targeted using a higher dose and/or number of applications, while infestations that are farther away from the infested area, but also more often of lower density, are targeted with a lower dose but with multiple applications (Fig. 5.4). This is because infestations farther away from the infested area contribute proportionally more to the rate of spread under stratified dispersal (see Chapter 3) than those that are closer.

A lesser used tactic under the Slow the Spread Program is the use of *Ld*MNPV, formulated as Gypchek® (Reardon *et al.* 1996). Because it is a naturally-occurring, highly host specific virus of the gypsy moth but availability is limited, Gypchek® is solely used in environmentally-sensitive habitats. Since 2000, approximately 17,401 ha have been treated under the STS Program using Gypchek®.

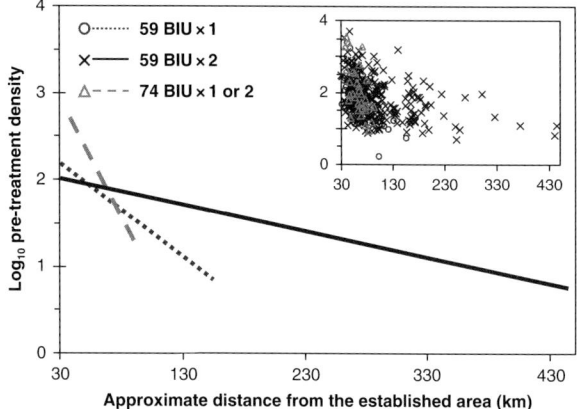

Fig. 5.4 Use of *Btk* in the gypsy moth Slow the Spread Program and for targeting populations outside of the transition zone (the area between the *dashed lines* in Fig. 5.3 is the transition zone; i.e. > 170 km from the established area), 2000–2007 (Tobin *et al.* 2004). Within the transition zone, higher density populations close to areas where gypsy moth is established tend to be targeted with higher doses, and for eradicating populations farther away, 2 applications are used even when densities are lower

5.3.5 Smaller Eradication Efforts

On a regular basis, gypsy moth is detected across the US in areas where it is not already established. Eradication programs have been conducted in 24 US states between 1980 and 2007 (Fig. 5.5A); while some of these states were near the leading edge of the gypsy moth population front in 1980 and 2007 (the thickened lines), many were not. States vary in the frequency of years that they have used *Btk* in eradication programs, although many of the states with the most experience with eradication programs (i.e. more years of eradication programs) have frequently used *Btk*. Figure 5.5B shows that frequently when states had large areas being treated in eradication programs they opted to rely totally on *Btk*.

In recent years, there have been two smaller eradication programs in Oregon that are of particular interest. Gypsy moth is no stranger to Oregon where large eradication programs against EGM occurred in 1984–1985 (Youngs 1985) and where AGM was eradicated in the early 1990s. In fact, between 1980 and 2006, 23 eradication programs were conducted in Oregon. In the early 2000s, gypsy moth trap catches were relatively low for a couple of years. However, in 2006, trap catches jumped from 9 (in 2005) to 66 (Oregon Department of Agriculture 2006). Of the 66 male moths trapped, 57 were EGM that had been trapped at one location near Bend, Oregon. Survey technicians discovered the source: in 2005 a property owner had purchased a 1967 Chevrolet over the internet auction site eBay® from a seller in Connecticut, a northeastern state where gypsy moth has been established since at least 1914 (Burgess 1930, Liebhold *et al.* 1992), and the vehicle contained viable gypsy moth egg masses. Egg masses were also detected on trees on the property,

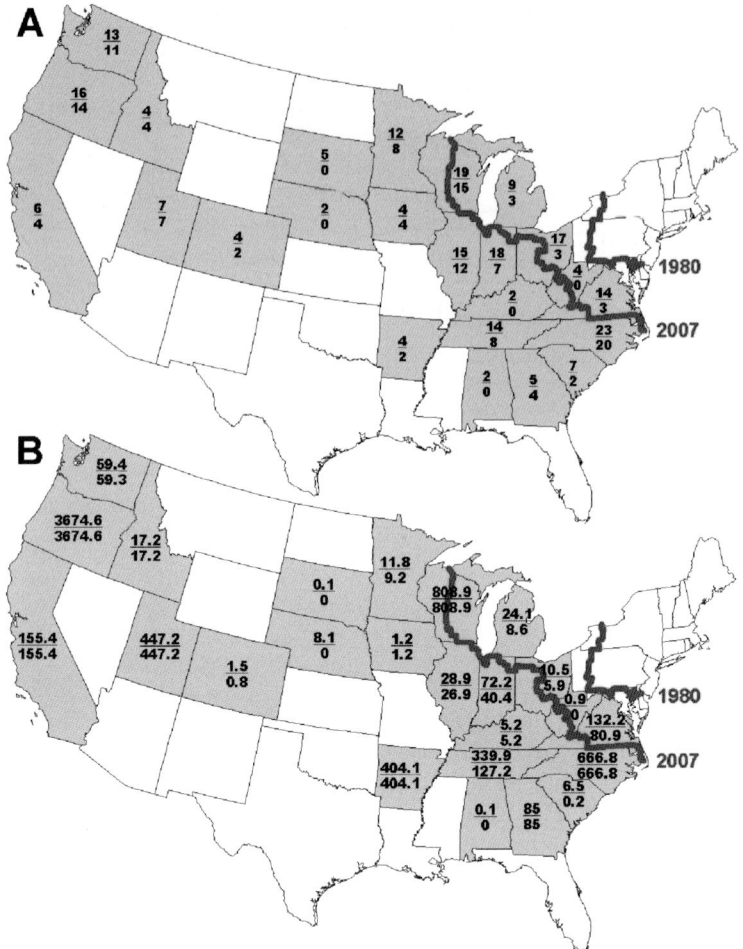

Fig. 5.5 Eradication efforts against the gypsy moth in the United States, 1980–2007 (*shaded states*). (**A**). The number of years (numerator) in which each state had an eradication project and the number of years in which *Btk* was used in an eradication effort (denominator). (**B**) The greatest amount (km²) of treated area in any year (numerator) and the greatest amount (km²) treated using *Btk* in any year (denominator). In each map, the thick grey lines indicate the gypsy moth infested area in 1980 and 2007, based upon county quarantine records (Tobin *et al.* 2007)

confirming that a reproducing population had successfully established. Another case involved detection of one AGM male that was trapped along the Columbia River, a large river where cargo ships arrive from Asia. Both of these infestations were treated with *Btk* following standard application protocols (three applications of *Btk* at each site; 259 ha for the area with the single AGM male and 216 ha for the EGM infestation) (Johnson *et al.* 2007). As a result, no moths were trapped at either site in 2007; 2 more years of no AGM catches and 1 more year of no EGM catches at the treated sites will be necessary to declare eradication. Additionally, in 2006, traps at other locations contained from 1 to 3 EGM males but these areas were not treated, demonstrating the differential threshold for treating populations of EGM relative to AGM (Oregon Department of Agriculture 2006).

5.4 Conclusions

Slowing the spread of EGM along its population front in the US and eradicating new gypsy moth infestations outside of the areas where EGM is established are taken very seriously in the US and Canada. Of even greater concern is AGM: new AGM infestations are treated much more aggressively than new infestations of EGM, e.g. the presence of even one AGM can lead to multiple aerial applications of *Btk*. It is a testament to the effectiveness, low non-target effects, and availability of *Btk* that this biopesticide has transitioned from being a minor component to the primary material used for eradication programs. In addition, *Ld*MNPV is used to eradicate gypsy moth populations in environmentally sensitive areas; it is generally the case that each year, more requests for *Ld*MNPV product are made than can be filled.

Invasive species often arrive where people live and work because they are frequently transported through human activities; thus, control activities must often be conducted in urban/suburban areas. Such is the case with gypsy moth. There has been public dissent regarding applications of *Btk* during some eradication programs conducted in urban areas (Myers *et al.* 2000, also see Chapter 4). A recently published book written on the first eradication effort against gypsy moth (1890–1901) was motivated by the author's disapproval of eradication efforts in Maryland in the 1990s (Spear 2005). Eradication has principally been criticized by the public due to concern about the potential for health effects as well as non-target effects from exposure to *Btk* (see Chapters 16, 17 and 18). The protest groups opposing use of *Btk* have questioned whether eradication programs are really necessary. However, if, for example, articles from Canada destined to be sent to the US are potentially contaminated by gypsy moth egg masses, the US might not accept them and this would clearly have a strong impact on trade. Other major costs of a gypsy moth infestation are tree mortality (Doane and McManus 1981, Gansner & Herrick 1984, Herrick & Gansner 1987), nuisance in urban and recreational areas (Payne *et al*. 1973, Moeller *et al*. 1977, Webb *et al*. 1991) and potential allergic reactions to the urticating hairs (Tuthill *et al*. 1984) associated with gypsy moth outbreaks.

Still, there is debate regarding the necessity for eradication programs; Myers (2003) suggested that if gypsy moth were to become established in British Columbia, environmental effects might not be as extensive as suggested by government agencies. She suggests that the expensive eradication programs are triggered by trade issues rather than concerns for nature and these priorities are not supported by all of the public. However, an economic analysis conducted in 1993 estimated that the benefits of detection and eradication in British Columbia at that time outweighed the costs of an annual suppression program by ten to one (Oliver 1994). A similar benefit-to-cost-analysis conducted on the Slow the Spread Program revealed a corresponding 4-to-1 ratio, with the primary benefit due to delaying the costs and losses due to gypsy moth in residential areas (Leuschner et al. 1996).

Authorities have instituted programs to prevent the arrival of AGM at US ports, as well as to minimize the potential for the transport of gypsy moth from overseas ports. However, until there are further improvements in our management of invasion pathways, gypsy moth, and other non-native species, will continue to be introduced through global trade and travel (Work et al. 2005, Liebhold et al. 2006, McCullough et al. 2006). Because of its effectiveness, availability and relatively low non-target effects, *Btk* will likely remain a primary management strategy in gypsy moth eradication efforts, at least for the near future.

Acknowledgments Results from eradication campaigns are usually not summarized in the peer-reviewed literature. We regret that many individual programs for eradicating gypsy moth are not included in this chapter. This chapter presents only selected examples from the many eradication programs that have been conducted. We thank many people for information, including Donna Leonard, Robert Fusco, Stephen Nicholson, Leah Bauer, Barry Bai, Timothy Marasco and Donald Eggen and we thank Laura Blackburn for her excellent assistance with manuscript preparation. The mention and use of product and trade names does not constitute endorsement by the United States Department of Agriculture.

References

Alexander DG (1992) Washington State Department of Agriculture 1992 Asian Gypsy Moth program. In: Proc 1992 Annual Gypsy Moth Review, 2–5 Nov 1994, Indianapolis, Indiana. pp 108–110
Baranchikov YN (1989) Ecological basis of the evolution of host relationships in Eurasian gypsy moth populations. In: Wallner WE, McManus KA (eds) Proc Lymantriidae: A comparison of features of new and old world tussock moths. USDA Forest Service, Gen Tech Rpt NE-123, Broomall, PA. pp 319–338
Baranchikov YN, Montgomery ME (1994) Tree suitability for Asian, European and American populations of gypsy moth. USDA Forest Service, Gen Tech Rpt NE-188:4
Bell J (1994) British Columbia report. In: Proc 1994 Annual Gypsy Moth Review, 30 Oct – 2 Nov 1994, Portland, Oregon. pp 153–155
Beroza M, Knipling EF (1972) Gypsy moth control with the sex attractant pheromone. Science 177:19–27
Bierl BA, Beroza M, Collier CW (1970) Potent sex attractant of the gypsy moth: its isolation, identification and synthesis. Science 170:87–89
Bogdanowicz SM, Wallner WE, Bell J, Odell TM, Harrison RG (1993) Asian gypsy moth (Lepidoptera: Lymantriidae) in North America: Evidence from molecular data. Ann Entomol Soc Am 86:710–715
British Columbia Ministry of Forest & Range (2007) History of gypsy moth infestations in British Columbia. http://www.for.gov.bc.ca/hfp/gypsymoth/history.htm [accessed 18 February 2008]
Burgess AF (1930) The gipsy [sic] moth and the brown-tail moth. United States Department of Agriculture Farmers' Bulletin 1623
Campbell RW, Sloan RJ (1977) Forest stand responses to defoliation by the gypsy moth. For Sci Monogr 19:1–34

Dahlsten DL, Garcia R, Lorraine H (1989) Eradication as a pest management tool: Concepts and contexts. In: Dahlsten DL, Garcia R (eds) Eradication of exotic pests: Analysis with case histories. Yale Univ Press, New Haven, CT. pp 3–15

de Amorim GV, Whittome B, Shore B, Levin DB (2001) Identification of *Bacillus thuringiensis* subspecies *kurstaki* strain HD1-like bacteria from environmental and human samples after aerial spraying of Victoria, British Columbia, Canada with Foray 48B. Appl Environ Microbiol 67:1035–1043

Doane CC, McManus ME (1981) The gypsy moth: Research toward integrated pest management. US Dept Agric For Serv Gen Tech Bull 1584

Elkinton JS, Liebhold AM (1990) Population dynamics of gypsy moth in North America. Annu Rev Entomol 35:571–596

Forbush EH, Fernald CH (1896) The gypsy moth. Wright and Potter, Boston, MA

Gansner DA, Herrick OW (1984) Guides for estimating forest stand losses to gypsy moth. Northern J Appl For 1:21–23

Garcia LE (1993) Exotic gypsy moth introduction at Sunny Point, North Carolina. Proc 1993 Annual Gypsy Moth Review, Nov, Harrisburg, Pennsylvania. pp 39–43

Garner KJ, Slavicek JM (1996) Identification and characterization of a RAPD-PCR marker for distinguishing Asian and North American gypsy moths. Ins Mol Biol 5:81–91

Hengeveld R (1989) Dynamics of biological invasions. Chapman and Hall, London

Herrick OW, Gansner DA (1987) Mortality risks for forest trees threatened with gypsy moth infestation. USDA Forest Service, Res Note NE-338

Johnson KJR, Bai BB, Rogg H (2007) Gypsy moth detection and eradication programs in Oregon – 2007. In: Proc 2007 Annual Gypsy Moth Review, 29 Oct–1 Nov, Sheperdstown, West Virginia

Keena M (1996) Comparison of the hatch of *Lymantria dispar* (Lepidoptera: Lymantriidae) eggs from Russia and the United States after exposure to different temperatures and durations of low temperature. Ann Entomol Soc Amer 89:564–572

Leuschner WA, Young JA, Walden SA, Ravlin FW (1996) Potential benefits of slowing the gypsy moth's spread. South J Appl For 20:65–73

Liebhold AM, Bascompte J (2003) The Allee effect, stochastic dynamics and the eradication of alien species. Ecol Lett 6:133–140

Liebhold AM, Halverson JA, Elmes GA (1992) Gypsy moth invasion in North America: a quantitative analysis. J Biogeogr 19:513–520

Liebhold AM, Work TT, McCullough DG, Cavey JF (2006) Airline baggage as a pathway for alien insect species entering the United States. Amer Entomol 52:48–54

Ludsin SA, Wolfe AD (2001) Biological invasion theory: Darwin's contributions from The Origin of Species. Bioscience 51:780–789

McCullough DG, Work TT, Cavey JF, Liebhold AM, Marshall D (2006) Interceptions of nonindigenous plant pests at U.S. ports of entry and border crossings over a 17 year period. Biol Inv 8:611–630

McFadden MW, McManus ME (1991) An insect out of control? The potential for spread and establishment of the gypsy moth in new forest areas in the United States. In: Baranchikov YN, Mattson WJ, Hain FP, Payne TL (eds) Forest insect guilds: patterns of interaction with host trees. USDA Forest Service, Gen Tech Rpt NE-153. pp 172–186

Moeller GH, Marler RL, McCay RE, White WB (1977) Economic analysis of the gypsy moth problem in the Northeast. III. Impacts on homeowners and managers of recreation areas. USDA Forest Service, Res Paper NE-360

Myers JH (2003) Eradication: Is it ecologically, financially, environmentally, and realistically possible? In: Rapport DJ, Lasley WL, Rolston DE, Nielsen NO, Qualset CO, Damania AB (eds) Managing for healthy ecosystems. Lewis Publ., Boca Raton. pp 533–539

Myers JH, Rothman LD (1995) Field experiments to study regulation of fluctuating populations. In: Cappuccino N, Price PW (eds) Population dynamics: New approaches and synthesis. Academic Press, San Diego, CA. pp 229–250

Myers JH, Savoie A, Van Randen E (1998) Eradication and pest management. Annu Rev Entomol 43:471–491

Myers JH, Simberloff D, Kuris AM, Carey JR (2000) Eradication revisited: Dealing with exotic species. Tr Ecol Evol 15:316–320

Oliver A (1994) "Do something" In: Proc 1994 Annual Gypsy Moth Review, 30 Oct-2 Nov 1994, Portland, Oregon. pp 150–152

Oregon Department of Agriculture (2006) Gypsy moth detections increase in Oregon this year. http://www.oregon.gov/ODA/news/061025gypsy_moth.shtml [accessed 19 February 2008]

Payne BR, White WB, McCay RE, McNichols RR (1973) Economic analysis of the gypsy moth problem in the Northeast. II. Applied to residential property. USDA Forest Service, Res Paper NE-285

Pogue MG, Schaefer PW (2007) A review of selected species of *Lymantria* (Hübner) [1819] (Lepidoptera: Noctuidae: Lymantriinae) from subtropical and temperate regions of Asia, including three new species, some potentially invasive to North America. USDA Forest Service FHTET-2006-07

Prasher D, Mastro VC (1994) Genotype analyses of 1994 port specimens. Proc 1994 Annual Gypsy Moth Review, 30 Oct–2 Nov, Portland, Oregon. pp 61–63

Reardon R, Dubois N, McLane W (1994) *Bacillus thuringiensis* for managing gypsy moth: a review. USDA Forest Service, National Center of Forest Health Management. FHM-NC-01-94

Reardon RC, Podgwaite J, Zerillo R (1996) Gypchek – the gypsy moth nucleopolyhedrosis virus product. USDA Forest Health Technology Enterprise Team, FHTET-96-16

Redman AM, Scriber JM (2000) Competition between the gypsy moth, *Lymantria dispar*, and the northern tiger swallowtail, *Papilio canadensis*: interactions mediated by host plant chemistry, pathogens, and parasitoids. Oecologia 125:218–228

Reineke A, Zebitz CPW (1999) Suitability of polymerase chain reaction-based approaches for identification of different gypsy moth (Lepidoptera: Lymantriidae) genotypes in Central Europe. Ann Entomol Soc Amer 92:737–741

Rejmánek M, Pitcairn J (2002) When is eradication of exotic pest plants a realistic goal? In: Veitch D, Clout M (eds) Turning the tide: the eradication of invasive species. IUCN SSC Invasive Species Specialist Group, pp 249–253

Riley CV, Vasey G (1870) Imported insects and native American insects. Amer Entomol 2:110–112

Roberts EA (2008) Decision-support system for the gypsy moth Slow-the-Spread Program. http://da.ento.vt.edu/[accessed 5 March 2008]

Schreiber DE, Garner KJ, Slavicek JM (1997) Identification of three randomly amplified polymorphic DNA-polymerase chain reaction markers for distinguishing Asian and North American gypsy moths (Lepidoptera: Lymantriidae). Ann Entomol Soc Amer 90:667–674

Sharov AA (2004) The bioeconomics of managing the spread of exotic pest species with barrier zones. Risk Anal 24:879–892

Sharov AA, Liebhold AM (1998) Model of slowing the spread of gypsy moth (Lepidoptera: Lymantriidae) with a barrier zone. Ecol Appl 8:1170–1179

Sharov AA, Leonard DS, Liebhold AM, Clemens NS (2002) Evaluation of preventive treatments in low-density gypsy moth populations. J Econ Entomol 95:1205–1215

Shigesada N, Kawasaki K, Takeda Y (1995) Modeling stratified diffusion in biological invasions. Am Nat 146:229–251

Simberloff D, Gibbons L. (2004) Now you see them, now. you don't! Biol Invasions 6:161–172

Spear RJ (2005) The great gypsy moth war: a history of the first campaign in Massachusetts to eradicate the gypsy moth, 1890–1901. University of Massachusetts Press, Amherst, MA

South M (1994) North Carolina gypsy moth program: 1994 overview. In: Proc 1994 Annual Gypsy Moth Review, 30 Oct-2 Nov 1994, Portland, Oregon. pp 304–306

Thorpe K, Reardon R, Tcheslavskaia K, Leonard D, Mastro V (2006) A review of the use of mating disruption to manage gypsy moth, *Lymantria dispar* (L.). US Dept Agric For Health Tech Ent Team 2006-13, Morgantown, WV. 76 pp

Tobin PC, Blackburn LM (2007) Slow the spread: a national program to manage the gypsy moth. USDA Forest Service, Gen Tech Rep NRS-6, Newtown Square, PA. 109 p

Tobin PC, Liebhold AM, Roberts EA (2007) Comparison of methods for estimating the spread of a non-indigenous species. J Biogeography 34:305–312

Tobin PC, Sharov AA, Liebhold AM, Leonard DS, Roberts EA, Learn MR (2004) Management of the gypsy moth through a decision algorithm under the Slow the Spread Project. Amer Entomol 50:200–209

Tuthill RW, Canada AT, Wilcock K, Etkind PH, O'Dell TM, Shama SK (1984) An epidemiological study of gypsy moth rash. Amer J Pub Health 74:799–803

USDA APHIS (Animal Plant Health Inspection Service) (2000) Importation of gypsy moth host material from Canada. Fed Reg 65:38171–38177

USDA APHIS (Animal Plant Health Inspection Service) (2001) Gypsy moth program manual. http://www.aphis.usda.gov/import_export/plants/manuals/domestic/[accessed 5 March 2008]

USDA APHIS (Animal Plant Health Inspection Service) (2003) Asian gypsy moth. http://www.aphis.usda.gov/lpa/pubs/fsheet_faq_notice/fs_phasiangm.html [accessed 18 February 2008]

USDA Forest Service (1995) Gypsy moth management in the United States: a cooperative approach. Final Environmental Impact Statement

USDA Forest Service (1996) 1996 Forest Insect and Disease Conditions for the Southern Region. http://www.digitalarborist.com/96conditions/[accessed 26 March 2008]

USDA Forest Service (1997) 1997 Forest Insect and Disease Conditions for the Southern Region. http://www.digitalarborist.com/97conditions/[accessed 26 March 2008]

USDA Forest Service (2008) Gypsy Moth Digest. http://na.fs.fed.us/fhp/gm [accessed 3 March 2008]

Valent Biosciences Corporation (1998) Foray® Dipel®: Forestry technical manual. Valent Biosciences Corp., Libertyville, IL

Wallner WE, Humble LM, Levin RE, Baranchikov YN, Carde RT (1995) Response of adult Lymantriid moths to illumination devices in the Russian Far East. J Econ Entomol 88:337–342

Washington State Dept Agriculture (2007) History of Asian gypsy moth in Washington State. http://agr.wa.gov/PlantsInsects/InsectPests/GypsyMoth/History/docs/HistoryOfAsianGM.pdf [accessed 18 February 2008]

Washington State Dept. Health (Environmental Health Programs) (2001) Report of Health Surveillance Activities: Aerial Spraying for Asian Gypsy Moth – May 2000 Seattle, WA. Office of Environmental Health and Safety, Olympia, WA. http://www.doh.wa.gov/ehp/Pest/AsianGypsyMothReport. PDF [accessed 18 February 2008]

Webb RE, Ridgway RL, Thorpe KW, Tatman KM, Wieber AM, Venables L (1991) Development of a specialized gypsy moth (Lepidoptera: Lymantriidae) management program for suburban parks. J Econ Entomol 84:1320–1328

Whitmire SL, Tobin PC (2006) Persistence of invading gypsy moth colonies in the United States. Oecologia 147:230–237

Williamson M, Fitter A (1996) The varying success of invaders. Ecology 77:1661–1666

Work TT, McCullough DG, Cavey JF, Komsa R (2005) Arrival rate of nonindigenous insect species into the United States through foreign trade. Biol Inv 7:323–332

Youngs L (1985) Panel discussion: Eradication – *B. t.* In: Proc 1985 Annual Gypsy Moth Review, 18–21 Nov 1985, Columbus, Ohio. pp 119–120

Zlotina MA, Mastro VC, Elkinton JS, Leonard DE (1999) Dispersal tendencies of neonate larvae of *Lymantria mathura* and the Asian form of *Lymantria dispar* (Lepidoptera: Lymantriidae). Environ Entomol 28:240–245

Part IV
Control

Chapter 6
Exotic Aphid Control with Pathogens

Charlotte Nielsen and Stephen P. Wraight

Abstract Exotic aphids are invading ecosystems worldwide. The principal factors favoring establishment of these invasive pests are their small size, parthenogenetic reproduction, short generation time, ability for long distance dispersal by winged morphs, and their explosive population dynamics. Attention has mainly been focused on invasive aphid pests of economic importance to agriculture, horticulture, and forestry. More recently, however, concerns have also concentrated on potential impacts of aphids on biodiversity, especially with respect to endangered native plants. Fungi are the most prevalent pathogens of aphids, and consequently fungi have been studied and used for biological control of invasive aphid species. Entomopathogenic fungi with high epizootic potential have been used in classical biological control programs, fungi have been mass produced and used for augmentation biological control (both inoculative and inundative), and crop and pest management practices have been modified to preserve/promote activity of naturally occurring fungi in conservation biological control programs. In this chapter we will review the various strategies that have been developed for control of invasive aphid species.

6.1 Introduction

Aphids (Hemiptera: Sternorrhyncha: Aphididae) are small, usually 1.5–3.5 mm, soft-bodied, phloem-feeding insects. Worldwide more than 4000 aphid species have been described (Dixon 1998). Aphids are one of the most economically important groups of insect pests and, while they are more significant in temperate climates, they can also cause substantial damage in the tropics. Their life cycles are diverse and complex, often including several morphs within the same species, and some species may have generations of individuals with high morphological variability. In some species, more than ten morphs may develop; including winged morphs

---

C. Nielsen
Department of Entomology, Comstock Hall, Cornell University, Ithaca, New York 14853-2601, USA; Department of Ecology, University of Copenhagen, Faculty of Life Sciences, Thorvaldsensvej 40, DK-1871 Frederiksberg C, Denmark
e-mail: chni@life.ku.dk

specialized for dispersal. In most cases, however, fewer morphs are required for completion of a life cycle (Heie 1980). Some species have sexually reproducing generations (holocyclic), while others do not reproduce sexually (anholocyclic) and consist only of viviparous parthenogenetic females.

In general, aphids have exceedingly short generation times due to their capacity for viviparous parthenogenetic reproduction and the fact that the offspring at birth contain embryos that also contain embryos (telescoping generations). Some holocyclic aphid species are dioecious, which means they exhibit host alternation between woody plants (the primary hosts, on which fertilised eggs hibernate and hatch in spring) and herbaceous plants (the secondary hosts or summer host). Other species are monoecious and do not alternate between hosts (Heie 1980, Dixon 1998).

Some aphid species are monophagous, having a very restricted plant host range. For example, the green spruce aphid, *Elatobium abietinum* (Walker), is an invasive aphid species in Iceland that feeds exclusively on trees in the genus *Picea* (spruce) (Blackman & Eastop 1994, Carter & Halldórsson 1998). In contrast, other aphid species are polyphagous, having much broader plant host ranges. One of the most polyphagous invasive aphids is the cotton aphid, *Aphis gossypii* Glover (also called the melon aphid), feeding on hosts from over 50 plant families (Blackman & Eastop 2000, Messing *et al.* 2007).

Aphids cause direct damage by sucking phloem sap and indirect damage by production of honeydew in which saprophytic fungi may grow. In addition, aphids can vector plant diseases with the potential to greatly increase crop losses (Minks & Harrewijn 1988). To date, most research on invasive aphids has focused on aphids as economic pests of agricultural, horticultural, and forestry crops. More recently, however, concerns have also concentrated on both the potential damage invasive aphids can inflict on native plants (especially endangered species) and their ecological effects on plant and insect communities (Teulon & Stufkens 2002, Messing *et al.* 2007, Andersen *et al.* 2008).

Colonization success of any invader is highly dependent on its ability to survive Allee effects (bottlenecks of a founder population). In the case of aphids the capacity for parthenogenesis and anholocycly lower the Allee population threshold considerably, making aphids more likely to establish in new areas (Mattson *et al.* 2007). Furthermore, these small insects can be very difficult to detect and have the ability for long distance dispersal as winged morphs. Given these attributes and favorable circumstances (presence of host plants, absence of predators, and suitable weather), very few individuals need to be introduced to result in establishment (Dixon 1987, 1998). For these reasons, aphids account for a substantial proportion of introductions of exotic insects worldwide (Teulon & Stufkens 2002, Messing *et al.* 2007).

Aphids generally cannot survive for very long when not feeding on plants, and thus, successful invasions are most likely to result from introduction of reproductive females on their host plant. Alternatively, invasions may derive from eggs, as their small size and relatively long dormancy allows them to survive long periods without sustenance (Teulon & Stufkens 2002).

The majority of the plants used in agriculture, horticulture, and forestry did not evolve in the areas where they are cultivated, but have also been introduced, and introduction of aphids along with crop plants has occurred with extraordinary frequency worldwide. For example, in New Zealand approximately 90% of the 120 aphid species present are not indigenous, and included among these are all of the recognized pest species on cultivated plants (Teulon & Stufkens 2002). In Hawaii, where no aphids are native, nearly 100 aphids species are documented to occur in the islands; every one an inadvertently introduced species (Messing *et al.* 2007).

In many cases, aphids were introduced with crops long ago and have come to be regarded as part of the established pest complex of a region; e.g. the pea aphid, *Acyrthosiphon pisum* (Harris), is native to Europe but has been established in the US since 1929 (Cameron & Milner 1981). Aphid invasions of major significance still occur with alarming frequency, however. For example, the soybean aphid, *Aphis glycines* Matsumura, was identified in North America in 2000 (although, it was probably present in the US in 1995) (Ragsdale *et al.* 2004). Regardless of whether an introduction is recent or not, the introduced species can validly be referred to as an 'invasive'. In this chapter, however, we will focus on relatively recent invasives, whose control is still regarded as a relatively recent problem.

As aphids are hemipteran pests with a piercing-sucking mode of feeding, work toward microbial control has focused on the fungi, the only non-aquatic microbial control agents capable of initiating infection via direct penetration of the aphid integument. All of the most prevalent and widely encountered fungi infecting aphids belong to the Order Entomophthorales (Phylum Zygomycota) (Dean & Wilding 1973, Milner *et al.* 1980, Dedryver 1983, Feng *et al.* 1991, Wraight *et al.* 1993, Steinkraus *et al.* 1995, Hatting *et al.* 2000, Nielsen & Hajek 2005, Scorsetti *et al.* 2007). Prevalence of infection may in some periods exceed 80%, indicating the possibility of utilising entomophthoralean fungi for microbial control of aphids (e.g. Latgé & Papierok 1988, Pell *et al.* 2001, Nielsen 2002, Shah & Pell 2003). The species of entomophthoralean fungi identified from aphids belong to eight genera: *Batkoa, Conidiobolus, Entomophaga, Entomophthora, Pandora, Erynia, Neozygites* and *Zoophthora* (Latgé & Papierok 1988, Humber 1989, 1992, 1997, Bałazy 1993, Keller 2006). Altogether 29 species of Entomophthorales have been described infecting aphids and, with few exceptions, these have host ranges restricted to the family Aphididae (Keller 2006). Nevertheless, *Zoophthora radicans*, has been recorded from numerous insect groups, but the literature suggests that individual isolates are best adapted to infect taxonomically related insect hosts (suggesting that *Z. radicans* comprises a species complex) (Milner & Mahon 1985, Keller 2006). Some species have a nearly cosmopolitan occurrence; e.g. *Pandora neoaphidis* (Remaudière et Hennebert) Humber is known from all continents except Antarctica. In contrast, others have a more restricted distribution: *Zoophthora anhuiensis* (Li) Humber, for example, is known only from its type locality, Anhui Province, China (Keller 2006). Certain aphid species seem particularly susceptible to certain fungi; for example, the cotton aphid is most frequently attacked by *Neozy-*

gites fresenii (Nowakowski) Batko (Steinkraus *et al.* 1995); the spotted alfalfa aphid, *Therioaphis trifolli* (Monell) f. *maculata*, an invasive in Australia and New Zealand, by certain isolates of *Z. radicans* (Milner 1997); the Russian wheat aphid, *Diuraphis noxia* (Mordvilko), an invasive of North America and South Africa, by *P. neoaphidis* (Feng *et al.* 1991, Wraight *et al.* 1993, Hatting *et al.* 2000); and the blue-green aphid, *Acyrthosiphon kondoi* Shinji & Kondo, an invasive of North America, Australia and New Zealand, by *Pandora kondoiensis* (Milner) Humber (Milner *et al.* 1983). Cameron & Milner (1981) showed that sympatric Australian populations of the exotic blue-green aphid and the exotic pea aphid were predominantly infected by different entomophthoralean species but that cross-infection was possible under laboratory conditions.

Despite the taxonomic diversity of entomophthoralean fungi infecting aphids, the different species exhibit the same overall pattern of life cycle and ecology (Brobyn & Wilding 1977, Tanada & Kaya 1993, Hajek & St. Leger 1994). From the infected aphids, primary conidia are forcibly discharged. Primary conidia produce secondary conidia, which like the primary conidia, are either forcibly discharged (ballistospores) or are borne on long, slender conidiophores (capilliconidia). Once the conidia land on a susceptible host under favourable conditions they produce germ tubes that directly penetrate the insect cuticle (or first produce appressoria that generate penetration hyphae). In some species, capilliconidia appear to be the principal infectious units (Glare *et al.* 1985, Pell *et al.* 2001). Once the fungus has penetrated the cuticle, it begins to multiply and, after a period, virtually all host tissues are invaded and the insect dies. The life cycle of the fungus can then follow one of three paths: (1) conidiophores emerge through the insect integument and conidia are formed, (2) zygo- or azygospores (resting spores) are formed or (3) in some cases dead hosts may support production of conidia (usually low numbers) in addition to resting spores (Brobyn & Wilding 1977, Zimmermann 1978, Tanada & Kaya 1993, Hajek & St. Leger 1994). Resting spores are not known from all entomophthoralean species (Nielsen 2002, Keller 2006, Nielsen *et al.* 2008). Survival structures are unknown in many species only because their basic biology has not been thoroughly studied. In other cases, e.g. *P. neoaphidis*, the pathogen-host interactions have been extensively studied over many years without any certain documentation of *in vivo* resting spores. It is generally believed that species lacking resting spores have other means of surviving unfavorable conditions (see review by Nielsen *et al.* 2008).

Entomophthoralean fungi are also characterized by extremely rapid modes of sporulation and infection, traits that enable them to exploit limited windows of time (e.g. overnight periods) when environmental conditions are favorable for their growth and development.

In addition to the entomophthoralean fungi, aphids have also been found naturally infected with *Beauveria bassiana* (Balsamo) Vuillemin, *Isaria fumosorosea* Wize (formerly assigned to the genus *Paecilomyces*), and *Lecanicillium* spp. (species previously identified as *Verticillium lecanii* (Zimmerm.) Viegas) (Hatting *et al.* 1999, 2000, Feng *et al.* 1990, Wraight *et al.* 1993, Nielsen & Hajek 2005). These fungi belong to the order Hypocreales (Ascomycota). Although fungi identified as

V. lecanii have been reported causing natural epizootics in aphid populations in the tropics and sub-tropics (Shah & Pell 2003), epizootics of Hypocreales have never been observed in aphids under natural conditions in the temperate zone. Like several of the entomophthoralean fungal species, Hypocreales also have a cosmopolitan distribution (Tanada & Kaya 1993, Inglis *et al.* 2001, Roberts & St. Leger 2004). Hypocreales routinely lack the teleomorphic stage (sexual stage) of their life cycle and exist only as anamorphic stages. Furthermore, these fungi are natural enemies of an extraordinarily broad range of insects and arachnids and none is restricted to the Aphididae (Tanada & Kaya 1993, Inglis *et al.* 2001, Roberts & St. Leger 2004). After infection, Hypocreales grow throughout the host and subsequently produce conidia externally on the cadavers. In contrast to the conidia of the Entomophthorales, these conidia are not actively discharged from the cadavers, although they can be wind dispersed. Compared to the Entomophthorales, successful infections by Hypocreales generally require larger numbers of conidia and longer periods of time under favourable conditions (more time is required for the processes of conidial germination and host penetration). In addition, because the conidia are not actively discharged, high rates of infection may be dependent upon the host moving into contact with concentrated sources of conidia (masses of conidia on cadavers or depositions of conidia on foliage or other substrates) (Milner 1997, Shah & Pell 2003). Besides the marked differences in life cycles, another significant difference between the Entomophthorales and Hypocreales is that the infectious propagules of many hypocrealean species are easily mass produced on artificial media (*in vitro*) and readily formulated as biopesticides (Tanada & Kaya 1993, Milner 1997, Shah & Pell 2003, Faria & Wraight 2007) in contrast to entomophthoralean fungi that are more difficult to mass produce. An important consequence of these numerous differences is that the use of these two groups of fungi for biological control has involved markedly different strategies. The Entomophthorales have been used primarily for classical and conservation biological and the Hypocreales for augmentation biological control (e.g. Milner 1985, 1997, Shah & Pell 2003). Examples of the implementation of these various strategies against invasive aphid species will be presented later in this chapter.

6.2 Natural Occurring Pathogens Found on Invasive Aphids

Reports on surveys of natural occurrence of pathogens on invasive aphids are numerous from all over the world (e.g. Milner *et al.* 1980, Feng *et al.* 1991, Wraight *et al.* 1993, Steinkraus *et al.* 1995, Hatting *et al.* 2000, Nielsen *et al.* 2001, Nielsen & Hajek 2005). When an insect species invades a new area, it is rarely accompanied by all the natural enemies that regulated its populations in its native range (certainly not the complex of natural enemies that is most often associated with aphids). Thus the exotic aphid may escape those natural enemies that are most highly adapted to it. This was, for example, the case when the spotted alfalfa aphid was introduced into Australia; surveys documented the absence of infections by *Z. radicans*, one of the most important fungal pathogens of this pest in both North America and Israel (Milner *et al.* 1980).

In other cases the complex of fungal entomopathogens infecting an exotic aphid in a new region is similar to the complex known from the native range of the aphid. Such is the case of the Russian wheat aphid in the US and South Africa (Wraight *et al.* 1993, Hatting *et al.* 2000, and see review by Poprawski & Wraight 1998). In such cases, it is unlikely that the exotic aphid was accompanied by the pathogens from its native region, but rather that the aphid has come under attack by virulent local strains of the pathogens. The true origins of such strains have rarely been investigated, but studies of the Russian wheat aphid in the US support this hypothesis, since *Entomophthora chromaphidis* Burger & Swain frequently infects several native aphids as well as the Russian wheat aphid in the US (Feng *et al.* 1991, Wraight *et al.* 1993), but is not known from Eurasia, the native area of the Russian wheat aphid. It is noteworthy that in some cases, results from field surveys suggest that exotic aphid species have more natural enemies in their expanded ranges than in their regions of origin (e.g. see Poprawski & Wraight 1998, Wu *et al.* 2004, Nielsen & Hajek 2005); however, this may be merely a result of sporadic surveys in low aphid populations in the countries of origin.

6.2.1 Case Study: The Russian Wheat Aphid, **Diuraphis noxia,** in the US and South Africa

The Russian wheat aphid is a cyclical pest of wheat and other small grains throughout its native range in Central Asia. Outbreak populations in Asia and in many invaded regions of the world have caused great economic damage since the late 1800s. Following devastating outbreaks in South Africa in the late 1970s and in the US and Canada in the mid 1980s, investigations were initiated to identify potential means for biological control. In addition to searches for Russian wheat aphid parasites and predators, surveys were also conducted to identify pathogens.

Surveys of cereal aphid populations worldwide have consistently identified large complexes of fungal pathogens, and Russian wheat aphid has generally been found susceptible to infection by many of the most common species. Pathogens from both of the most prominent groups of entomopathogenic fungi, the Hypocreales and Entomophthorales, have been identified as pathogenic to the Russian wheat aphid; however, only entomophthoralean species have been observed at epizootic levels in the field.

The entomophthoralean fungus *P. neoaphidis* was generally identified as the single most important fungal pathogen of cereal aphids, including the Russian wheat aphid. This fungus has been frequently identified from high-level epizootics that decimate populations of the Russian wheat aphid. Other common aphid-pathogenic Entomophthorales, including *Entomophthora, Conidiobolus, Neozygites,* and *Zoophthora* spp., have also been isolated with varying frequency from Russian wheat aphid populations (see review by Poprawski & Wraight 1998).

6.2.2 Case Study: The Soybean Aphid, Aphis glycines, *in the US*

The invasive soybean aphid is native to China but has been present in the US since at least 2000, when it was first found in the midwest. Soybean aphid spread very quickly across the Midwest and by the summer of 2001 it was also found in northeastern states (e.g. Losey *et al.* 2002). By 2004, the soybean aphid had expanded its range to most of the soybean-growing regions in the US (21 states) and other parts of North America (Ragsdale *et al.* 2004). Soybean aphids can quickly increase to very high population densities that, during early soybean reproductive stages, can cause reduced pod set and reduced plant height associated with major yield losses (Heimpel & Shelly 2004).

The diversity, abundance, and potential of naturally occurring entomopathogenic fungi for controlling the soybean aphid in the northeastern US were studied by Nielsen & Hajek (2005). They found seven aphid-pathogenic species: *L. lecanii, Conidiobolus thromboides* Drechsler , *E. chromaphidis, P. neoaphidis, Pandora* sp., *Zoophthora occidentalis* (Thaxt.) Batko, and *N. fresenii*, the first belonging to the order Hypocreales, the latter six from the order Entomophthorales. *P. neoaphidis* was the most dominant of all species documented and accounted for more than 90% of mycosed aphids; it was also the only species that occurred at epizootic levels. Mycosis was strongly correlated with aphid density, especially when aphid populations were increasing. In agreement, epizootic levels of infection were associated with subsequent declines in aphid populations. Although impacts of entomopathogenic fungi have not been quantified, reports from other states in the US and Canada support the findings that entomophthoralean fungal infections are common in populations of the soybean aphid (Rutledge *et al.* 2004). In contrast to the situation in the US, soybean aphids are only sporadic pests in China and South East Asia and rarely reach pest status in Japan. In China and Indonesia, natural enemies can maintain soybean aphid populations at low densities when insecticides are not applied (Heimpel *et al.* 2004). At present it is still too early to predict if the status of the soybean aphid in the US will, over time, decline to that of a sporadic pest as a result of effective control by naturally occurring entomophthoralean fungal pathogens acting in concert with other natural enemies.

6.2.3 Case Study: The Green Spruce Aphid, Elatobium abietinum, *in Iceland*

Anholocyclic populations of the green spruce aphid were accidentally introduced into Iceland, probably on Norway spruce, *Picea abies* (L.) Karst., imported from Denmark. The first individuals of the green spruce aphid were observed in Reykjavik in 1959 on Sitka spruce (*Picea sitchensis* (Bong.) Carr.), a non-native tree in Iceland, and have since then spread throughout Iceland. In the first three decades after the introduction, the green spruce aphid colonized spruce plantations in the southern and the eastern parts of Iceland, but from the 1990s the aphid was also found in plantations in the west and north (Carter & Halldórsson 1998). Sigurdsson *et al.* (1999) found that the green spruce aphid in Iceland consisted of two genetically distinct populations, one in the eastern and the other in the western part of the country; these populations most probably originate from two distinct introductions.

The natural occurrence of entomophthoralean fungi infecting the green spruce aphid in Iceland was studied from 1994 to 1999 by Austarå *et al.* (1997) and Nielsen *et al.* (2001). Only two species in the Entomophthorales, *N. fresenii* and *Entomophthora planchoniana* Cornu, were recorded in populations of the green spruce aphid. Later studies from 2003 to 2005 documented sporadic infection with *P. neoaphidis* and *Entomophaga pyriformis* Bałazy in a limited number of sites (J Eilenberg personal communication). These findings contrast with those from Europe, where six species of entomophthoralean fungi have been recorded from this host aphid (Austarå *et al.* 1997, Nielsen *et al.* 2000). Based on these Icelandic studies conducted from 1994 to 2005, *E. planchoniana* was the only species documented in the western part of Iceland while *N. fresenii* was the only species found in the eastern part. The geographical distributions of the two entomophthoralean fungi were identical to the known distributions of the two populations of the green spruce aphid earlier documented in Iceland (Sigurdsson *et al.* 1999). It is therefore possible that at each introduction the green spruce aphid carried only one of the six species of entomophthoralean fungi occurring in the rest of Europe. If this hypothesis is true, the dispersal of the fungi has apparently followed that of the host aphid population.

In addition, Nielsen *et al.* (2001) studied the ecological and physiological host range of entomophthoralean fungi on a range of native and non-native aphid species in Iceland. On aphid species occurring in the areas around Sitka spruce (excluding *E. abietinum*), four species of entomophthoralean fungi were documented throughout the country: *N. fresenii, E. planchoniana, P. neoaphidis* and *Conidiobolus obscurus* (Hall & Dunn) Remaudière & Keller. This strongly indicates that there are few or no interactions between the green spruce aphid and other aphids (native as well as exotic species) with regard to the entomophthoralean fungi attacking them. We believe that this is mainly due to a separation in time and space rather than a lack of susceptibility to specific fungal isolates. In Iceland peak populations of the green spruce aphid occur over the winter period, whereas most other aphid species in Iceland peak over the summer period. Furthermore, the spread of the green spruce aphid, and consequently, of its fungal diseases has probably been slow because most spruce plantations in Iceland are small and isolated.

6.3 Classical Biological Control

Classical biological control relies on introduction of exotic, usually co-evolved, biological control agents for permanent establishment and long-term pest control. The goal of the introduction is to release a natural enemy in a new area for permanent establishment so it creates a self-maintaining system (Carruthers *et al.* 1991, Eilenberg *et al.* 2001, Pell *et al.* 2001). At present only three attempts to establish entomophthoralean fungi for classical biological control of aphids have been recorded. This includes the release of *Z. radicans* to control the spotted alfalfa aphid in Australia (Milner *et al.* 1982), release of *Z. radicans* to control the Russian wheat aphid in the US (Poprawski & Wraight 1998), and *N. fresenii* has been released in areas with cotton aphid infestations in California (Pell *et al.* 2001, Steinkraus *et al.* 2002).

6.3.1 *The Spotted Alfalfa Aphid,* **Therioaphis trifolii** *f.* **maculata** *in Australia*

The spotted alfalfa aphid became an important pest of pasture and legume when it was first introduced to Australia in 1977. Initially, spotted alfalfa aphids were sampled in Australia in order to determine the natural occurrence of fungal infections. Although six species of entomopathogenic fungi were documented, the prevalence was very low, and *Z. radicans* was never recorded, even though this fungus seemed to be the most significant pathogen of the spotted alfalfa aphid in the US and Israel (Milner & Soper 1981). Therefore, *Z. radicans* was considered to have a great potential for classical biological control, and one isolate originating from an infected alfalfa aphid in Israel was released in Australia, either by placing sporulating *in vitro* cultures on plants with aphids or by releasing laboratory-infected aphids (Milner *et al.* 1982). Although the infection level was low the first few weeks after the release, an epizootic was in progress 41 days later, with up to 88% infection 3 m from the release point and 22% infection 14 m from the release point (Milner *et al.* 1982). No rain was recorded during the study but disease transmission was correlated with periods of high humidity. After release, *Z. radicans* became established in the area and apparently persisted during unfavorable periods as resting spores (Milner *et al.* 1982, Milner & Lutton 1983, Milner 1985). In 1985 Milner reported that the introduced pathogen was widely distributed over New South Wales and Queensland, causing epizootics in late summer/autumn.

6.3.2 *Case study: The Russian Wheat Aphid,* **Diuraphis noxia,** *in the US*

Based on the previous success with introduction of *Z. radicans* against spotted alfalfa aphid, this pathogen species was selected in the early 1990s for introduction against Russian wheat aphid in the western US. Two isolates of *Z. radicans* were identified as pathogenic to the Russian wheat aphid in laboratory bioassays, and one of these, a Yugoslavian isolate from the cicadellid *Empoasca vitis* Gothe, was released into *D. noxia*-infested wheat fields near Parma, Idaho, US, in the spring (mid-May) of 1992 (Poprawski & Wraight 1998). The introductions were made at two locations (4 × 4 m plots) within a 1.2 ha research field; the field was irrigated prior to the experiment to create conditions favorable for fungal infection. The fungus was released using two markedly different methods. In the first, adults of the aphid parasitoid *Aphelinus asychis* (Walker) were inoculated with a high dose of *Z. radicans* conidia and released into the field. The second method involved cutting small pieces of actively sporulating mycelium from pure cultures of the fungus and inserting these into the rolled flag leaves of wheat tillers harboring large colonies of the Russian wheat aphid. Sampling of aphids at the parasitoid release site revealed high levels of parasitism, but no detectable fungal infections. Aphids from colonies directly exposed to the sporulating mycelium were successfully infected, but only at a very low level (5%), and samples collected over the course of the field season failed to detect spread of the fungus to untreated plants. Dissections of aphid cadavers indicated that the fungus produced resting spores rather than conidia (the infectious propagules involved with epizootic spread), and this was ultimately attributed to the cool weather conditions that prevailed at the time of the release. The study was concluded at the end of the season, and no surveys for *Z. radicans* were conducted during subsequent years.

6.3.3 *The cotton aphid,* Aphis gossypii, *in California*

Classical biological control introductions of *N. fresenii* against the cotton aphid were made in 1994 and 1995 in Californian cotton fields. The pest status of this aphid increased during the nineties, but there was no evidence of *N. fresenii* infection, even though this fungus commonly causes epizootics in *A. gossypii* populations in other cotton growing regions of the US (Steinkraus *et al.* 1995, 2002). Isolates of the fungus originating from infected cadavers collected in Arkansas were released in cotton fields in California. After the release, *N. fresenii* persisted and spread in the aphid population over at least a few months. The control provided was reported to be moderately successful. However, whether the pathogen was able to persist in California and cause long-term suppression of the cotton aphid populations is still unknown (Pell *et al.* 2001, Steinkraus *et al.* 2002).

6.3.4 *Future Potential*

The future potential of classical biological control of aphids is generally regarded as limited. In all cases a careful investigation of the natural occurrence of fungi in the system in question is needed before classical biological control can be used. The highest potential will definitely be in situations where significant host specific pathogens are missing from the natural enemy complex in the new area of infestation (as was the case for *Z. radicans* on the spotted alfalfa aphid), or in cases where the new invasive aphid pest is separated in time and space from local entomophthoralean fungi. We suggest that classical biological control of the green spruce aphid in Iceland using entomophthoralean fungi would be an obvious possibility. Both *E. planchoniana* and *N. fresenii* cause epizootics in populations of the green spruce aphid in Iceland, but *E. planchoniana* was exclusively found in the western part of Iceland, while *N. fresenii* was found only in the eastern part of the country (Nielsen *et al*. 2001). A possibility for classical biological control would then be to release *E. planchoniana* into the eastern population of the green spruce aphid and *N. fresenii* into the western population to test whether these species can establish, proliferate and regulate the aphid populations. Six other species of entomophthoralean fungi have been recorded from the green spruce aphid in the northwestern part of Europe (Austarå *et al*. 1997, Nielsen *et al*. 2000). A release of one or several of these pathogens may lead to establishment and increased control of the green spruce aphid in Iceland.

6.4 Conservation Biological Control

The goal of conservation biological control is to enhance the naturally occurring enemy populations, e.g. by irrigation, by establishment of persistent herbaceous field margins, by reduction of pesticide application, or by reduced plowing or soil excavation (Eilenberg *et al*. 2001). Nevertheless, only very few attempts at habitat manipulation for conservation biological control of aphid have been reported in the literature. However, an excellent example of conservation biological control is provided by a service to predict epizootics of *N. fresenii* in cotton aphid populations in the mid-south and southeastern US and thereby prevent unnecessary insecticide applications. While this aphid is not considered native to the US, it has been established in the southeastern US for so many years that it is not considered a new problem and therefore, we recommend several reviews describing this program (e.g. Steinkraus *et al*. 1996, 1998).

6.4.1 Case Study: The Russian Wheat Aphid, **Diuraphis noxia,** *in the US and South Africa*

Field studies have revealed a strong correlation between moisture conditions and prevalence of the Russian wheat aphid fungal pathogens (Wraight *et al*. 1993, Feng *et al*. 1992). These observations suggest that manipulation of moisture conditions via irrigation may be a useful approach toward maximizing efficacy of fungal pathogens for biological control. This strategy, however, has not been rigorously investigated and will require a greater knowledge of the moisture requirements necessary for occurrence of fungal epizootics.

It has also been hypothesized that modification of the microhabitat (leaf rolling) induced by feeding of the Russian wheat aphid effectively shields this pest from naturally-disseminated (or inundatively-applied fungal propagules) (Feng *et al*. 1991, Knudsen & Wang 1998, Vandenberg *et al*. 2001) and that resistant wheat varieties exhibiting limited leaf rolling might support greater fungal activity (Vandenberg *et al*. 2001). This hypothesis, however, has not been supported by results from field tests of susceptible versus resistant varieties (Vandenberg *et al*.

2001, Hatting et al. 2004). Hatting et al. (2004) suggested that greater exposure of aphids to fungal inoculum and loss of favorable microhabitats due to reduced leaf roll could have offsetting effects.

6.4.2 Future Potential

No further examples of conservation biological control of invasive aphids have been reported in the literature. However, some of the basic biological surveys performed over the years suggest where efforts might be directed in order to improve the use of this strategy. For example, active inocula of *P. neoaphidis* and *C. obscurus* were present in soil in early spring, even though a winter had passed since aphids were present on plants above the soil (Nielsen et al. 2003). Soil is probably an important reservoir of inoculum needed for initiating infections in aphid populations in spring, especially since large numbers of aphids fall to the ground each day where they could contact inoculum (Sopp et al. 1987). The impact of this inoculum resource in the soil might be enhanced by reduced tillage of fields or avoidance of pesticides during critical time periods for epizootic development. To date, this has not been demonstrated for entomophthoralean fungi; however, Klingen et al. (2002) showed a significantly higher occurrence of insect pathogenic hypocrealean fungi in soils from organically managed fields versus conventionally managed fields.

The impact of aphid hosts feeding on weeds in crop fields and in field margins on the development of epizootics in populations of aphid pests varies among systems. Keller & Suter (1980) and Keller & Duelli (1998) suggested that epizootics in aphid populations in crop fields are initiated from aphid populations in weeds. Feng et al. (1991) also suggested that the early development of epizootics in populations of the Russian wheat aphid depends on the presence of other aphid species in the surrounding areas. Thus, management of field borders to maintain non-pest aphid populations alongside crops, could be investigated for enhancing development of epizootics in aphid populations in crops.

Interactions of aphid-pathogenic fungi with aphid predators and parasitoids may be both positive and negative. Predators and parasitoids may assist in the transmission of entomophthoralean fungi as passive vectors (Pell et al. 1997, Poprawski & Wraight 1998, Roy et al. 1998). Another positive effect of predators might be that the aphids drop to the soil more frequently in the presence of predators and thus predators may promote the likelihood of aphids contacting fungal inoculum in the soil or spreading inoculum to neighbouring colonies during epizootics.

The conservation biological control strategy is very attractive, since no fungal agent is released for control. However, as discussed above, it is also a very challenging strategy. The strategy relies heavily on a detailed understanding of the ecology of both the natural enemy and its host insect as well as management practices (Hajek 1997).

6.5 Inoculation Biological Control

Inoculative biological control relies on the release of a pathogen with the expectation that it will proliferate. Permanent pest control is not expected, and additional releases are therefore necessary, e.g. each season (Eilenberg *et al.* 2001). Usually the biological control agent is introduced before the damaging pest is anticipated so that it may build up to levels required for pest control (Carruthers & Hural 1990, Carruthers *et al.* 1991).

Several attempts to utilise fungi for inoculation biological control of aphids have been reported in the literature, both for invasive as well as native aphid species (Dedryver 1979, Wilding 1981a, b, Latteur & Godefroid 1983, Wilding *et al.* 1986, Silvie *et al.* 1990, Poprawski & Wraight 1998).

6.5.1 Case Study The Russian Wheat Aphid, **Diuraphis noxia,** *in the US*

As indicated previously, the fungus *P. neoaphidis* is an important natural regulator of cereal aphid populations worldwide. However, the frequently spectacular epizootics this fungus produces usually occur late in the growing season, after pest populations have reached damaging levels (Feng *et al.* 1991, 1992, Wraight *et al.* 1993). This is particularly problematic in light of the high fecundity and rapid population growth potential of aphid pests. Thus, in conjunction with the above-described *Z. radicans* studies, Poprawski & Wraight (1998) also investigated the potential for early-season releases of *P. neoaphidis* to substantially augment rates of infection and initiate epizootics earlier in the growing season. Using the method of direct inoculation of sporulating fungal mycelium into rolled flag leaves, these researchers introduced a large amount of infectious inoculum (conidia) of a locally collected isolate of *P. neoaphidis* into a small plot near the center of an infested wheat field near Parma, Idaho, on 29 May 1992. A localized epizootic was successfully initiated and rates of infection at the treatment site increased to 44% within 3 weeks. Spread of the pathogen into the surrounding areas of the field was exceedingly slow, however, and the pathogen did not begin to significantly impact the aphid populations until the pest had reached a highly damaging level.

Development of entomophthoralean epizootics in cereal aphid populations has been correlated with host density (Feng *et al.* 1992); however, Poprawski & Wraight (1998) concluded that the slow rate of epizootic development in this case was most likely attributable to unfavorable weather conditions. During the early part of the 1992 season, temperatures and wind speeds fluctuated greatly, and daily and nightly moisture levels were extremely erratic. Numerous studies have documented the exceptional dependence of many entomophthoralean pathogens on regular (daily) periods of high moisture and moderate temperature conditions (Missonnier *et al.* 1970, Milner & Bourne 1983, Galaini-Wraight *et al.* 1991). Overnight periods of high moisture (especially in the form of dew) have been shown to support epizootics during rain-free, apparently dry periods (Galaini-Wraight *et al.* 1991). Results of these studies indicate that successful inoculative augmentation of entomophthoralean pathogens will likely also require manipulation of environmental conditions, e.g. irrigation of fields when *P. neoaphidis* is released could provide the moisture necessary for epizootics to occur.

An obvious solution to the problem of slow pathogen spread is broadcast application of fungal materials, either in the form of infectious spores or granules of stabilized mycelium capable of producing infectious units post-application. As fresh mycelium has a very limited shelf life, successful stabilization is dependent upon desiccation tolerance. Unfortunately, the infectious units of most entomophthoralean fungi are thin-walled, short-lived conidia with limited potential for formulation as biopesticides. In addition, stabilization and formulation of entomophthoralean mycelia has also proven difficult. Substantial success has been achieved only with *Z. radicans* and to some extent *P. neopahidis* (McCabe & Soper 1985, Leite 1991, Shah *et al.* 2000, Wraight *et al.* 2003), and mycelial formulations of these fungi have not yet been tested against any aphid pests of small grains. Use of this technology has thus largely been restricted to those entomopathogenic fungi that produce desiccation-tolerant mycelia. Principal among these are the hypocrealean fungi, including species from such well-known genera as *Beauveria, Lecanicillium* and *Isaria*. Species of these fungi are naturally-occurring pathogens of the Russian wheat aphid and other cereal aphids, but infection levels in the field are usually very low (Feng *et al.* 1990, Hatting *et al.* 1999).

In the 1990s, the potential for using pelletized mycelium of *B. bassiana* for inoculation biological control of the Russian wheat aphid was investigated in a series of studies conducted at the University of Idaho (Knudsen *et al.* 1990, Knudsen *et al.* 1991, Knudsen & Wang 1998). Dried mycelial pellets of *B. bassiana* (formulated with alginate as a humectant) were applied against the Russian wheat aphid infesting wheat or barley in field cages. Results from tests conducted over two field seasons were disappointing in that resulting rates of infection did not exceeded 18% (Knudsen & Wang 1998). It was concluded that the foliar microenvironment was too dry to support *B. bassiana* epizootics.

6.5.2 Future Potential

Due to the epizootic potential of most entomophthoralean fungi inoculation biological control is usually regarded as having significant potential for this fungal order. Furthermore, inoculation biological control has the advantage that only a limited amount of inoculum is released at each treatment. All of the studies referred to above encourage further investigations with fungi for inoculation biological control of aphid pests. In particular high value crops may provide great potential, since such crops are less sensitive to increased treatment cost. Future trials aiming at controlling aphid pests should especially focus on a range of both abiotic and biotic parameters critical to the biology and ecology of fungal pathogens and aphid hosts. In particular, knowledge is needed on impact of other insects as vectors of aphid pathogenic fungi, optimisations of application time and fungal preparations as well as impact of humidity and temperature. In addition, molecular methods enabling tracing of the released isolates should be developed for further studies.

6.6 Inundative Biological Control

Inundative biological control relies on the use of living organisms to control pests when control is achieved exclusively by the released organisms themselves (Eilenberg *et al.* 2001). By this approach, the pathogen is introduced in large quantities into the pest population, often several times each growing season. In inundation biological control the entomopathogens are often referred to as 'biopesticides' or 'mycopesticides' reflecting that the strategy, in many respects, resembles the chemical control paradigm of insect control (Hajek 1997).

6.6.1 Case Study: The Russian Wheat Aphid, Diuraphis noxia, in the US and South Africa

Despite its generally low epizootic potential in crop canopies, *B. bassiana* is well known as one of the most successful fungal agents developed for inundation biological. Conidia of this pathogen have been formulated as numerous experimental and commercial biopesticides (Faria & Wraight 2007), and availability of these products has stimulated evaluations of their efficacy against a broad range of insect pests. *B. bassiana* strain GHA is a prominent commercial isolate available as emulsifiable oil dispersion (OD) and wettable powder (WP) formulations (Faria & Wraight 2007). These materials have been tested against field populations of *D. noxia* in the US and South Africa.

In the US, both the OD and WP formulations were applied against the Russian wheat aphid in Idaho over four field seasons at rates of 2.5–5×10^{13} conidia/ha to small plots of irrigated spring and winter wheat using a backpack hydraulic sprayer (Vandenberg *et al.* 2001). The high-rate applications in most cases produced substantial (>70%) reductions in Russian wheat aphid. However, the residual efficacy of fungal treatments was limited, suggesting that multiple applications would be required for effective control. Considering the economic constraints to biopesticide applications in small grains, Vandenberg *et al.* (2001) pursued an alternative approach of applying the fungus through an overhead irrigation system. Single applications of the above-described high rate in large field plots caused reductions in numbers of the Russian wheat aphid similar to those achieved with the hydraulic spray applications and 25–78% reductions in aphid infestation rates within 2 weeks after application. Vandenberg *et al.* (2001) ultimately concluded that additional study would be necessary to optimize rate and timing of applications if *B. bassiana* were to be effectively used for management of pests on low-value-per-hectare crops such as small grains.

Optimal timing of *B. bassiana* applications was investigated by Hatting *et al.* (2004) in dry-land wheat in South Africa. Weekly applications of the above-described OD formulation of strain GHA produced population reductions comparable to those reported by Vandenberg *et al.* (2001); however, the observed control was determined to have resulted almost exclusively from the application made at the time of aphid colonization of the emerging flag leaf. Hatting *et al.* (2004) hypothesized that aphids migrating to the flag leaves were directly impacted by the spray application or acquired lethal doses of conidia via secondary pick-up from the treated foliage. These results suggest that aphid movement might be exploited to enhance efficacy of mycoinsecticide applications. Aphid movement behavior was also identified by Knudsen *et al.* (1994) and Knudsen & Schotzko (1999) as an important factor in development of fungal epizootics.

In addition to *B. bassiana*, the fungus *I. fumosorosea* has been investigated for inundation biological control of Russian wheat aphid. Vandenberg *et al.* (2001) included this pathogen in two of the small-plot tests of *B. bassiana* in Idaho (see above). Results with *I. fumosorosea* were inconsistent; in one test, efficacy of *I. fumosorosea* formulated as either conidia or blastospores and applied at a rate of 5×10^{13} spores/ha, was equivalent to that of *B. bassiana*. In a second trial, application of *I. fumosorosea* blastospores was not associated with decreased populations of the Russian wheat aphid.

6.6.2 Future Potential

Mass production/formulation of the infectious units (conidia) of the most important natural pathogens of aphids, the entomophthoralean fungi, remains a difficult challenge, and until this problem is overcome, the potential of these agents for inundation biological control will remain unrealized. Greater potential exists for production of mycelial formulations of these fungi for use in inoculation biological control. While hypocrealean pathogens such as *B. bassiana* are more readily mass produced, they exhibit markedly slower action than the Entomophthorales and thus are less well adapted for exploiting the often limited or punctuated periods of favorable environmental conditions that characterize crop canopies. Efficacy of these fungi is especially limited in semi-arid environments and against pests like aphids that exhibit extraordinarily high rates of reproduction and development. Consequently, high doses and/or multiple, frequent applications of these fungi are usually required to achieve acceptable levels of control and the costs are generally prohibitive. Perhaps the greatest potential of the inundation approach for aphid control lies in its use in conjunction with conservation biological control approaches wherein environmental conditions are modified through irrigation or other mechanisms. There may be greater opportunities for manipulation of environmental conditions in the greenhouse than in the field (e.g. see Hall 1981). Exploitation of overhead irrigation systems to reduce the cost of broadcast application of fungal pathogens (e.g. as described by Vandenberg *et al.* 2001) is one approach that may offer potential for integration of these control agents into management programs for aphid pests in field crops.

6.7 Conclusion

Aphids are one of the most important groups of exotic insect pests. To date, several promising examples of biological control of exotic aphid pests have been reported. Generally it has been observed that successful outcomes of biological control programs targeting invasive aphids with entomopathogenic fungi require a comprehensive understanding of the host/pathogen/host plant system in question. In the future, we predict that a greater understanding of the epizootiology of fungal disease of aphids with the goal of prediction of prevalence of infection may be facilitated by the inclusion of molecular methods as well as mathematical models.

References

Andersen US, Córdova JPP, Nielsen UB, Olsen CS, Nielsen C, Sørensen M, Koolmann J (2008) Conservation through utilisation: a case study of the vulnerable *Abies guatemalensis* in Guatemala. Oryx 42: 206–213

Austarå Ø, Carter C, Eilenberg J, Halldórsson G, Harding S (1997) Natural enemies of the green spruce aphid in spruce plantations in maritime North West Europe. Búvísindi 11: 113–124

Bałazy S (1993) Flora of Poland. Polska Akademia Nauk, Kraków, Poland

Blackman RL, Eastop VF (1994) Aphids on the world's trees: An identification and information guide. CAB International, Wallingford, UK

Blackman RL, Eastop VF (2000) Aphids on the world's crops: An identification and information guide. John Wiley & Sons, New York

Brobyn P, Wilding N (1977) Invasive and developmental processes of *Entomophothora* species infecting aphids. Trans Br Mycol Soc 69:349–366

Cameron PJ, Milner RJ (1981) Incidence of *Entomophthora* spp. in sympatric populations of *Acyrthosiphon kondoi* and *Acyrthosiphon pisum*. NZ J Zool 8:441–446

Carruthers RI, Hural K (1990) Fungi as naturally occurring entomopathogens. In Baker R, Dunn P (eds) New directions in biological control. UCLA Symp Molec Cell Biol 112: 115–138

Carruthers RI, Sawyer AJ, Hural K (1991) Use of fungal pathogens for biological control of insect pests. In: Rice BJ (ed) Sustainable agriculture research and education in the field. National Academy Press, Washington, DC, USA. pp 336–372

Carter CI, Halldórsson G (1998) Origin and background to the green spruce aphid in Europe. In: Day KR, Halldórsson G, Harding S, Straw NA (eds) The Green Spruce Aphid in Western Europe; ecology, status, impacts and prospects for management. Forestry Commission, Technical paper 24:1–14

Dean GJ, Wilding N (1973) Infection of cereal aphids by the fungus *Entomophthora*. Ann Appl Biol 74:133–138

Dedryver CA (1979) Déclenchement en serre d'une epizootie à *Entomophthora fresenii* Nowak. sur *Aphis fabae* Scop. par introduction d'inoculum et régulation de l'humidite relative. Entomophaga 24:443–453

Dedryver CA (1983) Field pathogenesis of three species of Entomophthorales of cereal aphids in Western France. In: Cavalloro R (ed) Aphid antagonists. Commission of the European Communities, Rotterdam. pp 11–19

Dixon AFG (1987) Parthenogenetic reproduction and rate of increase in aphids. In: Minks AK, Harrewijn P (eds) Aphids their biology, natural enemies and control, Vol. 2A. Elsevier, Amsterdam, The Netherland. pp 269–286

Dixon AFG (1998) Aphid ecology. An optimization approach. 2nd edition. Chapman & Hall, London, UK

Eilenberg J, Hajek AE, Lomer C (2001) Suggestions for unifying the terminology in biological control. BioControl 46:387–400

Faria MR de, Wraight SP (2007) Mycoinsecticides and mycoacaricides: A comprehensive list with worldwide coverage and international classification of formulation types. Biol Control 43: 237–256

Feng MG, Johnson JB, Kish LP (1990) Survey of entomopathogenic fungi naturally infecting cereal aphids (Homoptera: Aphididae) of irrigated grain crops Southwestern Idaho. Environ Entomol 19:1535–1542

Feng MG, Johnson JB, Halbert SE (1991) Natural control of cereal aphids (Homoptera: Aphididae) by entomopathogenic fungi (Zygomycetes: Entomophthorales) and parasitoids (Hymenoptera: Braconidae and Encyrtidae) on irrigated spring wheat in Southwestern Idaho. Environ Entomol 20:1699–1710

Feng MG, Nowierski RM, Johnson JB, Poprawski TJ (1992) Epizootics caused by entomophthoralean fungi (Zygomycetes, Entomophthorales) in populations of cereal aphids (Hom., Aphididae) in irrigated small grains in southwestern Idaho, USA. J Appl Entomol 113:376–390

Galaini-Wraight S, Wraight SP, Carruthers RI, Magalhães BP, Roberts DW (1991) Description of a *Zoophthora radicans* (Zygomycetes: Entomophthoraceae) epizootic in a population of *Empoasca kraemeri* (Homoptera: Cicadellidae) on beans in Central Brazil. J Invertebr Pathol 58:311–326

Glare TR, Chilvers GA, Milner RJ (1985) Capilliconidia as infective spores in *Zoophthora phalloides*. Trans Br Mycol Soc 85:463–470

Hajek AE (1997) Ecology of terrestrial fungal entomopathogens. Adv Micr Ecol 15:193–249

Hajek AE, St. Leger RJ (1994) Interactions between fungal pathogens and insect hosts. Ann Review Entomol 39:293–322

Hall RA (1981) The fungus *Verticillium lecanii* as a microbial insecticide against aphids and scales. In: Burges HD (ed) Microbial control of pests and plant diseases. Academic Press, New York. pp 483–498

Hatting JL, Humber RA, Poprawski TJ, Miller RM (1999) A survey of fungal pathogens of aphids from South Africa, with special reference to cereal aphids. Biol Control 16:1–12

Hatting JL, Poprawski TJ, Miller RM (2000) Prevalences of fungal pathogens and other natural enemies of cereal aphids (Homoptera: Aphididae) in wheat under dryland and irrigated conditions in South Africa. BioControl 45:179–199

Hatting JL, Wraight SP, Miller RM (2004) Efficacy of *Beauveria bassiana* (Hyphomycetes) for control of Russian wheat aphid (Homoptera: Aphididae) on resistant wheat under field conditions. Biocon Sci Technol 14:459–473

Heie OE (1980) The Aphidoidea (Hemiptera) of Fennoscandia and Denmark. I.General part. Fauna Entomologica Scandinavica, Vol. 9. Scandinavian Science Press, Klampenborg, Denmark

Heimpel GE, Ragsdale DW, Venette R, Hopper KR, O'Neil RJ, Rutledge CE, Wu Z (2004) Prospects for importation biological control of the soybean aphid: Anticipating potential costs and benefits. Ann Entomol Soc Am 97:249–258

Heimpel GE, Shelly TE (2004) The soybean aphid: A review of its biology and management. Ann Entomol Soc Am 97:203

Humber RA (1989) Synopsis of a revised classification for Entomophthorales (Zygomycotina). Mycotaxon 34:441–460

Humber RA (1992) Collection of entomopathogenic fungal cultures: Catalog of strains. US Dept of Agriculture, Agricultural Research Service, ARS-110. 177pp

Humber RA (1997) Fungi: Identification. In: Lacey, L (ed) Manual of techniques in insect pathology. Academic Press, San Diego, USA. pp 153–186

Inglis GD, Goettel MS, Butt TM, Strasser H (2001) Use of Hyphomycetous fungi for managing insect pests. In: Butt T, Jackson C, Magan N (eds) Fungal biocontrol agents: Progress, problems and potential. Kluwer, Dordrecht, Netherlands. pp 23–69

Keller S (2006) Entomophthorales attacking aphids with a description of two new species. Sydowia 58:38–74

Keller S, Duelli P (1998) Aphidophaga and the case for nature protection and biotype conservation. In: Niemczuk E, Dixon AFG (eds) Ecology and effectiveness of Aphidophaga. SPB Academic Publishing, The Hargue, The Netherlands. pp 95–97

Keller S, Suter H (1980) Epizootiologische Untersuchungen über das *Entomophthora* – Auftreten bei feldbaulich wichtigen Blattlausarten. Acta Æcologica 1:63–81

Klingen I, Eilenberg J, Meadow R (2002) Effects of farming system, field margins and bait insect on the occurrence of insect pathogenic fungi in soils. Agric Ecosyst Environ 91:191–198

Knudsen GR, Schotzko DJ (1999) Spatial simulation of epizootics caused by *Beauveria bassiana* in Russian wheat aphid populations. Biol Control 16:318–326

Knudsen GR, Wang ZG (1998) Microbial control of the Russian wheat aphid (Homoptera: Aphididae) with the entomopathogen *Beauveria bassiana*. In: Quisenberry SS, Peairs FB (eds) Response model for an introduced pest – The Russian wheat aphid. Entomological Society of America, Lanham, MD. pp 209–233

Knudsen GR, Johnson JB, Eschen DJ (1990) Alginate pellet formulation of a *Beauveria bassiana* (Fungi: Hyphomycetes) isolate pathogenic to cereal aphids. J Econ Entomol 83:2225–2228

Knudsen GR, Eschen DJ, Dandurand LM, Wang ZG (1991) Methods to enhance growth and sporulation of pelletized biocontrol fungi. Appl Environ Microbiol 57:2864–2867

Knudsen GR, Schotzko DJ, Krag CR (1994) Fungal entomopathogen effect on numbers and spatial patterns of the Russian wheat aphid (Homoptera: Aphididae) on preferred and nonpreferred host plants. Environ Entomol 23:1558–1567

Latteur G, Godefroid J (1983) Trial of field treatments against cereal aphids with mycelium of *Erynia neoaphidis* (Entomophthorales) produced *in vitro* In: Cavalloro R (ed) Aphid antagonists. AA Balkema, Rotterdam, The Netherlands. pp 2–10

Latgé JP, Papierok B (1988) Aphid pathogens. In: Minks AK, Harrewijn P (eds) Aphids their biology, natural enemies and control, Vol 2B. Elsevier, Amsterdam, Netherlands. pp 323–335

Leite LG (1991) Estudo de alguns fatores que afetam a epizootia de *Zoophthora radicans* e utilização do fungo para o controle de *Empoasca* sp. MS thesis, Univ. São Paulo, Piracicaba, Brazil

Losey JA, Waldron JK, Hoebeke ER, Macomber LE, Scott BN (2002) First record of the soybean aphid, *Aphis glycines* Matsumura (Hemiptera: Aphididae), in New York. Grt Lks Entomol 35:101–105

Mattson W, Vanhanen H, Veteli T, Sivonen S, Niemela P (2007) Few immigrant phytophagous insects on woody plants in Europe: legacy of the European crucible? Biol Invasions 9:957–974

McCabe D, Soper RS (1985) Preparation of an entomopathogenic fungal insect control agent. US Patent 4,530,834

Messing RH, Tremblay MN, Mondor EB, Foottit RG, Pike KS (2007) Invasive aphids attack native Hawaiian plants. Biol Invasions 9:601–607

Milner RJ (1985) Pathogen importantation for biological control – Risk and benefits. In: Gibbs AJ, Meischke HRC (eds) Pest and parasite as migrants. Cambridge University Press, Cambridge, UK. pp 115–121

Milner RJ (1997) Prospects for biopesticides for aphid control. Entomophaga 42:227–239

Milner RJ, Bourne J (1983) Influence of temperature and duration of leaf wetness on infection of *Acyrthosiphon kondoi* with *Erynia neoaphidis*. Ann Appl Biol 102:19–27

Milner RJ, Lutton GG (1983) Effect of temperature on *Zoophthora radicans* (Brefeld) Batko: an introduced microbial control agent of the spotted alfalfa aphid, *Theriooaphis trifolii* (Monell) f. *maculata*. J Aus Entomol Soc 22:167–173

Milner RJ, Mahon RJ (1985) Strain variation in *Zoophthora radicans*, a pathogen on a variety of insect hosts in Australia. J Aus Entomol Soc 24:195–198

Milner RJ, Soper RS (1981) Bioassay of entomophthoralean fungi against the spotted alfalfa aphid *Therioaphis trifolii* f. *maculata*. J Invertebr Pathol 37:168–173

Milner RJ, Teakle RE, Lutton GG, Dare FM (1980) Pathogens (Phycomycetes, Entomophthoraceae) of the blue green aphid *Acyrthosiphon kondoi* Shinji and other aphids in Australia. Austral J Bot 28:601–619

Milner RJ, Soper RS, Lutton GG (1982) Field release of an Israeli strain of the fungus *Zoophthora radicans* (Brefeld) Batko for biological control of *Therioaphis trifoli* (Monell) f. *maculata* Spotted alfalfa aphid, pest of leguminous pastures. J Austral Entomol Soc 21:113–118

Milner RJ, Mahon RJ, Brown WV (1983) A taxonomic study of the *Erynia neoaphidis* Remaudière and Hennebert (Zygomycetes: Entomophthoraceae) group of insect pathogenic fungi, together with a description of the new species *Erynia kondoiensis*. Austral J Bot 31:173–188

Minks AK, Harrewijn P (1988) Aphid pathogens. In: Minks AK, Harrewijn P (eds) Aphids their biology, natural enemies and control, Vol. 2B. Elsevier, Amsterdam, Netherlands. pp V–VII

Missonnier J, Robert Y, Thoizon G (1970) Circonstances épidémiologiques semblant favoriser le développement des mycoses à Entomophthorales chez trios aphides, *Aphis fabae* Scop. *Capitophorus horni* Börner et *Myzus persicae* (Sulz.). Entomophaga 15:169–190

Nielsen C (2002) Interactions between aphids and entomophthoralean fungi: Characterisation, epizootiology and potential for microbial control. Ph.D. thesis. The Royal Veterinary and Agricultural University, Denmark

Nielsen C, Hajek AE (2005) Control of invasive soybean aphid, *Aphis glycines* (Hemiptera: Aphididae), populations by existing natural enemies in New York State, with emphasis on entomopathogenic fungi. Environ Entomol 34:1036–1047

Nielsen C, Eilenberg J, Harding S, Oddsdottir E, Halldórsson G (2000) Entomophthoralean fungi infecting the green spruce aphid (*Elatobium abietinum*) in the north-western part of Europe. IOBC/WPRS Bull 23:131–134

Nielsen C, Eilenberg J, Harding S, Oddsdottir E, Halldórsson G (2001) Geographical distribution and host range of Entomophthorales infecting the green spruce aphid *Elatobium abietinum* Walker in Iceland. J Invertebr Pathol 78:72–80

Nielsen C, Hajek AE, Humber RA, Bresciani J, Eilenberg J (2003) Soil as an environment for winter survival of aphid-pathogenic Entomophthorales. Biol Control 28:92–100

Nielsen C, Jensen AB, Eilenberg J (2008) Survival of entomophthoralean fungi infecting aphids and higher flies during unfavorable conditions and implications for conservation biological control. In: Ekesi S, Maniania NK (eds) Use of entomopathogenic fungi in biological pest management. Research SignPost, Kerala, India. pp 13–38

Pell JK, Pluke R, Clark SJ, Kenward M, Alderson PG (1997) Interactions between two aphid natural enemies, the entomopathogenic fungus *Erynia neoaphidis* Remaudière & Hennebert (Zygomycetes: Entomophthorales) and the predatory beetle *Coccinella septempunctata* L. (Coleoptera: Coccinellidae). J Invertebr Pathol 69:261–268

Pell J, Steinkraus D, Eilenberg J, Hajek A (2001) Exploring the potential of Entomophthorales in integrated crop management. In: Butt T, Jackson C, Magan N (eds) Fungal biocontrol agents: Progress, problems and potential. Kluwer, Dordrecht, Netherlands. pp 71–167

Poprawski TJ, Wraight SP (1998) Fungal pathogens of Russian wheat aphid (Homoptera: Aphididae). In: Quisenberry SS, Peairs FB (eds) Response model for an introduced pest – The Russian wheat aphid. Entomological Society of America, Lanham, MD. pp 209–233

Ragsdale DW, Voegtlin DJ, O'Neil R J (2004) Soybean aphid biology in North America. Ann Entomol Soc Am 97:204–208

Roberts DW, St. Leger RJ (2004) *Metarhizium* spp., cosmopolitan insect pathogenic fungi: Mycological aspects. Adv Appl Microbiol 54:1–70

Roy HE, Pell JK, Clark SJ, Alderson PG (1998) Implications of predator foraging on aphid pathogen dynamics. J Ivertebr Pathol 71:236–247

Rutledge CE, O'Neil RJ, Fox TB, Landis DA (2004) Soybean aphid predators and their use in integrated pest management. Ann Entomol Soc Am 97:240–248

Scorsetti AC, Humber RA, Garcia JJ, Lastra CCL (2007) Natural occurrence of entomopathogenic fungi (Zygomycetes: Entomophthorales) of aphid (Hemiptera: Aphididae) pests of horticultural crops in Argentina. BioControl 52:641–655

Shah PA, Pell JP (2003) Entomopathogenic fungi as biological control agents. Appl Microbiol Biotechnol 61:413–423
Shah PA, Aebi M, Tuor U (2000) Infection of *Macrosiphum euphorbiae* with mycelial preparations of *Erynia neoaphidis* in a greenhouse trial. Mycol Res 104:645–652
Sigurdsson V, Halldorsson G, Sigurgeirsson A, Thorsson AT, Anamthawat-Jonsson K (1999) Genetic differentiation of the green spruce aphid (*Elatobium abietinum* Walker), a recent invader to Iceland. Agri Forest Entomol 1:157–163
Silvie P, Dedryver CA, Tanguy S (1990) Application expérimentale de mycélium d'*Erynia neoaphidis* (Zygomycètes: Entomophthorales) dans des populations de pucerons sur laitues en serre maraîchère: étude du suivi de l'inoculum par charactérisation enzymatiques. Entomophaga 35:375–384
Sopp PI, Sunderland KD, Coombes DS (1987) Observations on the number of cereal aphids on the soil in relation to aphid density in winter wheat. J Appl Biol 111:53–57
Steinkraus DC, Hollingsworth RG, Slaymaker PH (1995) Prevalence of *Neozygites fresenii* (Entomophthorales: Neozygitaceae) on cotton aphids (Homoptera: Aphididae) in Arkansas cotton. Environ Entomol. 24:465–474
Steinkraus DC, Boys GO, Hollingsworth RG, Bacheler JS, Durant JA, Freeman BL, Gaylor MJ, Harris FA, Knutson A, Lentz GL (1996) Multi-state sampling for *Neozygites fresenii* in cotton. Proc Beltwide Cotton Conf 1996, Memphis, Tenn. pp 888–889
Steinkraus DC, Boys GO, Bagwell RD, Johnson DR, Lorenz GM, Meyers H, Layton MB, O'Leary PF (1998) Expansion of extension-based aphid fungus sampling service to Louisiana and Mississippi. Proc Beltwide Cotton Conf 1998. San Diego, California pp 1239–1242
Steinkraus DC, Boys GO, Rosenheim JA (2002) Classical biological control of Aphis gossypii (Homoptera: Aphididae) with *Neozygites fresenii* (Entomophthorales: Neozygitaceae) in California cotton. Biol Control 25:297–304
Tanada Y, Kaya HK (1993) Insect pathology. Academic Press, London
Teulon DAJ, Stufkens MAW (2002) Biosecurity and aphids in New Zealand. NZ Plant Prot 55:12–17
Vandenberg JD, Sandvol LE, Jaronski ST, Jackson MA, Souza EJ, Halbert SE (2001) Efficacy of fungi for control of Russian wheat aphid (Homoptera: Aphididae) in irrigated wheat. Southwestern Entomol 26:73–85
Wilding N (1981a) Pest control by Entomophthorales. In: Burges HD (ed) Microbial control of pest and plant diseases 1970–1980. Academic Press, London. pp 539–554
Wilding, N. (1981b) The effect of introducing aphid pathogenic Entomophthoraceae into field populations of *Aphis fabae*. Ann Appl Biol 99:11–23
Wilding N, Mardell SK, Brobyn PJ, Wratten SD, Lomas J (1986) The effect of introducing the aphid-pathogenic fungus *Erynia neoaphidis* into populations of cereal aphids. Ann Appl Biol 117:683–691
Wraight SP, Poprawski TJ, Meyer WL, Peairs FB (1993) Natural enemies of Russian wheat aphid (Homoptera, Aphididae) and associated cereal aphid species in spring-planted wheat and barley in Colorado. Environ Entomol 22:1383–1391
Wraight SP, Galaini-Wraight S, Carruthers RI, Roberts DW (2003) *Zoophthora radicans* (Zygomycetes: Entomophthorales) conidia production from naturally-infected *Empoasca kraemeri* and dry-formulated mycelium under laboratory and field conditions. Biol Control 28:60–77
Wu ZS, Schenk-Hamlin D, Zhan WY, Ragsdale DW, Heimpel GE (2004) The soybean aphid in China: A historical review. Ann Entomol Soc Am 97:209–218
Zimmermann G (1978) Biological control of aphids by entomopathogenic fungi: Present state and prospects. In: Cavalloro R (ed) Aphid antagonists. Balkema, Rotterdam. pp 33–40

Chapter 7
Steinernema scapterisci as a Biological Control Agent of *Scapteriscus* Mole Crickets

J. Howard Frank

Abstract *Scapteriscus didactylus, Sc. abbreviatus, Sc. borellii* and *Sc. vicinus* are South American mole crickets (Orthoptera: Gryllotalpidae) that have arrived in the Caribbean and the southeastern United States, where they have become pests of turf- and pasture-grasses, vegetable seedlings and other crops. All four species have been targets of classical biological control. Three classical biological control agents have been introduced and established against them. A wasp, *Larra bicolor* (Hymenoptera: Sphecidae), was introduced into Puerto Rico in the late 1930s, and decades later into Florida, whence it spread to Georgia and Mississippi. A nematode, *Steinernema scapterisci* (Nematoda: Steinernematidae), was introduced into Florida in 1985 and into Puerto Rico in 2001. A fly, *Ormia depleta* (Diptera: Tachinidae), was introduced into Florida in 1988. *Steinernema scapterisci* has been mass-produced on artificial diets and marketed as a biopesticide with a difference, in that it can establish permanent populations in sandy, low-organic soils with *Scapteriscus* mole crickets. As such, it can readily be deployed where its services are needed. It is highly specialized to the genus *Scapteriscus*, without non-target effects, and it is highly tolerant of chemical insecticides, thus adapted for use in integrated pest management strategies. It functions well against *Scapteriscus* adults (just like the other two biological control agents), but has limited effect against nymphs.

7.1 *Scapteriscus* Mole Crickets

7.1.1 The Genus **Scapteriscus** *in South America and its Emigrant Species*

The genus *Scapteriscus* (Orthoptera: Gryllotalpidae) now contains 19 species native to South America (Nickle 2003). Hundreds of years ago, *Scapteriscus didactylus*

J.H. Frank
Department of Entomology and Nematology, University of Florida, Gainesville, Florida 32611–0630, USA
e-mail: jhfrank@ufl.edu

(Latreille) began to expand its range northward through the chain of islands of the Lesser Antilles. This probably was by flight, colonizing island after island, and reaching Puerto Rico at least by the late 18th century (Frank *et al.* 2007). In the 20th century, it was detected in New South Wales, Australia, undoubtedly having arrived at least by 1982 as a contaminant of sea or air cargo (Rentz 1996). Early reports of the presence of *Sc. didactylus* in Georgia (US), Cuba, St. Croix (US Virgin Islands) and Jamaica were due to misidentification: it was other species of *Scapteriscus* that had arrived in those places.

Scapteriscus abbreviatus Scudder, whose adults are micropterous and flightless, may have originated in inland southern South America. It also occupies scattered localities northward along the coast of Brazil (Fowler 1987), scattered localities on West Indian islands of Cuba, Guadeloupe, Hispaniola, Jamaica, New Providence (Bahamas), Puerto Rico and St. Croix (Frank & Walker 2003), and scattered coastal areas of peninsular Florida (Walker & Nickle 1981). Most likely it traveled as a contaminant of ship ballast. The first recorded specimen from the port of Tampa, FL, was collected in 1899 and West Indian specimens were collected later, but this may simply reflect lack of entomological effort in the West Indies. Inability of *Sc. abbreviatus* to fly may be why the areas it has colonized are restricted in size.

Scapteriscus borellii Giglio-Tos and *Scapteriscus vicinus* Scudder have long-winged adults that can fly, but are thought to have traveled to the US in ship ballast. Earliest preserved US specimens of *Sc. borellii* were from 1904 in Brunswick, Georgia, and then from other southern US ports. Earliest US specimens of *Sc. vicinus* were from 1899 in Brunswick, Georgia, perhaps a single point of arrival (Walker & Nickle 1981). Neither of these species has been detected in the West Indies.

7.1.2 Behavior of the Emigrant Mole Crickets

Individuals of these four species spend most of their lives underground. Mated females excavate underground egg chambers and deposit in each a clutch of eggs. Chambers are sealed and not guarded by the females. The eggs take about 3 weeks to incubate. Hatchlings grow and molt 6–9 times, taking at least 5 months to do so in warm weather, longer in cooler weather. In the northern hemisphere, *Sc. abbreviatus* has oviposited in all months of the year, *Sc. didactylus* in all months except December although with a maximum in spring, *Sc. borellii* in spring (and summer too in southern Florida), and *Sc. vicinus* only in spring. Comparable data seem unavailable from the southern hemisphere. Flights of large numbers of adults (mainly females) of the three winged species are numerically greatest a few months before the maximal oviposition period, when females (some of them already mated) disperse to new habitats. The diet, most appropriately determined by dissection of field-caught individuals, is mainly plant materials for *Sc. abbreviatus* and *Sc. vicinus*, but for *Sc. didactylus* and *Sc. borellii* the diet includes a large proportion of animal materials (parts of other insects are abundant in the crop). All four species are typically found in sandy soils and sandy loams. They are generally absent from heavier soils unless those soils are made friable by cultivation and are moist, as by irrigation. One of the species, *Sc. abbreviatus*, stands out as able to colonize salt-laden coastal sands.

Despite scattered reports, the damage that any of these species causes in South America is poorly documented, in part because of questionable identification. They have been confused with each other and with mole crickets of other genera. Nevertheless, there is evidence they have caused damage to market gardens in Brazil, potato crops in Argentina, home lawns in Brazil, and golf courses in Brazil and Venezuela. The damage that they cause would doubtless be worse if their populations were not at least partially regulated by native natural enemies. Changes made by humans to the environment may cause mole cricket populations to increase locally. For example, land preparation for new golf courses typically involves removal of existing topsoil and the trucking in of sand from elsewhere to prepare a new surface, followed by artificial planting of selected grasses; this may destroy the existing soil fauna, allow rapid buildup of pest mole crickets flying in from elsewhere, eliminate wildflowers that could provide nectar to beneficial parasitoid insects that might later fly in, and eliminate entomopathogenic nematodes. Golf course superintendents are likely to use chemical pesticides to control any and all pest insects, and such use may forestall the recolonization by beneficial fauna capable of controlling pest mole crickets.

7.1.3 Damage Caused by **Scapteriscus** *spp. in their Expanded Range as Invasive Species*

Arrival of *Sc. didactylus* in the West Indies unleashed an important pest of vegetable seedlings (eggplant, tomato, sweet pepper, cabbage, etc.), sugarcane, tobacco, coffee seedlings, and turf grasses. In New South Wales (Australia) only damage to turf grasses by *Sc. didactylus* has seemed noteworthy. Where *Sc. abbreviatus* has arrived and established populations of limited area, principal damage has been to turfgrasses. Arrival of *Sc. borellii* in several seaports in the southern US soon led to it ranging from Florida northward to North Carolina and westward to Texas as a pest of turf grasses, where its tunneling activity is seen as about as much of a problem as its feeding on roots of grasses. Damage caused by *Sc. vicinus* in its new range in the southeastern US was noted first to vegetable seedlings, then to turfgrasses, and finally to pasture grasses. So, in the US, *Sc. vicinus* was judged as the worst of these pests, and now its range is almost as broad as that of *Sc. borellii*. Florida has three invasive *Scapteriscus* species and almost everywhere Florida soils are sandy (aiding excavation by mole crickets), so it has suffered more damage than other states of the US.

Baits containing chemical toxins (arsenic, cyanide) were the early methods of control. After WWII they were replaced by chlorinated hydrocarbons, especially chlordane because of its low cost and persistence. The banning of chlordane led to use of carbamates, and then organophosphates, and most recently a phenylpyrazole. But such compounds have been used mainly on turf, because they are too toxic and/or too expensive to use on pastures. Further, the commercial growing of vegetables in Florida has widely employed the use of a methyl bromide/chloropicrin mixture as a soil fumigant under plastic mulch, eliminating need for other soil treatments.

By the late 20th century, organically grown vegetables, pastures and, increasingly with public awareness, school playing fields were in need of a non-chemical control method. Pastures needed an inexpensive control method. Golf course turf was not included because its management operates under a unique set of economics in which the more prestigious golf courses have zero tolerance for damage by pests and have annual budgets to achieve that ideal.

7.2 The Search for a Permanent Solution to the Problems (Classical Biological Control)

It was in Puerto Rico that the search began for a permanent solution to problems with *Scapteriscus* mole crickets, but the first result was not deliberate. The cane toad *Bufo marinus* L. is native to South America but had been imported into Barbados and Martinique before 1850 to combat scarab larvae damaging to sugarcane. In 1920, cane toads were imported to Puerto Rico from Barbados for the same purpose. By 1927, it was claimed that not only had cane toads reduced the damage by scarab larvae to sugarcane, but they had also reduced the population of *Sc. didactylus* (Wolcott 1950). Wolcott, however, seems to have thought the level of damage by *Sc. didactylus* should be reduced still more, because he and colleagues in the late 1930s introduced from Belém, Pará, Brazil, a population of the sphecid wasp *Larra bicolor* F., which became established (Wolcott 1941). Unlike *B. marinus*, this wasp is a specialist natural enemy of *Scapteriscus* mole crickets.

In the early 1940s, entomologists in Florida attempted to introduce the same wasp. In part because a war was in progress and travel was difficult, the attempt was unsuccessful. That project was abandoned when use of chlordane against pest mole crickets became commonplace. But, when chlordane was banned by the US Environmental Protection Agency, a new program against pest mole crickets began in 1979 at the urging of cattlemen. This was the University of Florida Institute of Food & Agricultural Sciences (UF/IFAS), Mole Cricket Research Program (Walker 1985). This time, a stock of *L. bicolor* obtained from Puerto Rico was established at Ft. Lauderdale in 1981. However, attempts to bring in more wasps from Puerto Rico and establish them at more northerly points were unsuccessful (Frank & Walker 2006).

The new program investigated taxonomy, behavior, physiology, toxicology, ecology and population estimation methods (as well as non-specialist natural enemies occurring in Florida) of the pest mole crickets (Walker 1985, Hudson *et al.* 1988). Its ultimate objective was a permanent control method, meaning classical biological control, unless plant breeding should hold any promise. The program concentrated its attempts on finding suitable classical biological control agents in southern South America, the homelands of Florida's pest mole crickets. In 1985 a nematode from Uruguay was released, in 1988 a fly, *Ormia depleta* (Wiedemann) (Diptera: Tachinidae), from Brazil, and in 1988/1989 *L. bicolor* (again, but this time from Bolivia). All three became widely established in Florida, and *Scapteriscus* populations declined by 2000 to about 5% of what they had been before release of the biological control agents (Frank & Walker 2006). The wasp has also been established in Georgia, and has spread of its own accord to Mississippi and consequently probably also to Alabama; ultimately it may be expected to spread to southern Texas and southern Louisiana. The northernmost limit of the fly population is in northern Florida, and it is considered unlikely that it will spread northward. The following account is of the third biological control agent, the nematode, which later was named *Steinernema scapterisci* Nguyen & Smart.

7.3 *Steinernema scapterisci*

7.3.1 Detection of **Steinernema scapterisci** *in South America*

By the early 1980s, with the help of USDA funding, the UF/IFAS program was employing H.G. Fowler as a postdoctoral entomologist researcher in Rio Claro (São Paulo, Brazil) and A. Silveira-Guido, a retired university entomologist, in Montevideo, Uruguay. Both were charged with the same task: to find natural enemies of *Scapteriscus* mole crickets that could be used in Florida as classical biological control agents. Both of these scientists reported nematodes emerging from trapped mole crickets, but their shipments to Florida, examined by G.C. Smart, failed to reveal entomopathogenic nematodes. Funds were raised to send Smart and his technician K.B. Nguyen to visit Fowler and Silveira-Guido. In Silveira-Guido's laboratory, they were shown nematodes emerged from *Scapteriscus* mole crickets, perceived these to be of interest, and brought them to quarantine in Florida. Nguyen became Smart's graduate student and worked on characterization of the nematode of principal interest.

7.3.2 *Taxonomy of* **Steinernema scapterisci**

The species of *Steinernema* detected in *Sc. vicinus* was at first believed to be a strain of *Neoaplectana carpocapsae* Weiser (now considered a synonym of *Steinernema feltiae* (Filipjev)). Only later, as it performed poorly in standardized tests against larvae of Lepidoptera, was sufficiently detailed attention given to its structure that differences became apparent, and a description of this new species, *Steinernema scapterisci* Nguyen & Smart, was prepared with differentiation from three other species (Nguyen & Smart 1990a). Additional illustrations were provided by Stock (1992) who had detected the species in *Sc. borellii* in Argentina, and by Nguyen & Smart (1992c). Since then, additional species of *Steinernema* have been described, but none of them from *Scapteriscus* mole crickets. Nguyen & Smart (1992b) provided a key to the ten species of *Steinernema* known to that time.

7.3.3 Life Cycle and Behavior of Steinernema scapterisci

There are four juvenile stages. Third stage (infective) juveniles lurk in the soil as "ambush predators" ready to attack passing mole crickets. They enter mole crickets through the mouth or spiracles, break into the hemocoel, and release a specialist bacterium (Nguyen & Smart 1991c). The bacterium causes sepsis of the host and its eventual death. The nematodes mature to adults, mate, and the females lay eggs. If the nutrient supply is sufficient, the eggs develop through the juvenile stages to adults of the second generation within the host, these adult females lay eggs that develop into infective stage juveniles that disperse into the soil. From entry into mole crickets to release of progeny into the soil takes 8–10 days. However, if nutrients are deficient, as when the host is invaded by too many nematodes, eggs produced by the first-developing adults develop only to infective-stage juveniles that disperse and entry to release takes 6–7 days (Nguyen & Smart 1992a). The symbiotic bacterium was named *Xenorhabdus innexi* (Lengyel *et al.* 2005) and is known from no other *Steinernema* species.

St. scapterisci has been reared in the laboratory in *Sc. borellii* and *Acheta domesticus* L. (house crickets) (Nguyen & Smart 1992a). *A. domesticus* is an Old World species that does not occur in the field in Florida. After application of the nematode in the field, it has subsequently been detected from pitfall-trapped *Sc. borellii, Sc. abbreviatus, Sc. vicinus* (Parkman *et al.* 1994), and *Sc. didactylus* (Vicente *et al.* 2007), but not *Neocurtilla hexadactyla* (Perty), a species native to the eastern US (Parkman *et al.* 1993a). *Gryllus* spp. crickets were slightly susceptible in laboratory trials, but lepidopterous larvae, honey bees, cockroaches, and earthworms were poor hosts (Nguyen & Smart 1991b)., *St. scapterisci*-killed mole cricket hosts infected experimentally in the laboratory produced an average of 50,000 infective-stage juveniles per host (Nguyen & Smart 1991b). It should be noted that rearing mole crickets in the laboratory is laborious and thus expensive.

Infective-stage juveniles buried experimentally in containers in field soils suffered a rapid decline in numbers over the first 2 weeks, but thereafter decline was less rapid and some 25% survived at 8 weeks and these were still infective (Nguyen & Smart 1990b). Such nematodes applied on the surface moved downward into the soil whereas nematodes inserted into the soil moved upward or downward (Nguyen & Smart 1991a), doubtless to reach the optimal microhabitat. Laboratory trials suggested that 200,000 nematodes applied per m^2 of soil surface led to as high a proportion of hosts (*Sc. borellii* and *Sc. vicinus*) infected as did 500,000 or 1,000,000 per m^2 (Hudson & Nguyen 1989). Various trials and field observations have shown that *Sc. vicinus* is more susceptible than *Sc. abbreviatus* but less susceptible than *Sc. borellii*, and that mole cricket nymphs are not or scarcely susceptible to the nematode. The greater locomotory activity of *Sc. borellii* than *Sc. vicinus* probably explains its greater susceptibility in that it is likely to encounter more ambushing nematodes.

7.3.4 Initial Field Trials in Florida

Laboratory-reared *St. scapterisci* were applied to small (50 m^2) plots in Bahiagrass pastures in Alachua County, Florida, beginning in June 1985. Application was made (a) by burying cadavers of nematode-producing mole crickets at 1 per m^2, or (b) in water by sprinkling can. At the center of each plot a pitfall trap was installed to collect mole crickets for determination of infection with *St. scapterisci*. Then, as now, the only practical way to determine presence of *St. scapterisci* in the field was to trap mole crickets and hold them for as long as 10 days, expecting juvenile nematodes to emerge from any infected mole crickets during that time. After 5 years, the trials were terminated because nematode-infected *Scapteriscus* were still being collected routinely. Furthermore, *St. scapterisci* were being obtained from mole crickets trapped elsewhere in the county, presumably having spread in newly-infected winged mole crickets (Parkman & Frank 1992, Parkman *et al.* 1993a). Some 200 native mole crickets, *Neocurtilla hexadactyla* (Perty), were also taken at the plots, but none of them showed infection by *St. scapterisci* (Parkman *et al.* 1993a). Instead, some of them were infected by a previously unknown nematode, described as *Steinernema neocurtillae* Nguyen & Smart (Nguyen & Smart 1992b). That *Scapteriscus* and *Neocurtilla* should have different entomopathogenic nematodes is not altogether surprising because they are classified in different tribes of Gryllotalpidae (mole crickets).

7.3.5 Patent Application and Release of Material to Industry

In 1989 G.C. Smart, Jr. applied for a patent for the use of *St. scapterisci* as a biological control method against pest Orthoptera. The patent was granted in 1992. Then, in 1994 he filed another patent application "Biological control of mole crickets (*Scapteriscus* spp.)", granted in 1995 as US patent 5,445,819. He and his former student K B. Nguyen were named as the inventors, with the University of Florida as assignee. At some time before the first patent application, he provided stock of *St. scapterisci* to Biosys (Palo Alto, CA) which, at that time, mass produced other entomopathogenic nematodes on artificial diets.

7.3.6 Trials Sponsored by Industry in Florida and Elsewhere in the Southeast

In 1988, Biosys provided *St. scapterisci* to the UF/IFAS program for trial applications on a Gainesville golf course for comparison with chemical insecticides. Between-plot intervals were too narrow, and nematodes invaded control plots, rendering the comparison invalid.

In August–September 1989, Biosys provided *St. scapterisci* formulated in a water-soluble gel to the UF/IFAS program for trials on Florida pastures. Applications were made in 1-ha plots in the center of large Bahiagrass pastures in six counties (Clay, Flagler, Highlands, Hillsborough, Osceola, and Pasco). Researchers used 200,000 nematodes per m^2 (2 billion/ha) applied subsurface by chisel-rig in 100 gal. (378 L) of water per plot, and applications were made in the late afternoon. Control plots were in each case at least 1.7 km distant. Control and application fields each had an array of 13 pitfall traps (within and outside the treated plot) to collect mole crickets for evaluation of their infection by nematodes and estimation of the rate of spread of the nematode. Monitoring of trapped mole crickets (to reveal percentage parasitism) and quadrat sampling (to reveal mole cricket tunnels and grass cover) continued for 2 years. A population of nematodes was established successfully at five of the six sites. Mean maximum cumulative distance dispersed outside the treated plots was 60 m within 21 months after application. Other results were not consistent, perhaps because assessments of grass cover and damage levels were made by >6 people using varying standards. Nematodes were later detected in one of the control fields probably due to flight of infected mole crickets (Parkman *et al.* 1993b).

In October 1989 and April–May 1990, Monsanto (St. Louis, MO) provided *St. scapterisci* formulated in particulate clay to the UF/IFAS program for trials on three golf courses in northern Florida (Alachua County) and three in southern Florida (Broward County). Applications were made on large plots on irregularly-shaped fairways, using 200,000 nematodes per m^2 applied by spray-boom preceded and followed by irrigation. Controls were fairways that the golf course management was asked to treat normally as required, with the usual chemicals. Pitfall traps were installed in adjoining roughs to capture mole crickets and evaluate their infection by nematodes. Even at their widest aperture, the nozzles of the spray rigs were prone to clogging with clay. Mean trap catch in the second year was reduced 68% on treated plots where the nematode persisted, 62% on control plots where infected mole crickets were collected, and 41% on control plots where the nematode was not detected. Damage ratings and numbers of mole crickets soap-flushed from treated plots were significantly reduced in the second year after treatment. *Sc. abbreviatus* was the prevalent pest in Broward County, where *Sc. borellii* was absent. *Sc. borellii* became significantly more highly infected than *Sc. vicinus* in Alachua County, where *Sc. abbreviatus* was absent (Parkman *et al.* 1994).

The data obtained on one of the courses in northern Florida were then provided to a student in economics for analysis of expected payoff and risk. The student also obtained and used data on costs and income for that course. Decision theory methods were used for analysis and stochastic dominance demonstrated that depending on the risk attitudes of course managers, high-value golf courses should prefer to use *St. scapterisci* as their method of mole cricket control. Conversely, courses that expected lower ideal revenues, such as public golf courses, should prefer chemical pesticides (Lombardo *et al.* 1999). However, the economic analysis assumed that application of either chemicals or *St. scapterisci* would be made annually, whereas biological data showed that one application of *St. scapterisci* provided control for 2 years. The results of the economic analysis are thus debatable.

In 1990, a group of 28 golf courses and one sod farm passed funds through the Florida Turfgrass Research Foundation to sponsor a project applying the knowledge that *St. scapterisci* can be spread by newly-infected flying mole crickets (Parkman & Frank 1992). A mere three million *St. scapterisci* reared by hand in artificial culture in Gainesville were applied per site because commercially-produced nematodes were not then available. An "infection station" at each site used a sound-emitter broadcasting the call of a mole cricket male to attract winged adults (Walker 1996), expose them to *St. scapterisci*, and then release them. After 3 months, the sound-emitters were operated again to attract flying mole crickets, and these mole crickets were held to test for emergence of *St. scapterisci*. The logic was that the initially-infected mole crickets would have died of the infection, but would probably have released nematodes into the soil at their place of death, setting up foci for infection of additional mole crickets. Thus a local epidemic could be promoted at the cost of only a few million nematodes. Ultimately, after a second year of effort, infected mole crickets were trapped at 9 of the 29 sites (Parkman & Frank 1993).

In 1992, comparative trials sponsored by Biosys were carried out on mole cricket-infected turf in Alabama, Florida, Georgia, Louisiana, Mississippi, North Carolina and South Carolina. Each used large plots and compared effects with standard chemical treatments (acephate and bendiocarb) against pest mole crickets. Evaluations were made just before application, and at 2 and 4 weeks post-treatment. The nematodes (1 billion per acre, $\approx 250,000$ per m^2) performed nearly or about as well as did the chemicals under these conditions (WR Martin, Biosys, 1993 personal communication). Viewed from that limited perspective, the only advantage of the nematode to golf course management would be the lack of a waiting period before re-entry to the treated turf. It must also be remembered that head-to-head comparisons with chemicals suffer from at least one drawback: that chemicals are typically applied against small nymphs in summer, whereas *St. scapterisci* is best applied against adults in spring or autumn.

Public interest in mole cricket research waned in Florida in the early 1990s. Funding to the UF/IFAS program was almost eliminated, and no new research on *St. scapterisci* was conducted, in large part because the nematode was available commercially. After all, research on the wasp and the fly had been set aside to concentrate on the nematode, so now was the time to concentrate the very limited remaining resources on research on the other biological control agents. The program's objective had been complete classical biological control of *Scapteriscus* mole crickets in Florida, not to develop a biopesticidal nematode. The program, by intent of other parties, received no royalties from sales of the nematode.

7.3.7 Commercial Sales as a Biopesticide

Steinernema scapterisci was first marketed as Vector MC^{TM}, a biopesticide, by Biosys in the early 1990s, albeit without license to do so from the University of Florida. The University of Florida's Office of Technology Licensing (OTL) required

Biosys to desist, citing the pending patent. Biosys substituted *Steinernema riobrave* Cabanillas, Poinar & Raulston in its Vector MC^{TM} product. OTL then negotiated with a start-up company called BioControl, Inc. (Tampa, FL, later of Alachua, FL) the right to market *St. scapterisci*. BioControl obtained its supply of *St. scapterisci* from Ecogen Australia Pty Ltd. BioControl marketed the product in Florida for a few years, at first under the name Proact^{TM} and later as Proactant^{TM}, until Ecogen ceased production of all entomopathogenic nematodes. BioControl, with limited facilities and expertise tried to produce its own *St. scapterisci* for some months, but then foundered. By 1996, *St. scapterisci* had become unavailable commercially. By chance, that was the time when cattlemen in southwestern Florida perceived that the problem with pest mole crickets was worse and they wanted a solution, either chemical or biological. By then, there was no realistic chemical solution. Pressure built and, under new direction, the University of Florida's Office of Technology Licensing was persuaded to negotiate with Micro-Bio (UK) and achieved a marketing agreement just as Micro-Bio merged into Becker Underwood (Ames, IA). *St. scapterisci* is now marketed in the US as Nematac S^{TM} by Becker Underwood. Despite production of *St. scapterisci* in Australia, the nematode has never, to my knowledge, been marketed or even applied experimentally in Australia against *Sc. didactylus*, a supposed scourge of Australia's cricket (the sport) pitches. As noted below, Nematac S^{TM} has been applied experimentally in Puerto Rico, but I am unaware of any current marketing strategy for *St. scapterisci* in Puerto Rico or any other West Indian island or any South American country.

Sc. didactylus may more readily expand its population northward than southward from New South Wales. The (more tropical) Australian state to the north is Queensland, which has an invasive population of *B. marinus*, originally imported there in 1935 to control scarab beetles damaging sugarcane. This presents a natural test of Wolcott's (1950) idea that it was *B. marinus* that controlled *Sc. didactylus* in Puerto Rico. If Wolcott was correct, and if *B. marinus* populations in Queensland are as invasive as has been advertised, then *Sc. didactylus* should never become a problem in Queensland (unless somehow Australia's CSIRO succeeds in controlling *B. marinus*, a species that is now viewed as undesirable).

7.3.8 Resurrection of Interest in Steinernema scapterisci and New Trials

Interest in control of *Scapteriscus* mole crickets in pastures increased again late in 1996. The only one of the three biological control agents that could be applied inundatively was *St. scapterisci*, and only if it were again available commercially. So it was eventually brought to market, and the interested parties wanted to repeat some of the trials that had already been performed, but as demonstration projects involving the Agricultural Extension Service and audiences of cattlemen or turf managers. The fly and the wasp attracted little or no interest.

An immediate question was whether the progeny of nematodes applied to pastures in 1989 still persisted in those pastures. In 1997, pitfall traps were reinstalled in four of the six pastures where nematodes were applied in 1989 (Clay, Flagler, Osceola and Pasco Counties). Collaborating extension agents were asked to ship trapped mole crickets to Gainesville until 100 had been shipped from each site. All four sites produced nematode-infected mole crickets, showing persistence for 8 years; after 8 years, high proportions of trapped mole crickets were still found to be infected (Frank et al. 1999).

Graduate student K.A. Barbara was involved in new trials of *St. scapterisci* (Nematac S^{TM}) on two Gainesville golf courses in 2001. Before application, she checked trapped mole crickets to discover whether some might already be infected. She found that both courses still had a resident population of *St. scapterisci*, with 10–15% of mole crickets infected at 12 or 13 years after initial application, despite subsequent use of chemical pesticides (Frank et al. 2002, Barbara & Buss 2006). New applications temporarily boosted the soil inoculum (Barbara & Buss 2006). Her other work demonstrated a very high level of immunity of *St. scapterisci* to chemical insecticides (Barbara & Buss 2005) and the potential for using dilute permethrin instead of soap solution to flush living mole crickets from the soil to evaluate their infection by *St. scapterisci* (Barbara & Buss 2004).

A major effort was undertaken in application of *St. scapterisci* to Bahiagrass pastures. Nematodes were purchased from Ecogrow Australia Pty Ltd. Instead of using a chisel-rig as in 1989, a much more expensive device, a seed-drill, was adapted and modified to deliver nematodes in water. With this device, *St. scapterisci* was applied sub-surface in a pasture in September 2000 in one additional county (Polk) in peninsular Florida (Adjei et al. 2006). Spread of *St. scapterisci* outward from 1-ha treated plots had been observed in the 1989 trials. The 2000 trial took advantage of this spread by treating plots in swaths (strips) covering 50, 25 or 12.5% of the area of each plot to reduce costs of materials. After three years, there was no difference in outcome between these treatments, but there had been a reduction of 79% in the mole cricket population, and grass cover had increased from 33 to 96%. Thus, ranchers prepared to wait 3 years for full results can reduce the cost of treatment by 87.5% (Adjei et al. 2006). Then, applications of *St. scapterisci* (Nematac S^{TM}) were made in 2001 by spray-boom followed by irrigation in pastures in five additional counties (and also in some counties where treatments had been made in 1989). At the end of this time, one or more pastures in each of 12 counties had been treated successfully, with documentation that a population had been established. In addition, commercial treatments and demonstration projects by the Extension Service were made elsewhere.

In 2001, a study of mole crickets in Puerto Rico began with the intent to improve classical biological control where warranted. The major pest, widely distributed, was *Sc. didactylus*. It damaged turf and crops in sands and irrigated loams, but it was not detected in unirrigated loams. Infestations result from temporary immigration into newly planted and irrigated crops, but presence is permanent in irrigated turf. *Sc. abbreviatus* was detected only in sands, and it damaged crops planted in them. Applications of *St. scapterisci* were made on a golf course and a sod farm in western Puerto Rico in November 2001, by hand-held spray-boom attached to a small, truck-mounted pressurized tank containing water and nematodes, followed by irrigation. On large plots in loam, applications began late in the day as sunlight waned, and continued after dark. Pitfall traps in both areas collected mole crickets, which were held for emergence of nematodes. All mole crickets trapped were *Sc. didactylus*. Both applications successfully established populations of *St. scapterisci*, although the nematodes on the sod farm disappeared a month or so later when a flooding creek inundated the entire area. Pitfall trapping of mole crickets revealed nematode-infected mole crickets over the following 8 months, after which few mole crickets were trapped and none of them were infected. In May 2002, applications were made by sprinkling can on a highly organic sandy soil of an organic garden infested with *S. abbreviatus* and *S. didactylus*, but no infected mole crickets were trapped. A repeat of that application in August 2003 again failed to produce any nematode-infected mole crickets. A laboratory test of survival of nematodes in sterilized and unsterilized soil from the organic garden and from the golf course revealed poorer survival in the unsterilized soil (worse in the organic sand) raising the possibility of soil antagonists (Vicente *et al.* 2007). *Sc. didactylus* is susceptible to *St. scapterisci*, and there is reason to believe that the nematode's survival in at least some Puerto Rican soils is not as assured as in the sandy, low-organic soils typical of Florida.

7.3.9 The Concept of Area-Wide Biological Control and Role of Steinernema scapterisci

Area-wide biological control is just another expression coined to explain how classical biological control functions. Property owners and managers are familiar with applying a chemical pesticide to the property that they own or manage to control "their" pests. A golf course superintendent might realize that flying mole crickets invade the golf course every year from neighboring pastures. His response is to treat the golf course with chemicals. It may not be obvious to him that a classical biological control agent established on the neighboring pastures could reduce or eliminate the mole crickets flying in to his golf course, and that the classical biological control agent could spread to "his" course. *L. bicolor* wasps have been detected in good numbers on three Gainesville area golf courses by entomologists working there at other tasks. These numbers were measured by percentage parasitism of trapped mole crickets on two courses (RG Coler & KA Barbara personal communication), and by observation of numerous wasps hunting mole crickets on a third (T-I Huang personal communication). However, none of the three golf course superintendents has questioned their presence, and may not even have noticed them. *St. scapterisci*, being microscopic, is far less likely to be noticed. There is a huge educational task to be accomplished to gain public acceptance and demand for use of biological control on golf courses.

7.3.10 Interactions Between Parasitoids of Mole Crickets

Interactions between *St. scapterisci* and *O. depleta* within hosts were examined by exposing mole cricket hosts to each natural enemy alone, both natural enemies simultaneously, or to one natural enemy 3 days before exposure to the other. Because *O. depleta* appeared to be somewhat disadvantaged within multiparasitized hosts, and larvae and adult flies were susceptible to infection when exposed to the nematode in sand, the potential exists for field populations of the fly to be adversely affected in areas where *St. scapterisci* is abundant (Parkman & Frank 2002). Interactions between *St. scapterisci* and *L. bicolor*, and between *O. depleta* and *L. bicolor*, have not been evaluated.

7.3.11 The Role of Steinernema scapterisci *in IPM*

It is most informative that long-term monitoring showed that populations of *Sc. borellii* and *Sc. vicinus* had declined by 2000 to 5% of what they were in the 1980s, before *L. bicolor* and *St. scapterisci* reached the monitoring stations in northern Florida (Frank & Walker 2006). That indicates area-wide biological control is operational in at least part of Florida. Such a level of control should spread throughout Florida as populations of *L. bicolor* continue spreading and as *St. scapterisci* is spread by commercial sales and, to some extent, in newly-infected mole crickets. In central and southern Florida, *O. depleta* is contributing to area-wide biological control. Much the same should happen in southern areas of the Gulf Coast states Alabama, Mississippi, Louisiana and Texas due to continuing spread of *L. bicolor* and if *St. scapterisci* should be marketed in those states. In short, free classical biological control by *L. bicolor* and *St. scapterisci* together eventually achieves better control of *Scapteriscus* mole crickets than does any single application of 1980s-vintage chemical insecticides!

Provided that *St. scapterisci* and *L. bicolor* eliminate almost all *Scapteriscus* mole crickets so that damage is trivial, area-wide biological control will have done its job. The exception to this scenario is on those golf courses that strive for totally unblemished turf. Management of such golf courses will undoubtedly apply such chemical insecticides as are available and best suited for killing their target pests. Such insecticides will surely kill any *L. bicolor* adults or larvae that they contact, and will kill the mole cricket hosts of *L. bicolor* eggs. Only *L. bicolor* pupae may be immune, although this remains to be tested. Much the same is likely to happen to *O. depleta*. The biological control agents could be expected to survive in untreated roughs of golf courses (Frank & Parkman 1999). However, *St. scapterisci* seems highly immune to most chemical insecticides (Barbara & Buss 2005). Golf course management would be well advised to apply *St. scapterisci* if it is not known to be present, and then to apply chemical insecticides only to areas (tees and greens) needing maximal protection, and then only if there is evidence of damage by mole crickets (not prophylactically); this should save them money.

7.4 Conclusion

Steinernema scapterisci is highly successful as a classical biological control agent against *Scapteriscus* mole crickets in northern Florida in combination with the wasp *L. bicolor* and, in central and southern Florida with *L. bicolor* and the fly *O. depleta* (Parkman *et al.* 1996, Frank & Walker 2006). It is established also in southern Georgia together with *L. bicolor*. It has been applied as a biopesticide on turf in other southern states, but its permanent establishment there has not been reported.

St. scapterisci performed poorly in standardized tests against larvae of *Galleria mellonella* (L.) compared with other entomopathogenic nematodes, suggesting that it would not be an effective biopesticide for general use (Nguyen & Smart 1990a, Ricci *et al.* 1996). Its appropriate and best biopesticidal use is against *Scapteriscus* mole crickets, and it has the great advantage over other entomopathogenic nematodes that it can reproduce in those mole crickets, release progeny into the soil, and establish a continuing population.

The targets of use of *St. scapterisci* in the southeastern US are *Sc. vicinus*, *Sc. borellii*, and *Sc. abbreviatus*. Although *Sc. borellii* is more susceptible than *Sc. vicinus*, both can be controlled. Nymphs of these mole crickets are less susceptible than adults and their susceptibility decreases with decreasing size, which may be a function of the size of spiracular openings that are common points of entry. In Puerto Rico, *Sc. didactylus* has proven about as susceptible as *Sc. borellii* in Florida. However, *Sc. abbreviatus*, for unknown reasons, seems less susceptible.

Large-scale successful field applications of *St. scapterisci* have been made in August–October, and April–May in Florida, predicated on presence and activity of large nymphs and adults of *Sc. vicinus* and knowing that small to medium-sized nymphs are not, or only slightly, susceptible. Nevertheless, the June 1985 initial applications successfully established populations. It may be best to restrict applications to August–November or February–April, avoiding December–January when mole crickets may be deep underground due to cold temperatures, and May–July when the most susceptible stage (adults) become uncommon. Such restricted application times should certainly be applied if the objective is immediate kill of adults. There are two advantages of this approach: greatest immediate knockdown of target mole crickets, and greatest amplification of the nematode population in susceptible hosts.

The life cycle of *Sc. didactylus* assures the presence of adults throughout the year in Puerto Rico and other West Indian islands (perhaps also in Australia) so application could be contemplated in any month. It remains to be determined whether antagonists really are present in Puerto Rican soils.

Sampling for the presence of *St. scapterisci* anywhere is best performed by trapping living adult mole crickets and holding these under conditions that will support their life for 10 days, during which time many thousands of *St. scapterisci* should emerge from them if they are infected. The difficult logistics of making collections in every Florida county explain lack of detailed knowledge of the *S. scapterisci* distribution in Florida.

Data on percent infection of field-trapped mole crickets by *St. scapterisci* are informative of mortality levels at the moment of collection. Such mortality varies during the year but is continuous due to established populations of *St. scapterisci*. Mathematical models are required to calculate total generational mortality inflicted on mole crickets, which obviously vastly exceeds the percent mortality in any one collection of mole crickets. If 10% were the percent infection measured throughout a year, and 10 days were the survival time of infected mole crickets, then annual mortality during a generation (365 days) might be calculated using an amortization schedule. This shows that even "a mere" 10% infection could lead to 97% annual losses in a mole cricket population. Percent infection measured in pastures 8 years after application of *St. scapterisci* ranged from 37 to 50% (Frank *et al.* 1999) and, at this level, mole cricket populations would decline much more rapidly. It is reasonable to suppose that the soil inoculum of infective-stage *St. scapterisci* adjusts in time to the level that can be supported by the resident mole cricket population, although further studies are necessary to test this concept.

Acknowledgments I thank my colleague T.J. Walker for critical review of a draft of this chapter.

References

Adjei MB, Smart Jr GC, Frank JH, Leppla NC (2006) Control of pest mole crickets (Orthoptera: Gryllotalpidae) on pasture with the nematode *Steinernema scapterisci* (Rhabditida: Steinernematidae). Fla Entomol 89:532–535

Barbara KA, Buss EA (2004) Survival and infectivity of *Steinernema scapterisci* (Nematoda: Steinernematidae) after contact with soil drench solutions. Fla Entomol 87:300–305.

Barbara KA, Buss EA (2005) Integration of insect parasitic nematodes (Rhabditida Steinernematidae) with insecticides for control of pest mole crickets (Orthoptera: Gryllotalpidae: *Scapteriscus* spp.). J Econ Entomol 98:689–693

Barbara KA, Buss EA (2006) Augmentative applications of *Steinernema scapterisci* (Nematoda: Steinernematidae) for mole cricket (Orthoptera: Gryllotalpidae) control on golf courses. Fla Entomol 89:257–262

Fowler HG (1987) Man, insects and littoral ecosystems in Brazil: The role of trade, history, and chance in shaping biogeographical distributions. Anais Simp Ecosist Costa Sul Sudeste (Acad Ciênc Estado São Paulo):175–182

Frank JH, Parkman JP (1999) Integrated pest management of pest mole crickets with emphasis on the southeastern USA. Integr Pest Manage Rev 4:39–52

Frank JH, Walker TJ (2003) Mole crickets (Orthoptera: Gryllotalpidae) in Jamaica. Fla Entomol 86:484–485

Frank JH, Walker TJ (2006) Permanent control of pest mole crickets (Orthoptera: Gryllotalpidae: *Scapteriscus*) in Florida. Am Entomol 52:138–144

Frank JH, Buss EA, Barbara K (2002) Beneficial nematodes in turf: Good for how many years against pest mole crickets? Fla Turf Digest 19(4):48–50

Frank JH, Grissom C, Williams C, Jennings E, Lippi C, Zerba R (1999) A beneficial nematode is killing pest mole crickets in some Florida pastures and is spreading. Fla Cattleman 63(7): 31–32

Frank JH, Vicente NE, Leppla NC (2007) A history of mole crickets (Orthoptera: Gryllotalpidae) in Puerto Rico. Insecta Mundi (2007) 0004:1–10

Hudson WG, Nguyen KB (1989) Effects of soil moisture, exposure time, nematode age, and nematode density on laboratory infection of *Scapteriscus vicinus* and *S. acletus* (Orthoptera: Gryllotalpidae) by *Neoaplectana* sp. (Rhabditida: Steinernematidae). Environ Entomol 18:719–722

Hudson WG, Frank JH, Castner JL (1988) Biological control of *Scapteriscus* mole crickets. Bull Entomol Soc Am 34:182–198

Lengyel K, Lang E, Fodor A, Szállás E, Schumann P, Stackebrandt E (2005) Description of four novel species of *Xenorhabdus*, family Enterobacteriaceae: *Xenorhabdus budapestensis* sp. nov., *Xenorhabdus ehlersii, Xenorhabdus innexi* sp. nov., and *Xenorhabdus szentirmaii* sp. nov. System Appl Microbiol 28:115–122

Lombardo J, Weldon RN, Frank JH (1999) Quantification of risk for integrated pest management strategies as a decision making aid with applications for turfgrass in the Southeast. Univ Florida, Food & Resource Economics Dept., Staff Paper SP 99–1

Nguyen KB, Smart GC (1990a) *Steinernema scapterisci* n. sp. (Rhabditida: Steinernematidae). J Nematol 22:187–199

Nguyen KB, Smart GC (1990b) Preliminary studies on survival of *Steinernema scapterisci* in soil. Proc Soil Crop Sci Soc Fla 49:230–233

Nguyen KB, Smart GC (1991a) Vertical distribution of *Steinernema scapterisci*. J Nematol 22:574–578

Nguyen KB, Smart GC (1991b) Pathogenicity of *Steinernema scapterisci* to selected invertebrates. n. sp. (Rhabditida: Steinernematidae). J Nematol 23:7–11

Nguyen KB, Smart GC (1991c) Mode of entry and site of development of *Steinernema scapterisci* in mole crickets. J Nematol 23:267–268

Nguyen KB, Smart GC (1992a) Life cycle of *Steinernema scapterisci* n. sp. (Rhabditida: Steinernematidae). J Nematol 24:160–169

Nguyen KB, Smart GC (1992b) *Steinernema neocurtillis* n. sp. (Rhabditida: Steinernematidae) and a key to the species of the genus *Steinernema*. J Nematol 24:463–467

Nguyen KB, Smart GC (1992c) Addendum to the morphology of *Steinernema scapterisci*. J Nematol 24:478–481

Nickle DA (2003) A revision of the mole cricket genus *Scapteriscus* with the description of a morphologically similar new genus (Orthoptera: Gryllotalpidae: Scapteriscinae). Trans Am Entomol Soc 129:411–485

Parkman JP, Frank JH (1992) Infection of sound-trapped mole crickets, *Scapteriscus* spp., by *Steinernema scapterisci*. Fla Entomol 75:163–165

Parkman JP, Frank JH (1993) Use of a sound trap to inoculate *Steinernema scapterisci* (Rhabditida: Steinernematidae) into pest mole cricket populations (Orthoptera: Gryllotalpidae). Fla Entomol 76:75–82

Parkman JP, Frank JH (2002) Interactions between *Ormia depleta* (Diptera: Tachinidae) and *Steinernema scapterisci* (Nematoda: Steinernematidae), natural enemies of pest mole crickets (Orthoptera: Gryllotalpidae). Environ Entomol 31:1226–1230

Parkman JP, Hudson WG, Frank JH, Nguyen KB, Smart GC (1993a) Establishment and persistence of *Steinernema scapterisci* (Rhabditida: Steinernematidae) in field populations of *Scapteriscus* mole crickets (Orthoptera: Gryllotalpidae). J Entomol Sci 28:182–190

Parkman JP, Frank JH, Nguyen KB, Smart GC (1993b) Dispersal of *Steinernema scapterisci* (Rhabditida: Steinernematidae) after inoculative applications for mole cricket (Orthoptera: Gryllotalpidae) control in pastures. Biol Control 3:226–232

Parkman JP, Frank JH, Nguyen KB, Smart GC (1994) Inoculative release of *Steinernema scapterisci* (Rhabditida: Steinernematidae) to suppress pest mole crickets on golf courses. Environ Entomol 23:1331–1337

Parkman JP, Frank JH, Walker TJ, Schuster DJ (1996) Classical biological control of *Scapteriscus* spp. (Orthoptera: Gryllotalpidae) in Florida. Environ Entomol 25:1415–1420

Rentz DCF (1996) The changa mole cricket, *Scapteriscus didactylus* (Latreille), a New World pest established in Australia. J Austral Entomol Soc 34:303–306

Ricci M, Glazer I, Campbell JF, Gaugler R (1996) Comparison of bioassays to measure virulence of different entomopathogenic nematodes. Biocontr Sci Technol 6:235–243

Stock SP (1992) Presence of *Steinernema scapterisci* Nguyen et Smart, parasitizing the mole cricket *Scapteriscus borellii* in Argentina. Nematol Medit 20:7–9

Vicente NE, Frank JH, Leppla NC (2007) Use of a beneficial nematode against pest mole crickets in Puerto Rico. Proc Carib Food Crops Soc 42(2):180–186

Walker TJ (ed) (1985) Mole crickets in Florida. Univ Fla Agric Exp Sta Bull 846 (1984)

Walker TJ (1996) Acoustic methods of monitoring and manipulating insect pests and their natural enemies. In: Rosen D, Bennett FD, Capinera JL (eds) Pest management in the subtropics: Integrated pest management-a Florida perspective. Intercept, Andover, UK. pp 245–257

Walker TJ, Nickle DA (1981) Introduction and spread of pest mole crickets: *Scapteriscus vicinus* and *S. acletus* reexamined. Ann Entomol Soc Am 74:158–163

Wolcott GN (1941) The establishment in Puerto Rico of *Larra americana* Saussure. J Econ Entomol 34:53–56 [*L. americana* is a synonym of *L. bicolor*]

Wolcott GN (1950) The rise and fall of the white grub in Puerto Rico. Am Nat 84:183–193

Chapter 8
The Use of *Oryctes* Virus for Control of Rhinoceros Beetle in the Pacific Islands

Trevor A. Jackson

Abstract The rhinoceros beetle (*Oryctes rhinoceros*) was accidentally introduced into Samoa in 1909 from where it spread to many islands in the south-west Pacific. A novel virus pathogenic to the beetle, originally designated as *Rhabdionvirus oryctes* and later *Oryctes* virus, was isolated from Malaysia and introduced into Samoa in 1963. Later releases took place in Tonga, Fiji and other Pacific Islands. The virus rapidly established and caused high levels of infection that spread as epizootics through the beetle populations. The virus killed larvae in breeding sites and caused adult beetles to cease feeding, leading to reduced damage and a decline in fecundity of the pest population. Researchers reported spectacular declines in the treated populations within 1–3 years of application. Reapplication has proven effective where there has been a resurgence of beetle damage. After 40 years from the initial releases, high palm damage has been reported from some areas suggesting a breakdown of control. Selection of more virulent strains and improved methods of application could overcome these problems.

8.1 Introduction

The rhinoceros beetle *Oryctes rhinoceros* (Scarabaeidae; Dynastinae) is a large horned beetle endemic to the Asia/West Pacific region. Its biology has been reviewed by Catley (1969) and Bedford (1980), and quantitative ecological data was reviewed and incorporated into a model by Hochberg & Waage (1991). The adult beetles feed and aggregate in the crowns of palm trees (Fig. 8.1A). The beetles enter the leaf axil and feed by boring into the unfurled tissue at the meristem of the palm. Male beetles produce an aggregation pheromone, attractive to both males and females, which results in a patchy distribution of the beetles within a stand of palms. Beetle presence is evidenced through notching, fanning and breaking of the emergent palm fronds (Fig. 8.1B). In cases of high beetle feeding pressure, the

---

T.A. Jackson
AgResearch Ltd., Biocontrol, Biosecurity & Bioprocessing, Lincoln Science Centre, Christchurch 8140, New Zealand
e-mail: trevor.jackson@agresearch.co.nz

Fig. 8.1 **A**. Adult rhinoceros beetle. **B**. Young coconut palm with extensive leaf damage caused by adult rhinoceros beetle feeding (Photos courtesy of Sada Nand Lal)

growing tip of the palm will be killed leaving a dead standing stump. After mating, female beetles will fly to accumulations of dead organic matter, heaps of fronds, fallen logs and even standing dead palms, for oviposition, laying approximately 60 eggs into the decaying organic matter. The first instar larva will emerge from the egg after two weeks and begin to feed on the surrounding organic matter. Development of the three larval instars will take approximately 6 months in tropical conditions before transformation into the pupal stage, which lasts for about 1 month prior to emergence of the neonate adult. The adult beetles can live for about 6 months and are strong fliers over relatively short distances. A sibling pest species, *Oryctes monoceros*, is known from Africa and Islands of the Indian Ocean and the massive *Scapanes australis* occupies a similar niche damaging coconut palms in Asia and Papua New Guinea. Within its endemic zone the rhinoceros beetle can be a pest of coconut palm (*Cocos nucifera*) and oil palm (*Elaeis guineensis*), especially when there is an abundance of decaying organic matter suitable for breeding. Young palms on replant sites are particularly susceptible.

8.2 Rhinoceros Beetle in the Pacific

In the early part of the 20th century the major economic activity for most Pacific Islands was production of copra from coconuts in natural or managed plantations. Rhinoceros beetle is believed to have entered the Pacific concealed in rubber plant seedlings from Ceylon in 1909 (Catley 1969). The insect established rapidly in Samoa and subsequently spread to Tonga (1921) (since successfully eradicated); Wallis Island (1931); Palau, New Britain and West New Guinea (1942); Vavua, Tonga and New Ireland (1952); Pak Island and Manus Island, New Guinea (1960); Tongatapu, Tonga (1961) and the Tokelau Islands (1963). The beetle was first reported from Fiji in 1953 and, despite intensive attempts to implement quarantine procedures, had spread through most of the Fiji Island group by 1971 (Bedford 1976). Initial outbreaks of the pest were devastating; on Palau 50% of palms were killed within 10 years of *Oryctes* introduction (Gressit 1953). Establishment of quarantine systems, coupled with a reduction in beetle numbers through control measures, appears to have reduced the rate of spread of the insect with few new outbreaks reported since 1970. However, recent reports indicate that the beetle has been discovered on Guam and damage has been reported from Raratonga (SN Lal personal communication).

8.3 The Search for Controls

Damage by rhinoceros beetle, the effect on villagers' livelihoods and the threat to economic viability of the island communities, prompted regional bodies to investigate possible solutions. Initial responses included cultural control through destruction of breeding sites and treating palms with chemical pesticides, which had limited success (Catley 1969). Biological control was attempted and gained momentum with the initiation of the UNDP/SPC Project for Research on the Control of the Coconut Rhinoceros Beetle in 1965 (Young 1986), resulting in the release of a large number of natural enemies of rhinoceros beetle into the Pacific Islands. Most releases of parasites and predators had little success (Waterhouse & Norris 1987) but the search for pathogens in the region of origin of the pest was to prove more successful. Dr Alois Huger of the Institute for Biological Control, Darmstadt, Germany, was contracted to search for pathogens of the beetle by UNDP/SPC (Huger 2005). He concentrated his survey studies on the Malay peninsular working closely with staff of the oil palm plantations where large numbers of beetles were found in decaying logs in the replanting areas. Unusual larvae were identified displaying a lethargic, translucent condition indicative of dissolution of the fat body. Feeding macerates of putative diseased larvae to healthy larvae from a laboratory colony resulted in similar symptoms and strong indications of an infective agent, which was revealed as a virus by histological and electron microscopy studies.

8.4 The *Oryctes* Virus

The *Oryctes* virus is a non-occluded dsDNA virus that was first described as *Rhabdionvirus oryctes* (Huger 1966) and later defined as *Oryctes* virus, the type species of Subgroup C of the Baculoviridae by the International Committee on Taxonomy of Viruses (ICTV). It has recently been suggested that *Oryctes* virus be incorporated into a new virus genus and designated as *Oryctes rhinoceros* nudivirus (OrNV) (Wang *et al.* 2007). Characteristics of the virus and pathology of the infected beetles are described by Huger (1966, 2005) and Huger & Krieg (1991). Following ingestion by larvae, the virus will invade the midgut epithelium and migrate into other tissues. The abdomen becomes turgid and glassy and internal turgor may increase to cause prolapse of the rectum. The virus also infects the adult beetle with initial infection of the midgut epithelial cells causing a proliferation of cells, swelling of the gut and release of infected nuclei into the gut lumen. Infected adults will defecate large amounts of virus during the early stages of infection and contaminate their surrounding habitat. Infection leads to cessation of feeding, reduces the fecundity of females and decreases the life span of infected beetles (Zelazny 1973a).

8.4.1 Release of the Virus for Rhinoceros Beetle Control

The first field releases of the virus were made in 1967 in Samoa. Infected rhinoceros beetle larvae were shipped from the BBA laboratory, Darmstadt, to Samoa for multiplication of the virus by feeding healthy larvae with rotting sawdust contaminated by macerates of diseased larvae (Huger 1972, 2005). Dead infected larvae were distributed and applied to breeding sites on the Samoan islands of Manono and Savaii in March and April 1967. By October 1968, infected larvae were recovered from the field and subsequently virus infection was found to be widespread throughout the Samoan islands, even on Upolu where no releases had taken place (Marschall 1970). In the site of original release on Manono, Marschall (1970) reported "the beetle had almost disappeared" and noted a conspicuous decrease in damage in areas where the virus was well established.

Following the success of the initial virus releases in Samoa, further releases were made on other rhinoceros beetle infested islands of the Pacific (Bedford 1981). *Oryctes* virus was introduced into Fiji from Samoa in 1970 and released on multiple sites by capturing healthy adult beetles, infecting them in the laboratory and

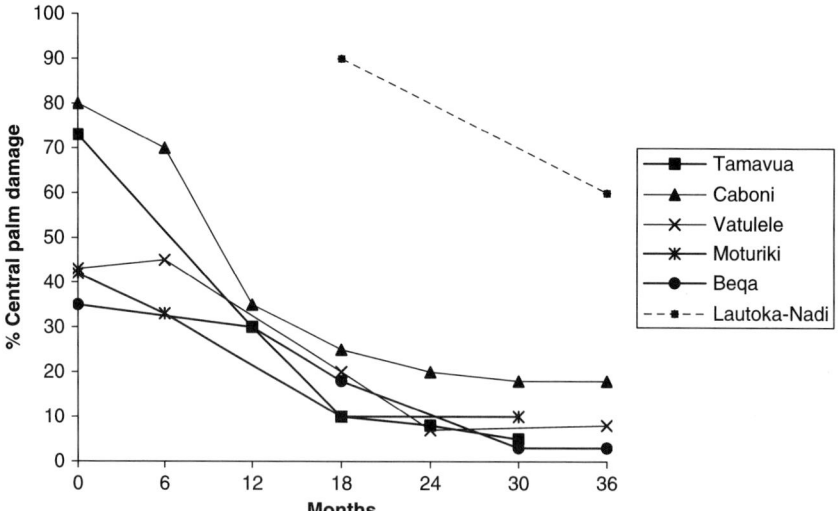

Fig. 8.2 Reduction in palm damage recorded over 36 months after release of *Oryctes* virus on five sites in the Fiji Islands from 1970–1972. No virus was released in the Lautoka area where damage remained high 18 months after the start of the programme but natural incidence of disease was recorded in the area after 36 months coinciding with a decline in visible damage (adapted from Bedford 1981)

releasing them into the field at selected sites (Bedford 1976). Between 1970 and 1974 several thousand infected beetles were released to badly infested areas and impact monitored by disease spread and estimates of damage to the palms. Several untreated sites were also monitored to provide controls for impact assessment. The results were dramatic, with 50–70% of trapped beetles infected and palm damage, which ranged from 35–85% before treatment, dropping to less that 10% on most sites within 3 years of release (Fig. 8.2). Dramatic photographic evidence of the recovery of palms post treatment is provided by Bedford (1976, 1981).

On Tongatapu, Tonga, virus was released by pouring macerate of larvae infected in Samoa onto heaps of rotting sawdust artificially infested with beetle larvae. More than 50 breeding heaps were treated on the western end of the island in 1970 and the resulting epizootic was monitored and estimated to have spread across the island at a rate of 3 km/month (Young 1974) with a peak of 40% larvae infected at the height of the epizootic. At the release site, number of palms with central crown damage dropped from 28 to 5%, 15 months from treatment.

The virus has also been released on other Pacific Islands including Tokelau (1967), Palau (1970, 1983), Wallis Island (1970–1971), American Samoa (1972) and Papua New Guinea (1978–1979) (Hajek et al. 2005) as well as releases within the Maldives and India. The virus has also been released for control of *Oryctes monoceros* in the Seychelles and Tanzania (Hajek et al. 2005).

8.4.2 Effect of Multiple Releases of Virus

Young (1974) indicated that disease incidence fell behind the "front" of the epizootic, which would be expected for the normal delayed-density-dependent activity of a pathogen following pest density reduction (Barlow 1999). In Samoa, Marschall & Ioane (1982) noted that after the spectacular decrease associated with the first release of the virus, beetles were increasing in number and palm damage was once again increasing, especially in areas where there was bush or plantation clearing and a high number of fresh breeding sites. A reinfection programme was established and monitored by trapping adults with the attractant ethyl chrysanthemumate (Maddison et al. 1973). Releases were made at several sites and beetle populations and disease monitored for up to 3 years with the authors concluding that damage could be reduced through implementation of this process. Increased levels of infection and reduced palm damage have been reported in a number of areas following regular release of virus including the Philippines (Alfiler 1992), Malaysia (Tuck 1996; Ramle et al. 2005) and India (Babjan et al. 1995).

8.4.3 Assessment Methods

While introduction of the virus into a new outbreak area where the rhinoceros beetle population is free of the disease is relatively simple to manage, ongoing management through repeated introductions will be more complex and depend on some simple decision rules. The decision to re-release is generally made on the basis of need – when beetle numbers seem to be on the rise. However, a rational programme would be based on absence or low incidence of the virus from the beetle population. Diagnosis of disease and determination of disease incidence is not always simple. Several methods have been developed for diagnosis of *Oryctes* virus and Zelazny (1978) reports on visual diagnosis, bioassay and gut content smear and microscopic examination as alternative methods for assessment. He concluded that direct visual diagnosis was the easiest method of diagnosis, but was dependent on the experience of the observer. To develop a consistent method for monitoring the virus, Richards *et al.* (1999) designed a PCR primer for the virus and a monitoring system that has been tested in Malaysia and Samoa. While this method has proven effective in monitoring virus spread (Ramle *et al.* 2005), sensitivity of PCR means that care must be taken in sample collection and preparation to avoid cross contamination and overestimation of disease.

8.5 Discussion

The introduction of *Oryctes* virus into the Pacific Islands has been a major success in reducing extreme outbreak populations of rhinoceros beetle and resolving the critical problem of beetle damage to palms. The success of the virus can be attributed to its biology and mode of pathogenicity. As observed by Huger (1972) infected beetles act as "virus reservoirs" releasing large quantities of virus into the beetle habitat. Beetles aggregate at feeding and breeding sites, behaviour encouraged by the male attractant pheromone (Hallett *et al.* 1995). Not surprisingly, mature beetles caught in traps show very high levels of infection as the aggregation behaviour of the insects ensures cross contamination. Levels of infection are higher among mature beetles than neonates or larvae (Ramle *et al.* 2005) suggesting that the virus is predominantly horizontally transmitted between the adults. The virus is successful in protecting the palms by bringing about cessation of feeding and a reduction in egg laying rate by females, leading to reduced damage and population decline. The pathobiology of *Oryctes* virus also reveals its weakness as a plant protection agent and suggests that with a high density of emergent adults, the reduced level feeding will not be sufficient to prevent palm damage, especially to young establishing trees. Damage, even in the presence of virus in the population, can be observed where there is a high density of organic matter for beetle breeding such as after old palm clearing for replanting, hurricane damage or in the areas near to sawmills.

Release of virus into fresh healthy outbreak populations will remain an essential tool in rhinoceros beetle management but the question remains whether the system could be improved (Jackson *et al.* 2005). Crawford & Zelazny (1990) showed that the virus evolved rapidly after release in the Maldives and data have been collected from Malaysia (Ramle *et al.* 2005) suggesting that some of the natural isolates are only weakly virulent. Success of reintroduction programmes in Samoa and the Philippines suggests that there may be a benefit from release of highly virulent strains even where the virus is already established. Zelazny (1973b) showed that highly virulent strains can cause cessation of feeding within a few days with a consequent reduction in damage and fecundity in the population. Virus delivery can also be improved by development of stable formulations that could be combined with attractants for a lure-and-infect strategy. A combination of new strains and new delivery systems could enhance the performance of *Oryctes* virus and provide an excellent new technology for coconut palm protection in the Pacific.

Acknowledgments My thanks to scientists who worked on the discovery and implementation phase of the *Oryctes* virus programme for their thorough reports, especially Dr Alois Huger for his enthusiastic insights and stimulating discussions on the history of the programme. I would also like to acknowledge the researchers and field workers in the Pacific and Asia for their efforts that have contributed to the continued use of *Oryctes* virus for protection of the "tree of life".

References

Alfiler ARR (1992) Current status of the use of a baculovirus in *Oryctes rhinoceros* control in the Philippines. In: Jackson TA, Glare TR (eds) Use of pathogens in scarab pest management. Intercept, Andover, pp 261–267

Barlow ND (1999) Models in biological control: a field guide. In: Hawkins BA, Cornell HV (eds) Theoretical approaches to biological control. Cambridge University Press, Cambridge, pp 43–70

Babjan B, Sudha Devi K, Dangar TK, Sathiamma B (1995) Biological suppression of *Oryctes rhinoceros* by re- release of *Baculovirus oryctes* in an infected contiguous area. J Plantation Crops 23:62–63

Bedford GO (1976) Use of a virus against the coconut palm rhinoceros beetle in Fiji. Pest Articles News Summ (PANS) 22:11–25

Bedford GO (1980) Biology, ecology and control of palm rhinoceros beetles. Annu Rev Entomol 25:309–339

Bedford GO (1981) Control of rhinoceros beetles by baculovirus. In: Burges HD (ed) Microbial control of pests and plant diseases, 1970–1980. Academic Press, London, New York, pp 409–426

Catley A (1969) The coconut rhinoceros beetle *Oryctes rhinoceros* (L.). Pest Articles News Summ (PANS) 15:18–30.

Crawford AM, Zelazny B (1990) Evolution of *Oryctes* baculovirus: rate and types of genomic change. Virology (New York) 174:294–298

Gressit JL (1953) The coconut rhinoceros beetle (*Oryctes rhinoceros*) with particular reference to the Palau Islands. Bulletin Bernice P. Bishop Museum, No 212, 157 pp

Hajek AE, McManus ML, Delalibera Junior I (2005) Catalogue of introductions of pathogens and nematodes for classical biological control of insects and mites. USDA For Serv FHTET-2005-05. 59 pp

Hallett RH, Perez AL, Gries G, Gries R, Pierce HD Jr, Yue J-M, Oehlschlager AC, Gonzalez LM, Borden JH (1995) Aggregation pheromone on coconut rhinoceros beetle, *Oryctes rhinoceros* (L.) (Coleoptera: Scarabaeidae). J Chem Ecol 21:1549–1570

Hochberg ME, Waage JK (1991) A model for the control of *Oryctes rhinoceros* (Coleoptera: Scarabaeidae) by means of pathogens. J Appl Ecol 28:514–531

Huger AM (1966) A virus disease of the Indian rhinoceros beetle, *Oryctes rhinoceros* (Linnaeus), caused by a new type of insect virus, *Rhabdionvirus oryctes* gen. n., sp. n. J Invertebr Pathol 8:38–51

Huger AM (1972) Grundlagen zur biologischen Bekämpfung des Indischen Nashornkäfers, *Oryctes rhinoceros* (L.), mit *Rhabdionvirus oryctes*: Histopathologie der Virose bei Käfern. Z Angew Entomol 72:309–319

Huger AM (2005) The *Oryctes* virus: Its detection, identification, and implementation in biological control of the coconut palm rhinoceros beetle, *Oryctes rhinoceros* (Coleoptera: Scarabaeidae). J Invertebr Pathol 89:75–84

Huger AM, Krieg A (1991) Baculoviridae. Nonoccluded baculoviruses. In: Adams JR, Bonami JR (eds) Atlas of invertebrate viruses. CRC Press, Boca Raton. pp 287–319

Jackson TA, Crawford AM, Glare TR (2005) *Oryctes* virus – time for a new look at a useful biocontrol agent. J Invertebr Pathol 89:91–94

Maddison PA, Beroza M, McGovern TP (1973) Ethyl-chrysanthemumate as an attractant for the coconut rhinoceros beetle. J Econ Entomol 66:591–592

Marschall KJ (1970) Introduction of a new virus disease of the coconut rhinoceros beetle in Western Samoa. Nature 225:228

Marschall KJ, Ioane I (1982) The effect of re-release of *Oryctes rhinoceros* baculovirus in the biological control of rhinoceros beetles in Western Samoa. J Invertebr Pathol 39:267–276

Ramle M, Wahid MB, Norman K, Glare TR, Jackson TA (2005) The incidence and use of *Oryctes* virus for control of rhinoceros beetle in oil palm plantations in Malaysia. J Invertebr Pathol 89:85–90

Richards NK, Glare TR, Aloali'i I, Jackson TA (1999) Primers for the detection of *Oryctes* virus. Molec Ecol 8:1552–1553

Tuck HC (1996) The integrated management of *Oryctes rhinoceros* (L.) populations in the zero burning environment. Proc 1996 PORIM Internatl Palm Oil Congress. pp 336–368

Wang Y, van Oers MM, Crawford AM, Vlak JM, Jehle JA (2007) Genomic analysis of Oryctes rhinoceros virus reveals genetic relatedness to Heliothis zea virus 1. Arch Virol 152:519–531

Waterhouse DF, Norris KR (1987) *Oryctes rhinoceros* (Linnaeus). In: Biological control; Pacific prospects. CSIRO, Canberra, Australia. pp 101–117

Young EC (1974) The epizootiology of two pathogens of the coconut palm rhinoceros beetle. J Invertebr Pathol 24:82–92

Young EC (1986) The rhinoceros beetle research project: History and review of the research programme. Agric Ecosys Environ 15:149–166

Zelazny B (1973a) Studies on *Rhabdionvirus oryctes*. II. Effect on adults of *Oryctes rhinoceros*. J Invertebr Pathol 22:122–126

Zelazny B (1973b) Studies on *Rhabdionvirus oryctes*. III. Incidence of the *Oryctes rhinoceros* population in Western Samoa. J Invertebr Pathol 22:359–363

Zelazny B (1978) Methods of inoculating and diagnosing the baculovirus disease of *Oryctes rhinoceros*. FAO Plant Prot Bull 26:163–168

Chapter 9
Use of Microbes for Control of *Monochamus alternatus*, Vector of the Invasive Pinewood Nematode

Mitsuaki Shimazu

Abstract Pine wilt disease, the most serious problem in pine forests in Asia, is caused by the pinewood nematode, *Bursaphelenchus xylophilus*. This nematode is an invasive pest in Asia where it is vectored by adults of a native wood boring cerambycid, *Monochamus alternatus*, during their maturation feeding. Surveys of pathogens were carried out to develop microbial control of this insect, and a strain of *Beauveria bassiana* highly virulent to *M. alternatus* was selected for development. Since larvae of *M. alternatus* live under the bark of pine trees, they were rarely being infected with conidia applied on the outside of the bark. A band-formulation of the fungus using nonwoven material was developed to maintain conidia at a high density, and larval mortalities were improved when using this formulation. Effective control of pine wilt disease also entails preventing feeding by *M. alternatus* adults on pine twigs, thereby avoiding transmission of the pinewood nematode. It takes 10–15 days for band-formulations of *B. bassiana* to kill emerged adults of *M. alternatus*. However, since infected adults are very immobile and seldom feed on pine twigs, their effective transmission of *B. xylophilus* is very limited.

9.1 Introduction

As well as invasive insect pests, invasive plant pathogens often cause serious damage to plants. Major forest diseases such as chestnut blight (*Cryphonectria parasitica*), Dutch elm disease (*Ophiostoma ulmi*), and white pine blister rust (*Cronartium ribicola*) are caused by invasive pathogens. These pathogens cause heavy damage and are difficult to eradicate or even control. The pine wilt disease in East Asia (Mamiya 1988, Yang 2004) and Europe (Mota *et al.* 1999) is also an invasive forest disease comparable with the above diseases. Pine wilt disease has been destroying native pine species in Japan and is the most serious pest of pine forests in these areas. This

---

M. Shimazu
Forestry and Forest Products Research Institute, Department of Forest Entomology, Matsunosato 1, Tsukuba, Ibaraki Pref. 305-8687, Japan
e-mail: shimazu@ffpri.affrc.go.jp

disease is caused by the pinewood nematode, *Bursaphelenchus xylophilus* (Mamiya & Kiyohara 1972).

In East Asia, the nematode is mainly transmitted by the Japanese native *Monochamus alternatus* (Coleoptera: Cerambycidae) (common name: Japanese pine sawyer), and usually this vector is targeted for control of the pine wilt disease. Chemical insecticides and fumigants are normally used to control this beetle. Biological control agents for *M. alternatus* have also been studied; a coleopteran ectoparasite, *Dastarcus helophoroides* (Coleoptera: Colydiidae) (Urano 2003) and an entomopathogenic fungus *Beauveria bassiana* (Deuteromycota) were found to be promising biocontrol agents. The latter has now been registered as a commercial insecticide. In this chapter, the biologies of the nematode and of the vector of pine wilt disease, methods used for control of this disease, and development of microbial methods for controlling the vector are described.

9.2 The Pine Wilt Disease and Its Practical Control

In Japan, an outbreak of a wilting disease on pines was recorded in 1905 in Nagasaki Prefecture (Yano 1913). Descriptions of the symptoms were similar to those of today's pine wilt disease and this is believed to be the first record of pine wilt disease in Japan. Later, similar damage occurred and expanded in western Japan until the 1930s (Kishi 1995). Damage was especially serious in the disordered period during and after World War II, and Dr. R. L. Furniss (1950), an entomologist working for the General Headquarters of the Occupation Forces, made a recommendation to control the damage by felling, stripping and burning damaged pine trees. Although the cause of this epidemic was not clear at that time, he made this recommendation based on his speculation that it was associated with insects under pine bark and his recommendations resulted in a decrease in pine wilt damage during the following few decades. This suggested one or some wood boring insects were involved in the pine wilt. Tokushige and Kiyohara (1969) reported that a species of nematode of the genus *Bursaphelenchus* was found universally in dead pine trees and called this species pinewood nematode. Inoculation tests of the pinewood nematode revealed that this was the pathogen causing the wilting disease (Kiyohara 1970, Kiyohara & Tokushige 1971), and the pinewood nematode was described as *B. lignicolus* (Mamiya & Kiyohara 1972). Just after discovery of this nematode, the Japanese pine sawyer, *Monochamus alternatus* was found to be the vector of this nematode (Morimoto & Iwasaki 1971). *B. lignicolus* was later found to be the synonym of *B. xylophilus*, which is native in North America (Nickle *et al.* 1981).

After discovery of the pinewood nematode as the cause of this disease, it became clear that three organisms were involved; pine trees as the host, *B. xylophilus* as the pathogen, and *M. alternatus* as the vector. The nematodes are carried in tracheae of *M. alternatus* adults. Adults of *M. alternatus* feed on bark of young pine twigs during a period of sexual maturation (= maturation feeding). At the time of maturation feeding, the nematodes exit from the beetle and invade the host tree through feeding scars on the twigs. Pine trees affected by *B. xylophilus* are not distinguishable at the early stage of the disease by their appearance. However, their conductive tissues have already been interrupted by cavitation and the trees are physiologically wilted. Such pine trees emit alpha-pinene and ethanol and the sexually mature *M. alternatus* adults are attracted by these substances. Males and females meet on wilting pine trees, mate, and lay eggs. Larvae of *M. alternatus* feed on the inner bark of dead pine trees. When *M. alternatus* pupates in spring, the nematodes gather around the pupal chamber, and at the moment of adult eclosion, the nematodes enter into the tracheae of the beetle. Thus, *M. alternatus* itself is not capable of killing pine trees; it only reproduces in dead pines, which have been killed by *B. xylophilus* that was transmitted at the time when *M. alternatus* fed on the twigs.

Control measures against this disease have been systematically developed after the discovery that this beetle vectors the nematode causing the disease. Studies on breeding resistant pines and integration of disease management methods have also been carried out as a long-term approach to control. Conventional control methods are largely directed against the vector insect and the pathogen (Kishi 1995). Trunk injection is the only practical control method against the pathogen, *B. xylophilus*. It prevents the possible increase of the nematode in the injected trees, even when *M. alternatus* adults feed on twigs and nematodes enter the tree through the maturation feeding scars. Chemicals similar to anthelminthics, e.g. morantel tartrate (ex. Greenguard®), are used for this purpose. Trunk injection results in excellent disease prevention, few adverse environmental effects and is effective for the entire feeding period of adults. However, its high cost and the risk of decay following the drilling of tree trunks for injection of the material are major drawbacks. Injections must be repeated every 2–4 years. Because of the high cost of injection, this method can be used only for important or irreplaceable trees growing at famous temples, shrines, golf courses and the like, and use of this tactic in forests is not practical.

Two main approaches are used for controlling the vector, *M. alternatus*: preventive sprays of healthy pine trees and destruction of insect-infested trees. A preventive spray can be applied to avoid maturation feeding of *M. alternatus* adults and prevent transmission of the nematode to healthy host trees. Insecticides are sprayed onto the foliage of healthy pine trees. This is the only practical way to protect pine forests from pine wilt today. The disease is well controlled with this method, but it is difficult to use this method alone to eliminate the disease. A second approach for controlling *M. alternatus*, the destruction of insect-infested trees should also be done, because damaged trees must be completely disposed of to eliminate the source of the vector and pathogen in the following year. Several methods can be used to dispose of infested trees, including insecticide sprays, fumigation, burning, burying, and chipping. Subsequent disposal of these dead trees is necessary, but difficult.

9.3 Pathogens of the Vector *Monochamus alternatus*

Pathogens of *M. alternatus* were known before the discovery that this insect vectors the pinewood nematode. Hasegawa and Koyama (1937) reported *Isaria farinosa* as a pathogen of this insect. However, at that time in Japan, the scientific names of entomopathogenic fungi were confused and the name "*I. farinosa*" meant *B. bassiana* of today (Aoki 1971). Therefore, the fungus that Hasegawa and Koyama reported was probably *B. bassiana*. Apparently what we call *I. farinosa* today did not occur frequently on *M. alternatus* during Hasegawa and Koyama's studies (1937). To explore microbial natural enemies of *M. alternatus*, Shimazu and Katagiri (1981) collected cadavers of *M. alternatus* from all over Japan, and isolated fungal pathogens. They found fungi in the genera *Beauveria, Verticillium*, and *Isaria* (= *Paecilomyces*, Luangsa-Ard *et al.* 2005) as important fungal pathogens. Above all, *B. bassiana* was isolated from almost all localities. The prevalence of diseased *M. alternatus* was generally low in the field and no distinct epizootic attributable to any particular pathogen occurred on this insect. Even the natural infection rate of the most common pathogen, *B. bassiana*, was very low, based on *M. alternatus* cadavers that were found. Togashi (1989) also recorded *B. bassiana, Verticillium* species and a facultative pathogenic bacteria, *Serratia marcescens*, but their prevalence in cadavers was very low. Wang *et al.* (2004) reported that 1–5% of *M. alternatus* larvae in China were found to be infected by pathogens in nature; among the pathogen-infected larvae found, about 12–19% were infected by *B. bassiana*. *Metarhizium anisopliae* is also one of the most common entomopathogenic anamorphic fungi worldwide, although it rarely occurs on *M. alternatus*. Shimazu and Kushida (1983) reported that *M. anisopliae*, in addition to *B. bassiana*, was highly pathogenic to *M. alternatus*. Soper and Olson (1963) surveyed the pathogens associated with *Monochamus* species in Maine, USA, and recorded *B. bassiana, I. farinosa, Verticillium* sp., *Aspergillus flavus*, and *Fusarium* sp., the same genera as pathogens of the Japanese pine sawyer. Recently, Francardi *et al.* (2003) also studied the possibility of using *B. bassiana* for the control of *Monochamus galloprovincialis*, the pinewood nematode vector in Portugal.

9.4 Use of *Beauveria bassiana* for Control of *Monochamus alternatus*

9.4.1 Initial Approach: Spraying Conidial Suspensions

To select a microbial control agent of *M. alternatus* for further study and potential development for control, Shimazu and Kushida (1983) bioassayed many fungal isolates from *M. alternatus*, and selected an isolate of *B. bassiana*, F-263, from Kumamoto Prefecture, Japan. This isolate caused most mortality of *M. alternatus* larvae and adults when they were inoculated by dipping individuals in conidial suspensions. The median lethal concentration (LC_{50}) for *B. bassiana* F-263 against *M. alternatus* larvae was 1.1×10^3 conidia/ml (gross mortality: based on the total number of dead larvae) or 2.0×10^3 conidia/ml (net mortality: based on only the larvae with outer mycelial growth on cadavers) (Shimazu 1994). This isolate also produced conidia abundantly on artificial media.

At first, the fungus was applied by spraying conidial suspension, modifying application techniques used for applying conventional chemical insecticides. Dead pine trees were treated targeting larvae under the bark and adults emerging from the trees, and live trees were treated, targeting the adults feeding on them (Shimazu & Kushida 1980, Shimazu et al. 1982, 1983). The treatments aimed at adults resulted in low mortality and required a long period to death. This delay before death allowed the adults to transmit the nematodes and to lay eggs, and it was concluded that treatments targeting adults showed little promise at that time. Spraying the fungus on the surface of dead pine trees to target larvae under the bark resulted in greater mortality, about 75% (Shimazu & Kushida 1980), but the larval mortality was still insufficient to suppress pine wilt disease.

In these field experiments, concentrations of fungal suspensions were about 10,000-fold higher than the LC_{50} in the laboratory but the resulting mortality was too low. This may result from the fact that there was little chance for the fungus, which was sprayed onto the bark, to come into contact with the larvae under the bark. Larvae of *M. alternatus* live and feed in the inner bark of weakened or freshly dead pine trees, and bore into wood to make pupal chambers when they are matured in autumn. To increase mortality, further studies focused on trying to introduce the fungus directly under the bark, or to increase the fungal concentrations applied to the outer bark.

9.4.2 Novel Methods of Use: Introduction of Beauveria bassiana *under the Bark*

Nobuchi (1989) used a novel method to introduce *B. bassiana* under the bark. That is, he used the pine bark beetle, *Cryphalus fulvus* to carry the *B. bassiana* conidia. This insect bores and breeds only under the bark of dead pine trees, in a manner similar to that of *M. alternatus*. This insect does not have a diapause and is easy to rear artificially so it is well suited to the task. Its use as a carrier for *B. bassiana* was based on the idea that when adults of this insect were released in pine wilt-infested forests after they were contaminated with *B. bassiana* conidia, they would bore into dead pine trees, be killed by the fungus within several days and *M. alternatus* larvae inhabiting the same areas would come into contact with the conidia and become infected (Nobuchi 1989). Moreover, since larvae of *C. fulvus* grow in larval galleries connected with the egg gallery of their mother, an increase in fungal inoculum under the bark as the result of larval infection of *C. fulvus* was expected. Enda et al. (1989) confirmed this hypothesis in a large cage experiment. Field experiments releasing pine bark beetles contaminated with *B. bassiana* were also conducted at various sites in Japan, and mortality of *M. alternatus* by *B. bassiana* was again confirmed (Enda et al. 1989, 1991). Infection of *M. alternatus* with *B. bassiana* was found for several years in *Pinus thunbergii* forests in which *B. bassiana* had been introduced by this method (Forestry Agency 1993). This method has the advantage that it is not necessary to locate *M. alternatus* infestations and then select and fell dead trees before applying the fungus because *C. fulvus* feeds on almost the same part of the dead pine trees as *M. alternatus*. Some disadvantages of this method include the labor intensive mass production of the carrier *C. fulvus*, and the fact that the fungal conidia which are applied do not always enter under the bark of dead pines because

Fig. 9.1 Spawn chips and culture bags for introduction of *Beauveria bassiana* under pine bark

they are carried by free flying bark beetles. The problems with mass production and mass-releasing contaminated *C. fulvus* were subsequently solved by the development of an "automatic contaminating device" by Nobuchi (1993).

To increase the density of *B. bassiana* under the bark, Shimazu et al. (1992) developed *Beauveria* spawn chips. These spawn chips were made of 7×20 mm cylindrical, commercially available wheat bran pellets added to water, autoclaved and inoculated with *B. bassiana* to induce conidial production on the surface (Fig. 9.1). To apply the spawn chips, the trunks of pine trees infested with *M. alternatus* were drilled to ca 20 mm depth and the pellets inserted. Tests showed that about 80% of *M. alternatus* larvae were killed by *B. bassiana* when these pellets were applied in August to September at a rate of 12 pellets/m length of trees or logs. On average, 45–100% mortality was obtained in experiments carried out by the Forestry and Forest Products Research Institute or prefectural forestry research institutes, and higher mortality was obtained when applications were made earlier in the season (Shimazu 1993). When the pellets were applied to standing dead trees at the height of 1–3 m, infected larvae were most often found in regions where the pellets were applied, although some infected larvae were found in areas of the trees lower than 1 m, but seldom in the upper areas. Mortality increased considerably when spawn chips were used but again there were some limitations associated with this method, such as high labor input for drilling the holes, the observation that the treatment was most effective near where it was applied, and difficulties in treating small branches.

9.4.3 Use of Fungal Bands for Application of Beauveria bassiana

9.4.3.1 Use against Larvae

Another way to increase pine sawyer mortality under the bark of dead trees is to increase the amount of conidia applied to the outside of the bark, but there is a limit to the conidial density that can be applied in a suspension sprayed onto the outside of the bark. Therefore, the "solid state" application of conidia was tested.

The culture medium itself or wheat bran pellets could be used as the carrier of high numbers of conidia, but these materials are difficult to apply to the surfaces of dead pine trees. Substrates made of polyurethane foam or nonwoven cellulose fabric strips impregnated with cultures of the entomopathogenic fungus *Beauveria brongniartii* have been used for control of cerambycid pests in orchards (Kashio & Ujiye 1988, Hashimoto *et al.* 1991). Nitto Denko Co. has commercialized this type of formulation (referred to as "fungal bands") of *B. brongniartii* under the name "BiolisaKamikiri," as a control agent for *Anoplophora malasiaca* and *Psacothea hilaris*. Regrettably, *B. brongniartii* is only weakly pathogenic to *M. alternatus* (Shimazu 1994), and "BiolisaKamikiri" cannot be used to control *M. alternatus*. However, the substrates used for BiolisaKamikiri, i.e. non-woven fiber bands, are suitable for applying *B. bassiana* to dead pine trees. Shimazu *et al.* (1995) employed band-shaped fabric as the carrier for *B. bassiana* and used this application method against *M. alternatus* larvae. The substrate material used was the same as that used for "BiolisaKamikiri". *B. bassiana* was shake-cultured to obtain the seed culture, then mixed with the fresh medium and the nonwoven fabric band was soaked in the medium. The band materials were incubated in plastic baskets for 3 weeks at 25 °C to obtain maximum sporulation. After culture, ca 1.8×10^8 conidia/cm^2 had been produced on the material. The fungal bands covered with conidia were then applied to the bark of dead pine trunks and branches in various ways and stapled in place (Fig. 9.2). Shimazu *et al.* (1995) applied the fungal bands for control of *M. alternatus* larvae and the insecticidal effect was compared with spawn chips. The fungal bands were applied in two different ways; on top of the stacked logs (pile) or coiled around the top of standing logs (wrap), or the spawn chips were applied at a density of 1 spawn chip/600 cm^2 ($= \times 1$) or double this amount ($\times 2$) (Table 9.1).

Fig. 9.2 Application of fungus bands onto dead pine trunks to target *Monochamus alternatus* larvae. **A**. On a standing dead tree. **B**. On pine logs

Table 9.1 Number and mortality of *Monochamus alternatus* larvae in pine logs treated with *Beauveria bassiana* in various ways (Shimazu et al. 1995)

Treatment* and method	Month	Number of logs	Number of larvae (mean ± SE)/2 m log				Percent infected	
			Alive	*B. bassiana*	Other death	Total	Measured (mean ± SE)	Corrected
Control	July	4	6.8 (1.4)	1.0 (1.0)	0.5 (0.3)	8.3 (2.0)	7.69 (7.69)	0.00
Control	August	8	5.4 (1.1)	0.9 (0.5)	0.1 (0.1)	6.4 (1.4)	12.02 (4.96)	0.00
Band, wrap	July	7	0.1 (0.1)	2.7 (0.9)	0.0 (0.0)	2.9 (1.0)	96.00 (3.38)	95.67
Band, piled	July	8	2.8 (1.0)	3.5 (0.8)	0.1 (0.1)	6.4 (1.6)	66.14 (8.77)	63.32
Band, wrap	August	8	0.1 (0.1)	7.1 (2.5)	0.0 (0.0)	7.3 (2.5)	96.43 (3.34)	95.94
Band, piled	August	8	0.4 (0.3)	8.5 (2.0)	0.1 (0.1)	9.0 (2.0)	94.60 (3.59)	93.86
Band, piled, sunlight	August	5	1.2 (0.7)	4.8 (2.5)	0.0 (0.0)	6.0 (3.2)	78.84 (9.61)	75.95
Spawn chips × 1	July	4	5.5 (3.2)	6.3 (1.8)	0.3 (0.3)	12.0 (3.0)	58.74 (16.46)	55.31
Spawn chips × 2	July	4	0.8 (0.5)	7.8 (2.1)	0.0 (0.0)	8.5 (1.7)	87.50 (7.98)	86.46
Spawn chips × 1	August	8	1.5 (0.5)	4.9 (1.0)	0.0 (0.0)	6.4 (1.2)	78.23 (6.99)	75.25
Spawn chips × 2	August	8	1.1 (0.5)	7.0 (1.8)	0.1 (0.1)	8.3 (2.1)	87.52 (4.80)	85.81
Spawn chips × 2, sunlight	August	5	2.6 (1.6)	5.0 (1.0)	0.0 (0.0)	7.6 (1.7)	76.33 (14.50)	73.10

*See text.

Many *M. alternatus* larvae were infected, and their mortality was equivalent to or greater than that obtained by implanting spawn chips in the wood. There was a significant difference in mortalities between each treatment and the controls, but no difference was found among the treatments ($p < 0.05$, Tukey-Kramer's Post Hoc test). Fungal bands were superior to spawn chips because they are easier to apply and are highly effective. Moreover, fungal bands can be used on standing dead trees as well as piled dead branches. At present, this method is thought to be the most practical way to apply *B. bassiana* as a microbial insecticide against *M. alternatus*.

9.4.3.2 Use against Adults

As mentioned above, adults of *M. alternatus* are comparatively more resistant to *B. bassiana* than the larvae. Therefore, simple sprays of *B. bassiana* conidia on bark or foliage do not cause sufficient mortality of adults in a short time. However, recently, it was found that *B. bassiana* causes high mortality of *M. alternatus* adults in a shorter time when the adults were allowed to walk on the *B. bassiana* conidia on fungal bands (Okabe *et al.* 2001, Okitsu *et al.* 2000). The use of fungal bands to apply *B. bassiana* was studied through joint work by the Forestry and Forest Products Research Institute, universities, prefectural governments and a production company. In this project, virulence of *B. bassiana* against adults exposed to fungal bands, field experiments using bands applied to dead pine logs to control emerged adults, and population dynamics of the applied fungus in the field were studied. In this project, field treatments of damaged pine trees with fungal bands were carried out at seven different sites in Japan. The sawyer-infested boles were cut into logs, stacked, and 10 fungal bands were applied per cubic meter volume of stacked logs. The pile with bands was then covered with a polytarp. The experiments were carried out in large screen cages in the field, and the emerged adults were reared individually. In most prefectures, except Kagoshima, the southern-most prefecture, more than 90% of emerged adults were killed within 15 days (Table 9.2).

Using fungal bands for application, adults become infected by walking on the conidial mass; thus, the virulence of the fungus cannot be measured by the conventional method of dipping experimental insects into the conidial suspensions. Consequently, a novel bioassay technique was developed by mixing dead conidia with live conidia at different ratios to regulate inoculum density, and the virulence of dry

Table 9.2 An example of percent mortalities of *Monochamus alternatus* adults in field experiments within 15 days after exposure to *Beauveria bassiana* fungal bands

Experimental site	2003	2004	2005
Kumamoto	95	88	–
Kagoshima	52	61	–
Shiga (1)	94	95	100
Shiga (2)	97	83	71
Kanagawa	92	71	–
Tokyo	96	100	98
Akita	99	89	91

conidia could then be measured (Shimazu 2004b). Using this bioassay technique, young *M. alternatus* adults, within 4 days after emergence, were found to be more susceptible to *B. bassiana* than adults >10 days after emergence.

9.5 Development of the Fungal Band Formulation as a Commercial Insecticide: Safety and Efficacy

9.5.1 Safety of Beauveria bassiana Isolate F-263

B. bassiana is often used as a microbial control agent throughout the world because it is not pathogenic to mammals. Shimazu (2004a) showed that isolate F-263 did not grow at the body temperature of mammals. Inoculation tests of F-263 on several species of natural enemies and beneficial insects were conducted by the Forestry Agency (1999, 2000, 2001) to determine the pathogenicity of this isolate. This fungus caused only slight mortality of adult *Harmonia axyridis* (Coleoptera: Coccinellidae) and *Dastarcus helophoroides* (Coleoptera: Colydiidae). It was weakly pathogenic to *Orius sauteri* (Heteroptera: Anthocoridae) adults, and the LC_{50} was $2.5 \times 10^7 - 2.5 \times 10^9$/ml. It was moderately pathogenic to the honeybee, *Apis mellifera* (Hymenoptera: Apidae), with an LC_{50} of 6.1×10^6/ml, therefore, when applied commercially careful application is required to avoid direct contact with honeybees.

Phytotoxicity tests were carried out (Forestry Agency 1999, 2000) with direct inoculation of fungal bands with F-263 to dominant forest trees, undergrowth and agricultural crops. The fungus showed no effects on Japanese fir, *Abies firma* (Pinaceae); hinoki cypress, *Chamaecyparis obtusa* (Cupressaceae); sugi, *Cryptomeria japonica* (Taxodiaceae); eurya, *Eurya japonica* (Theaceae); Japanese beautyberry, *Callicarpa mollis* (Verbenaceae); *Hydrangea hirta* (Saxifragaceae); the soybean, *Glycine max* (Leguminosae); eggplant, *Solanum melongena* (Solanaceae); Japanese ginger, *Zingiber mioga* (Zingiberaceae); gold-banded lily, *Lilium auratum* (Liliaceae); a bamboo, *Pleioblastus chino* (Gramineae); and lily turf, *Liriope graminifolia* (Liliaceae). Conidia of F-263 were mixed into forest soil to determine the effect of *B. bassiana* on soil microorganisms, and the dynamics of soil bacteria, actinomycetes, and general fungi (Shimazu *et al.* 2002a). F-263 had no effect on densities of bacteria, actinomycetes, and soil fungi, as enumerated by selective media and germination rates. *B. bassiana* conidia did not germinate in forest soil, and over 12 months, numbers of *B. bassiana* gradually decreased to 1/10 of the initial number applied.

B. bassiana is also known to cause white muscardine on the silkworm, *Bombyx mori* (Lepidoptera: Bombycidae). The virulence of *B. bassiana* F-263 to silkworm larvae was determined by the Forestry Development Technological Institute (1990). The LC_{50} was 9.8×10^4 conidia/ml for 1st instar larvae and 8.9×10^5 conidia/ml for 5th instar larvae. These LC_{50} values were almost the same as those of other wild *B. bassiana* strains to the silkworm (Kawakami 1973).

Suzuki *et al.* (1991) looked at the risk of accidental contamination of mulberry leaves (i.e. food sources for *B. mori*) by releasing *Beauveria*-contaminated pine bark beetles, *C. fulvus*, and placing potted mulberry plants at 10–200 m from the release point. Mulberry leaves from plants at the *C. fulvus* release point caused a high rate of *B. bassiana* infection when fed to silkworm larvae, but plants further than 10 m from the release point caused almost no infection. It was concluded that the risk of releasing *C. fulvus* is limited to only a very small distance near the release point.

As *Beauveria*-inoculated fungal fabric bands carried more conidia than *C. fulvus* contaminated with conidia, greater dispersal was thought to be likely from the use of bands than from releasing *Beauveria*-contaminated bark beetles. Shimazu et al. (2002b) investigated conidial dispersal from experimental fungal bands with *B. bassiana* using selective medium plates to detect airborne *B. bassiana*. Conidial density more than 60 m from the bands was $5.2 \times 10^2/\text{m}^2/\text{day}$ at maximum, and we concluded that if conidia continued to disperse from bands at this same rate, it would take more than 40,000 days to achieve a conidial density on mulberry leaves lethal to silkworm larvae (i.e. 2.1×10^7 conidia/m^2) (Wada & Miyamoto 1997, 1999). It was concluded that contamination of mulberry leaves more than 60 m away from fungal bands would be extremely rare.

9.5.2 Use and Efficacy of the Commercial Band Formulation

The band formulation of *B. bassiana* F 263was thus developed and Nitto Denko Co. applied for its approval as a microbial insecticide under the name of "BiolisaMadara" (Fig. 9.3) (the department of Nitto Denko working on this product was moved to Idemitsu Kosan Co. in 2007). The formulation was registered with the Japanese government for use in Japan in February 2007. To use the fungal bands for control of *M. alternatus* adults emerging from dead pine trees, dead pine trees infested with *M. alternatus* are felled, and 2500 cm^2 of fungal bands per 1 m^3 of logs are placed on the logs, and then covered with a opaque plastic sheet (Fig. 9.4). The adult *M. alternatus* emerging from these logs move across the surfaces of the logs, toward the cut ends under the sheet, by phototaxis. During this movement, they contact fungal bands. It is recommended that fungal bands be placed near the

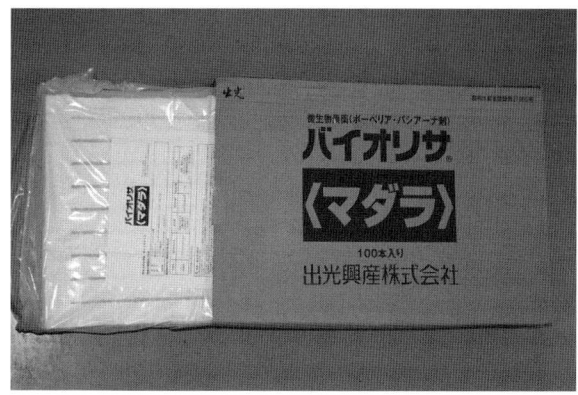

Fig. 9.3 "BiolisaMadara", a commercial formulation of *Beauveria bassiana* for control of *Monochamus alternatus*

Fig. 9.4 Application of a fungus band onto dead pine trunks targeting *Monochamus alternatus* adults

cut ends of logs or near the top of the log pile, to promote adults contacting bands. The purpose of the plastic sheet is to promote contact of walking adults with the fungal bands.

As stated before, using this method, more than 90% of emerged adults could be killed within 15 days. The pre-oviposition period and the peak of maturation feeding for *M. alternatus* is 2–3 weeks after emergence; therefore, high mortality during the pre-oviposition period could prevent both transmission of the nematodes and reproduction by *M. alternatus*. Both male and female *M. alternatus* adults were allowed to walk on this formulation immediately after emergence from pine logs and were then released in a large screen cage with fresh pine logs in the field. While non-treated adults laid eggs and the next generation of adults emerged, in the treatment cage, only a few eggs were laid and no adults of the next generation emerged (Okabe *et al*. 2002). Maehara *et al*. (2007) investigated the amount of maturation feeding and transmission of the pinewood nematodes by *M. alternatus* adults exposed to the *B. bassiana* formulation. Inoculated beetles fed less than the control beetles and ceased feeding several days before death. Among fungal-inoculated adult beetles carrying >1000 nematodes, some died before nematode transmission. Although the remaining heavily nematode-infested beetles lived until the beginning of nematode exit, they had stopped feeding, preventing the nematodes from entering pine twigs, i.e. preventing inoculation of host trees. From these experiments, it was concluded that fungal bands used in this way can be effective in preventing the transmission of the nematode to healthy pine trees as well as preventing the emergence of the next generation.

9.6 Issues and Future Prospects

In many cases, to utilize microbial control, a higher economic damage threshold must be tolerated because effects of pathogens are generally not immediate; forest insects are typically suitable targets for this reason. However, the situation is different for *M. alternatus*. This insect itself is not harmful to pine forests, and it causes problems by vectoring the pinewood nematode. If this were a simple defoliator, the damage would be well-suppressed by using the fungal band formulation of *B. bassiana*. However, for control of pine wilt disease, it is necessary to eliminate dead pine trees, prevent infection of healthy trees and isolate healthy trees from the surrounding nematode-infested pine forest. At Okinoerabu Island of Kagoshima Prefecture, Japan, the pine wilt disease was eradicated by complete disposal of infested wood and careful preventive spraying; land managers no longer need to conduct preventive insecticide spraying (Muramoto 1999). There are some other cases of eradication of this disease in Japan. However, use of chemical insecticides for preventive sprays has always been needed, and at the same time, effective means of disposing of the damaged pine trees (i.e. fumigation, incineration, burying, etc.) were also necessary (Yoshida 2006). These are the difficulties encountered when controlling the pine wilt disease.

The fungal band formulation is aimed at treatment of dead pine trees to kill emerging adults of *M. alternatus*, and the use of this formulation does not necessarily achieve eradication of pine wilt disease. It might be difficult for this formulation to compete with chemical insecticides with respect to cost. In agricultural areas on the other hand, microbial insecticides are used even though they are more expensive than chemicals. Recently, consumer concerns about safety of foods have been increasing, and in the cases of agricultural crops, organic products can be valuable although they are more expensive to produce. Similarly, there have been demands for reduction in the use of chemicals in control of the pine wilt disease, regardless of expense. Formerly, there was no weapon to answer these demands, but now the fungal band formulation is an available alternative. Moreover, fungal bands are easier to apply than fumigants. Another advantage of this formulation is ease of application. When fumigants are used for treatment of dead pine trees, pine logs must be covered with a plastic sheet, a ditch is created around the logs, and the edges of the plastic sheet must be buried to create an airtight enclosure. In contrast, fungal bands are placed on the pine logs and covered with a plastic sheet, without any need for airtight sealing.

At present, approved targets of the fungal band formulation are adults of *M. alternatus*. However, *B. bassiana* is, in fact, more pathogenic to larvae than the adults. Therefore, expanded use of fungal bands for *M. alternatus* larvae should be explored. To target larvae, the fungal band should be placed on dead pine trees as early as possible in the season, i.e. just after death of the tree, before the larvae bore into the heartwood where they do not come into contact with the conidia (Shimazu & Sato 2003).

B. bassiana has a relatively wide host range and this fungal band formulation could be utilized against other susceptible insect species in addition to *M. alternatus*. Since this formulation is band-shaped, insects overwintering on tree trunks could be targeted, providing a location for them to rest and hide between the band and the tree trunk, and, while hiding, they would become infected. Katagiri *et al.* (1983) attached fungus-cultured materials such as straw matting or corrugated fiberboard to the trunks of pines, and obtained infection of overwintering larvae of *Dendrolimus spectabilis* (Lepidoptera: Lasiocampidae) with *B. bassiana*. This idea was based on

Fig. 9.5 Traditional trapping method for the overwintering pine caterpillars by banding trees with straw mats

a traditional Japanese method for trapping pine caterpillars overwintering in old bark, by banding lower trunks with woven straw (Fig. 9.5). Similarly, infection of *D. spectabilis* with *B. bassiana* can be expected if the nonwoven fabric bands containing *B. bassiana* are used instead of woven straw. Moreover, if we are not particular about the exact shapes of fungal bands, further targets for application should be possible.

None of the pathogens of *M. alternatus* has been known to kill adults immediately; *B. bassiana* F-263 is the most promising agent against the adults, but still takes nearly 1 week on average to kill. The adult emergence of *M. alternatus* continues for about 2 months, and thus ages of the adults are uneven. Therefore, although young adults of *M. alternatus* do not often transmit the pinewood nematodes, adults attacking healthy pines have a high risk of transmitting the nematodes immediately. In programs for control of pine wilt disease, pathogens cannot be applied to healthy trees to prevent nematode transmission, but are necessarily limited to killing larvae and adults associated with dead pine logs. If we could find a pathogen that kills *M. alternatus* adults immediately (e.g. with a toxin), it could be sprayed onto foliage and twigs of healthy pine trees to kill adults when they are feeding, before nematodes leave them and before *M. alternatus* eggs are laid. However, for the present, availability of a product that kills *M. alternatus* before dispersal is an excellent environmentally safe alternative for control of this vector.

References

Aoki J (1971) *Beauveria bassiana* (Bals.) Vuill. isolated from some lepidopterous species in Japan (in Japanese with English summary). Jap J Appl Ent Zool 15:222–227

Enda N, Igarashi M, Fukuyama K, Nobuchi A (1989) Control of *Monochamus alternatus* Hope (Coleoptera; Cerambycidae) by the entomogenous fungus, *Beauveria bassiana* Vuillemin (Deuteromycotina; Hyphomycetes), carried by *Cryphalus fulvus* Niijima (Coleoptera; Scolytidae) (a preliminary report) (in Japanese). Trans 100th Mtg Jpn For Soc 579–580

Enda N, Gotoh T, Fukuyama K, Tsuchiya D (1991) Control of *Monochamus alternatus* Hope (Coleoptera; Cerambycidae) by the entomogenous fungus, *Beauveria bassiana* Vuillemin

(Deuteromycotina; Hyphomycetes) carried by *Cryphalus fulvus* Niijima (Coleoptera; Scolytidae) in Izu-Ohshima Island (in Japanese). Trans 102nd Mtg Jpn For Soc 281–282

Forest Development Technological Institute (1990) Reports on research commissioned for 1989. Biological control of pine wilt disease. Forestry Devel Technol Inst, Tokyo (in Japanese)

Forestry Agency (1993) Reports on research commissioned for 1992. Biological control of pine wilt disease. Forestry Devel Technol Inst, Tokyo (in Japanese)

Forestry Agency (1999) Reports on research commissioned for 1998. Biological control of pine wilt disease. Forestry Devel Technol Inst, Tokyo (in Japanese)

Forestry Agency (2000) Reports on research commissioned for 1999. Biological control of pine wilt disease. Forestry Devel Technol Inst, Tokyo (in Japanese)

Forestry Agency (2001) Reports on research commissioned for 2000. Biological control of pine wilt disease. Forestry Devel Technol Inst, Tokyo (in Japanese)

Francardi V, Rumine P, deSilva J (2003) On microbial control of *Monochamus galloprovincialis* (Olivier) (Coleoptera Cerambycidae) by means of *Beauveria bassiana* (Bals.) Vuillemin (Deuteromycotina Hyphomycetes). Redia 86:129–132

Furniss RL (1950) Recommendations for forest insect control in Japan. No. 536 GHQ, SCAP, National Resources Section

Hasegawa K, Koyama R (1937) Studies on pathogens of forest insects and their application (preliminary report) (in Japanese). Bull For Exp Sta Imperial Household 3:1–26

Hashimoto S, Sakaguchi N, Kashio T, Gyotoku Y, Kai I, Narahara M (1991) A carrier for cultures of the entomogenous fungus, *Beauveria brongniartii*, for the biological control of the whitespotted longicorn beetle, *Anoplophora malasiaca* (in Japanese). Proc Assoc Pl Prot Kyushu 37:170–174

Kashio T, Ujiye T (1988) Evaluation of the entomogenous fungus *Beauveria tenella*, isolated from the yellowspotted longicorn beetle, *Psacothea hilaris* for the biological control of the whitespotted longicorn beetle *Anoplophora malasiaca* (in Japanese with English summary). Proc Assoc Pl Prot Kyushu 34:190–193

Katagiri K, Shimazu M, Kushida T (1983) A trial of microbial control of *Dendrolimus spectabilis* using *Beauveria bassiana* without direct spraying of conidial suspension (in Japanese). Proc 94th Mtg Jpn For Soc 165

Kawakami K (1973) Studies on the disease of the silkworm *Bombyx mori* L., with special references to the invasion of causative fungi and pathological changes of infected larvae (in Japanese with English summary). Bull Seric Exp Sta 25:347–370

Kishi Y (1995) The pine wood nematode and the Japanese pine sawyer. Thomas Co Ltd, Tokyo

Kiyohara T (1970) Inoculation of the pine wood nematode to living Japanese black pine, *Pinus thunbergii* (in Japanese). Trans 24 Mtg Kyushu Branch Jpn For Soc :243–244

Kiyohara T, Tokushige Y (1971) Inoculation experiment of a nematode, *Bursaphelenchus* sp., onto pine trees (in Japanese with English summary). J Jap For Soc 53:210–218

Luangsa-ard JJ, Hywel-Jones NL, Manoch L (2005) On the relationships of *Paecilomyces* sect. *Isarioidea* species. Mycologia 96:773–780

Maehara N, He X, Shimazu M (2007) Maturation feeding and transmission of *Bursaphelenchus xylophilus* (Nematoda: Parasitaphelenchidae) by *Monochamus alternatus* (Coleoptera: Cerambycidae) inoculated with *Beauveria bassiana* (Deuteromycotina: Hyphomycetes). J Econ Entomol 100:49–53

Mamiya Y (1988) History of pine wilt disease in Japan. J Nematol 20:219–226

Mamiya Y, Kiyohara T (1972) Description of *Bursaphelenchus lignicolus* n.sp. (Nematoda: Aphelenchoididae) from pine wood and histopathology of nematode-infested trees. Nematologica 18:120–124

Morimoto K, Iwasaki A (1971) Transmission of the pine wood nematode by *Monochamus alternatus* (in Japanese). Trans 25 Mtg Kyushu Branch Jpn For Soc:165–166

Mota MM, Braasch H, Bravo MA, Penas AC, Burgermeister W, Metge K, Sousa H (1999) First report of *Bursaphelenchs xylophilus* in Portugal and in Europe. Nematology 1:727–734

Muramoto M (1999) Ending of pine wilt disease in Okinoerabu Island, Kagoshima Prefecture. Proc Intl Symp on Sustainability of Pine Forests in Relation to Pine Wilt and Decline, Tokyo, 27–28 Oct. 1998, pp 193–195

Nickle WR, Golden AM, Mamiya Y, Wergin WP (1981) On the taxonomy and morphology of the pine wood nematode, *Bursaphelenchus xylophilus* (Steiner & Buhrer 1934) Nickle 1970. J Nematol 13:385–392

Nobuchi A (1989) A trial of microbial control of *Monochamus alternatus* utilizing *Cryphalus fulvus* as a carrier of the pathogen (in Japanese). Forest Pests 38:133–137

Nobuchi A (1993) An automatic releasing equipment of *Beauveria*-contaminated bark beetle for microbial control of *Monochamus alternatus* (in Japanese). Forest Pests 42:213–217

Okabe T, Takai K, Suzuki T, Higuchi T (2002) Biological control of the Japanese pine sawyer, *Monochamus alternatus* by *Beauveria bassiana* (II) – Fecundity of the female adults contaminated with *Beauveria bassiana* (in Japanese). Kyushu J For Res 55:73–74

Okabe T, Nakashima K, Takai K, Suzuki T, Higuchi T (2001) Biological control of the Japanese pine sawyer, *Monochamus alternatus* by *Beauveria bassiana* (in Japanese). Trans 54 Mtg Kyushu Branch Jpn For Soc 115–116

Okitsu M, Kishi Y, Takagi Y (2000) Control of adults of *Monochamus alternatus* Hope (Coleoptera: Cerambycidae) by application of non-woven fabric strips containing *Beauveria bassiana* (Deuteromycotina: Hyphomycetes) on infested tree trunks (in Japanese with English summary). J Jpn For Soc 82:276–280

Shimazu M (1993) Control of *Monochamus alternatus* using a pathogen, *Beauveria bassiana* cultured on wheat-bran pellets (in Japanese). Forest Pests 42:232–236

Shimazu M (1994) Potential of the cerambycid-parasitic type of *Beauveria brongniartii* (Deuteromycotina: Hyphomycetes) for microbial control of *Monochamus alternatus* Hope (Coleoptera: Cerambycidae). Appl Entomol Zool 29:127–130

Shimazu M (2004a) Effects of temperature on growth of *Beauveria bassiana* F-263, a strain highly virulent to the Japanese pine sawyer, *Monochamus alternatus*, especially its tolerance to high temperatures. Appl Entomol Zool 39:469–475

Shimazu M (2004b) A novel technique to inoculate conidia of entomopathogenic fungi and its application for investigation of susceptibility of the Japanese pine sawyer, *Monochamus alternatus*, to *Beauveria bassiana*. Appl Entomol Zool 39:485–490

Shimazu M, Katagiri K (1981) Pathogens of the pine sawyer, *Monochamus alternatus* Hope, and possible utilization of them in a control program. Proc 17th IUFRO World Congress Div 2:291–295

Shimazu M, Kushida T (1980) Microbial control of *Monochamus alternatus* – Treatment of pathogens on the infested pine trees (in Japanese). Trans 32nd Mtg Kanto Branch Jpn For Soc 93–94

Shimazu M, Kushida T (1983) Virulences of the various isolates of entomogenous fungi to *Monochamus alternatus* Hope (in Japanese). Trans 35th Mtg Kanto Branch Jpn For Soc 165–166

Shimazu M, Kushida T, Katagiri K (1982) Microbial control of *Monochamus alternatus* -Spray of pathogens onto the infested pine trees just before adult emergence- (in Japanese). Trans 93rd Mtg Jpn For Soc 399–400

Shimazu M, Kushida T, Katagiri K (1983): Microbial control of *Monochamus alternatus* -Spraying of pathogens during maturation feeding. Trans 94th Mtg Jpn For Soc 485–486

Shimazu M, Sato H (2003) Effects of larval age on mortality of *Monochamus alternatus* Hope (Coleoptera: Cerambycidae) after application of nonwoven fabric strips with *Beauveria bassiana*. Appl Entomol Zool 38:1–5

Shimazu M, Maehara N, Sato H (2002a) Density dynamics of the entomopathogenic fungus, *Beauveria bassiana* Vuillemin (Deuteromycotina: Hyphomycetes) introduced into forest soil, and its influence on the other soil microorganisms. Appl Entomol Zool 37:263–269

Shimazu M, Sato H, Maehara N (2002b) Density of the entomopathogenic fungus, *Beauveria bassiana* Vuillemin (Deuteromycotina: Hyphomycetes) in forest air and soil. Appl Entomol Zool 37:19–26

Shimazu M, Kushida T, Tsuchiya D, Mitsuhashi W (1992) Microbial control of *Monochamus alternatus* Hope (Coleoptera: Cerambycidae) by implanting wheat-bran pellets with *Beauveria bassiana* in infested tree trunks. J Jpn For Soc 74:325–330

Shimazu M, Tsuchiya D, Sato H, Kushida T (1995) Microbial control of *Monochamus alternatus* Hope (Coleoptera: Cerambycidae) by application of nonwoven fabric strips with *Beauveria bassiana* (Deuteromycotina: Hyphomycetes) on infested tree trunks. Appl Entomol Zool 30:207–213

Soper RS, Olson RE (1963) Survey of biota associated with *Monochamus* (Coleoptera: Cerambycidae) in Maine. Can Entomol 95:83–95

Suzuki S, Makihara H, Fujioka H (1991) Effect of releasing pine bark beetle contaminated with *Beauveria bassiana* on silkworm (in Japanese). Tohoku Sanshi Kenkyu Hokoku 16:13–14

Togashi K (1989) Studies on population dynamics of *Monochamus alternatus* Hope (Coleoptera: Cerambycidae) and spread of pine wilt disease caused by *Bursaphelenchus xylophilus* (Nematoda: Aphelenchoididae) (in Japanese). Bull Ishikawa-ken For Expt Stn 20:1–142

Tokushige Y, Kiyohara T (1969) *Bursaphelenchus* sp. in the wood of dead pine trees. J Jap For Soc 51:193–195

Urano T (2003) Preliminary release experiments in laboratory and outdoor cages of *Dastarcus helophoroides* (Fairmaire) (Coleoptera: Bothrideridae) for biological control of *Monochamus alternatus* Hope (Coleoptera: Cerambycidae). Bull FFPRI 2:255–262

Wada S, Miyamoto K (1997) Effect of larval stage on the susceptibility of silkworm, *Bombyx mori*, to *Beauveria bassiana* isolated from cerambycid beetle, *Monochamus alternatus* (in Japanese). Abs 48th Ann Meet Jpn Soc Ser Sci Kanto Branch p2

Wada S, Miyamoto K (1999) Effects of *Beauveria bassiana* isolated from *Monochamus alternatus* on the silkworm (in Japanese). Abs 69th Ann Meet Jpn Ser Soc p79

Wang L, Xu F, Jiang L, Zhang P, Yang Z (2004) Pathogens of the pine sawyer, *Monochamus alternatus*, in China. Nematology Monographs and Perspectives, Vol. 1, EJ Brill, Leiden, Netherlands. pp 283–289

Yang BJ (2004) The history, dispersal and potential threat of pine wood nematode in China. Nematology Monographs and Perspectives, Vol. 1. EJ Brill, Leiden, Netherlands. pp 21–24

Yano M (1913) Investigations on the cause of dead pine in Nagasaki Prefecture. Sanrin-Koho 4 appendix 1–14 (in Japanese)

Yoshida N (2006) A strategy for controlling pine wilt disease and its application on-site (in Japanese with English summary). J Jpn For Soc 88:422–428

Chapter 10
Use of Entomopathogens against Invasive Wood Boring Beetles in North America

Ann E. Hajek and Leah S. Bauer

Abstract *Anoplophora glabripennis* and *Agrilus planipennis* are wood-boring beetles introduced from China to North America that are capable of killing healthy trees; *A. glabripennis* is polyphagous but attacks on maples (*Acer* spp.) are of major concern in North America, and *A. planipennis* attacks ash trees (*Fraxinus* spp.). Bioassays against *A. glabripennis* with entomopathogenic fungi identified that a strain of *Metarhizium anisopliae* (F 52) is virulent against adults. The primary deployment method investigated is propagation of the fungus within bands of non-woven fiber material. The fungal bands are then wrapped around tree trunks or branches, where wandering adults become contaminated with spores when walking across bands. Bands retain concentrations of viable conidia above the LC_{50} for >3 months. Infections also decrease reproduction before females die, resulting in fewer offspring. A strain of the entomopathogenic fungus *Beauveria bassiana* (GHA) sprayed on infested ash trees causes mortality of adult *A. planipennis* as they emerge from tree trunks. Cover sprays also result in fungal infections of *A. planipennis* larvae, pupae and adults that have not yet emerged, due to bark splits that form over the larval galleries providing points of entry for fungal inoculum under the tree bark. In addition, recent bioassays identified a strain of *Bacillus thuringiensis* virulent against *A. planipennis*. Development of this microbial control agent for aerial application is planned to target adult beetles that feed throughout their lives on ash foliage in the tree canopy.

10.1 Introduction

Twenty-five exotic species of bark- and wood-boring beetles were found to be established in the continental United States between 1985–2005, including two species of buprestids, five cerambycids, and 18 scolytids (Haack 2006). Invasive beetle species that bore in wood and are targets of microbial control agents belong to the families

---

A.E. Hajek
Department of Entomology, Comstock Hall, Cornell University, Ithaca, New York 14853-2601 USA
e-mail: aeh4@cornell.edu

Cerambycidae (longhorned beetles) and Buprestidae (jewel beetles). It is thought these species were initially accidentally introduced when beetle-infested wood was moved to areas where these species are not native, most probably through shipping. Here we discuss microbial control of two invasive species of beetles in North America, the cerambycid *Anoplophora glabripennis* (Asian longhorned beetle) and the buprestid *Agrilus planipennis* (emerald ash borer). While many species in these beetle families are associated with dead or dying trees, both of these species are known to be able to kill healthy trees.

Wood-boring beetles are difficult to control because the larvae live under tree bark or within wood. These beetles develop slowly and larval stages, that may be present in trees for long periods, are usually very difficult to detect. Once adults emerge, they are also difficult to detect because pheromones and host attractants for these beetle groups are poorly understood, so allelochemical lures are inefficient or unavailable. Adults typically disperse high into tree canopies to feed, and are therefore also difficult to control. Adding to difficulties in control, adults of these species emerge from wood asynchronously, over many months, and can be long-lived.

One factor driving the development of microbial control agents for management of wood-boring beetles is the limited efficacy of conventional insecticides. It is difficult or impossible to reach all larval or adult beetle feeding and resting sites when cover sprays are applied to tree canopies. The favored application method for control of wood borers is systemic uptake of insecticides through either trunk or soil injections. Systemic uptake, however, can result in uneven distribution of low insecticide concentrations within trees (Poland *et al*. 2006a). Moreover, systemic insecticides also travel up the tree through the xylem, resulting in limited efficacy against phloem feeders such as *Agrilus* spp. and early instar *A. glabripennis*.

There are numerous advantages to using microbial agents for control, particularly for wood-boring beetles. Due to lack of optimal control methods for wood-boring beetles, infested trees, once detected, are often removed and destroyed either to eradicate an invasive pest or, if the pest is established, reduce the population and remove hazardous trees. When possible, the removal of valuable urban and suburban trees can be avoided by individual treatment and use of microbials; this is especially true for trees in urban/suburban areas. In forests, all infested trees cannot be removed necessitating development of alternate methods for controlling these destructive beetles.

The fungi constitute the only insect pathogen group that has been exploited against invasive wood-boring beetles. The fungal species used are anamorphs of Hypocreales, e.g. *Metarhizium anisopliae* and *Beauveria bassiana* (Hajek & Bauer 2007). Appropriate application methods have been adapted for the different behaviors of targeted species. As will be described below, the development of microbial control against *A. planipennis* has focused on use of conidial sprays. In contrast, for *A. glabripennis*, non-woven fiber bands impregnated with fungal cultures have been emphasized. The "fungal band" application technology was originally developed in Japan for control of native cerambycids attacking orchards (Higuchi *et al*. 1997). Fungal bands are now utilized for management of the cerambycid vector of the pinewood nematode, *Monochamus alternatus*, in Japan (Chapter 9), and fungal bands in combination with allelochemicals are presently being developed against *M. alternatus* in China (Li *et al.* 2007).

The most widely used microbial control agents for all arthropods worldwide are formulated isolates of the insect-pathogenic bacterium *Bacillus thuringiensis* (*Bt*). Most registered products based on *Bt* are for control of lepidopteran, dipteran, or coleopteran pests (Glare & O'Callaghan 2000). A coleopteran-active *Bt* strain with high toxicity against *A. planipennis* adults is currently under development for possible use in an aerial spray program to suppress population densities of adults of this species, which feed in the ash canopy on foliage throughout their relatively long lives.

10.2 Development of Entomopathogenic Fungi for Control of *Anoplophora glabripennis*

Anoplophora glabripennis, the Asian longhorned beetle, is native to China and Korea but has been found in several areas of North America (Hajek 2007). It was first discovered in Brooklyn, New York, in 1996 and was subsequently also found in Chicago, New Jersey and Toronto. *A. glabripennis* can attack and kill relatively healthy trees of numerous hardwood species. This beetle species is a major killer of poplar (*Populus* spp.) and willow (*Salix* spp.) trees in China. In North America, maples (*Acer* spp.) are the hosts of greatest concern. Because this beetle has caused unparalleled tree mortality in China (Lingafelter & Hoebeke 2002), eradication is a priority in North America. Successful eradication seems possible because *A. glabripennis* is only present in a few areas of urban forest in North America, and is not known to occur in the native forest. Populations of *A. glabripennis* are relatively low, and the US and Canadian governments have been willing to support costly eradication campaigns. In the US alone, eradication programs cost $US 225 million from 1997–2006 (US GAO 2006) and >8,000 and >12,000 trees have been cut down in the US and Canada, respectively (see Poland *et al.* 2006a).

A. glabripennis generally has one generation per year in the US. Most adults emerge from trees from May–October and feed for 9–15 days before becoming sexually mature (see Hajek *et al.* 2008). Adults then mate, after which males guard females while egg niches are chewed by females in the bark and an egg is subsequently laid under the bark at the side of each niche. After eggs hatch, larvae tunnel in the sapwood under the bark for a few instars before moving into the heartwood. When initially attacking a tree, females often lay eggs near the tree crown base in the trunk and main branches but, with successive years of attack, eggs can be laid throughout much of the tree. Tunneling by larval *A. glabripennis* initially causes branch dieback and trees are eventually killed when there are high populations or when *A. glabripennis* reinfests the tree over successive years (Haack *et al.* 1997). Before tree death, larval tunneling weakens tree branches and trunks, which break more easily, leading to hazard trees.

The North American *A. glabripennis* eradication campaigns include detection, removal and then destruction of infested trees. Aside from removing infested trees, susceptible host trees occurring in a zone around the infestation are treated with preventive systemic applications of the neonicotinoid insecticide imidacloprid. Imidacloprid applications principally target adult beetles that feed on midribs of leaves, leaf petioles and the bark of twigs. However, the distribution of imidacloprid within trees is variable (Poland *et al.* 2006a) so adults ingest unpredictable dosages. Applications could affect newly hatched larvae in the phloem-cambial region but it is thought that larger larvae in the xylem might not be exposed to lethal doses of insecticide (Poland *et al.* 2006a). The extent to which imidacloprid applications kill adults or larvae in North America is unknown (populations are very low) although it is known that sublethal doses could lead to increased adult dispersal because imidacloprid acts as an antifeedant (Poland *et al.* 2006b). Therefore, while systemic imidacloprid certainly impacts some *A. glabripennis*, there is a need for additional eradication methodologies.

Development of a microbial control strategy was based on a commercial product used against several cerambycid pests of Japanese orchards, including *Anoplophora chinensis* (= *A. malasiaca*) (Higuchi *et al.* 1997). In Japan, cultures of *Beauveria brongniartii* are grown in non-woven fiber bands, producing lawns of conidia on band surfaces. The bands are attached around branches and tree trunks when cerambycids reach adulthood and emerge from trees. Cerambycids commonly walk on tree trunks and branches and, as they walk across the bands, they contaminate themselves with conidia. Optimal exposure to bands occurs during the preovipositional period when infection can prevent or reduce oviposition. Contaminated beetles can also transmit infective spores when mating (Tsutsumi 1998). Bands of *B. brongniartii* can be stored at 5°C for >1 year and remain active in the field for at least one month (Higuchi *et al.* 1997). In addition, bands produced in Japan are made from wood pulp so they are biodegradable and therefore do not have to be removed from trees.

10.2.1 Laboratory Trials

Development of fungal bands for control of *A. glabripennis* began with bioassays comparing twenty-two isolates of four species of entomopathogenic Hypocrealean fungi [*Beauveria bassiana*, *Beauveria brongniartii*, *Isaria farinosa*, and *Metarhizium anisopliae*] against adults and larvae in China and in a quarantine in Ithaca, New York. Preliminary studies demonstrated lower virulence of *B. brongniartii* and *B. bassiana* isolates against larvae compared with adults so further bioassays focused only on adults (Dubois 2003). Survival times for 50% of the beetles tested (ST_{50}) ranged from 5.0 days (*M. anisopliae* ARSEF 7234 and *B. brongniartii* ARSEF 6827) to 24.5 days (*I. farinosa* ARSEF 8411) days (Dubois *et al.* 2007). Screening studies initially included strains of *B. brongniartii*, which is registered as a microbial control agent in Europe, Asia and South America but not in North America (Faria & Wraight 2007). At that time, we could not confirm that this fungal species is native to North America, which added uncertainty regarding future registration for pest control in the US. Further laboratory bioassays identified three isolates of *M. anisopliae* killing adults in 5–6 days, with no difference in time to death between males and females. Therefore, our subsequent studies focused on one of these isolates (*M. anisopliae* F 52) that is already registered for control of various ticks and beetles, root weevils, flies, gnats and thrips in the US (US EPA 2003). Fungal bands were produced using similar procedures to those used in Japan (Shanley 2007). Adult beetles were exposed to squares of *M. anisopliae* F 52 fungal bands to establish the median lethal concentration of 6.8×10^6 conidia/cm^2 (Shanley 2007).

10.2.2 Field Trials with Fungal Bands

Field trials with *A. glabripennis* have been conducted in China because North American populations are very low and this species is targeted for eradication, which means that any infestations detected are quickly destroyed. Initial studies in 2000 used field-collected *A. glabripennis* adults caged within 1 m long window screen cages on tree trunks, with fungal bands encircling the trunks within some of the cages (Dubois *et al.* 2004a). Adults caged with *B. bassiana* and *B. brongniartii* bands died more quickly than controls and female oviposition decreased significantly in treated cages. For comparison, conidia were sprayed onto the surfaces of tree trunks within cages, and days to death for adults exposed to treated tree bark were the same as for adults in cages with fungal bands. However, 10 days after application, while conidial viability remained high on fungal bands, viability of conidia that had been sprayed onto tree trunks was drastically reduced. A similar study was conducted in 2001 but results were more variable. Researchers hypothesized that the variable results in 2001 were caused by high temperatures during the trial (range maxima: 30.2–38.5°C). This hypothesis was supported by loss of fungal band viability after 15 days on trees during 2001.

In 2001, uncaged trials compared areas with bands of two *B. brongniartii* isolates on each of 100 trees with nearby control areas without bands. Adult beetles were regularly collected and reared and oviposition was quantified (Dubois *et al.* 2004b). Some treatments resulted in decreased adult longevity compared with controls or decreased oviposition. In 2002, a similar study was conducted in a different site, large enough to allow additional replication and with higher populations of *A. glabripennis* (Hajek *et al.* 2006). Adults collected from fungal-treated plots 7–22 days after band placement died more quickly than controls, although results were more consistent for the *B. brongniartii* isolate tested than for the *M. anisopliae* isolate. Once again, oviposition was reduced in the treated plots compared with the control plots.

10.2.3 Longevity of Activity of Fungal Bands

To evaluate the length of time that fungal bands maintain viability, during the summers of 2001–2004, bands impregnated with strains of *M. anisopliae*, *B. bassiana*, and *B. brongniartii* were attached to tree trunks in localized sites in Queens, New York City (AE Hajek unpublished data). Bands were removed at varying intervals to quantify viable conidia and to conduct bioassays. Percent germination of conidia from bands did not decrease consistently with time in the field although total conidial density and density of viable conidia decreased with increasing time in the field (AE Hajek unpublished data).

For bioassays, death of quarantine-colony adults within 40 days of exposure to pieces of band from the field was used to calculate percent mortality. For bioassays of bands from 2001–2003, 100% mortality of treated beetles always occurred. In 2004, however, mortality dropped below 100% before the end of the trial with *M. anisopliae* F 52 (112 days after bands were placed in the field), although mortality still remained at > 50%. Bioassays conducted using samples of *M. anisopliae* F 52 bands taken from the field identified an LC_{50} of 7.22×10^6 conidia/cm^2, which was very similar to laboratory results. Although densities of viable conidia on bands decreased over time in 2004, densities of viable conidia never decreased below the LC_{50}. These results demonstrated that *M. anisopliae* F 52 bands retain virulence in the field for at least 112 days (AE Hajek unpublished data).

10.2.4 Indirect Effects of Fungal Bands

Studies were also conducted to evaluate the indirect effects of fungal bands to address whether only adults walking across bands are affected by treatments. Field studies had demonstrated an impact of fungal treatment on female fitness so assays were conducted using a quarantine colony to explore this further. Bioassays tested the effects of *M. anisopliae* infection on reproduction by adult female *A. glabripennis* and progeny survival (Hajek et al. 2008). The effect of infection on fecundity was evaluated for females already laying eggs and for newly eclosed females using two isolates of *M. anisopliae*. Both longevity and oviposition were significantly lower for females that were already laying eggs when exposed to *M. anisopliae* ARSEF 7234, when compared with controls. Newly eclosed females exposed to *M. anisopliae* ARSEF 7711 also displayed shortened longevity compared with controls (10.0 ± 0.7 days vs 74.3 ± 6.8 days for controls) and decreased oviposition (1.3 ± 0.7 eggs per ARSEF 7711-exposed female vs 97.2 ± 13.7 eggs per female for controls). Percentages of eggs that did not hatch were greater than controls for both age groups of fungal-treated females and unhatched eggs frequently displayed signs of fungal infection. The percentage of larvae dying within 9 weeks of oviposition was higher for progeny from sexually mature females exposed to ARSEF 7234 compared with controls, and dead larvae usually displayed signs of fungal infection. Thus, for both ages of females and both fungal isolates, fewer surviving larvae were produced after fungal inoculation of females, compared with controls. Infection with *M. anisopliae* affects female fitness by decreasing female longevity and fecundity, and through horizontal transmission of *M. anisopliae* to some offspring.

Japanese studies have shown that when adult yellow-spotted longicorn beetles (*Psacothea hilaris*) were exposed to a *B. brongniartii* band and then introduced to potential mates, horizontal transmission could occur (Tsutsumi 1998). During initial studies, female *A. glabripennis* were exposed to *M. anisopliae* bands and then caged with males. All exposed males died more quickly than controls but they did not die as quickly as the exposed females, which is consistent with receiving a lower dose (AE Hajek unpublished data).

In a related study, Shanley & Hajek (2008) investigated whether conidia from *M. anisopliae* bands could be dispersed to other parts of the environment by *A. glabripennis* adults that had contacted bands and whether *A. glabripennis* would become infected with exposure to tree bark contaminated with conidia in this way. One or five adult *A. glabripennis* were used to contaminate tree bark with conidia. All adults subsequently exposed to contaminated environments were killed by *M. anisopliae* infections. Beetles exposed to environments contaminated by five beetles died more quickly than beetles exposed to environments contaminated by one beetle. Beetles at both density treatments died in fewer days than beetles exposed to environments without *M. anisopliae* conidia.

A follow-up field study examined whether conidia from *M. anisopliae* bands spread naturally to other parts of the environment, and if these conidia are infective to *A. glabripennis*. In the field, bands containing *M. anisopliae* were hung on tree trunks at 3 m height (Shanley & Hajek 2008). Bark samples were taken 10–30 cm above and 10–60 cm below bands up to 9 days after band placement to quantify densities of viable conidia. More viable conidia were detected in samples below bands compared with samples above bands. A significant positive correlation was found between rainfall and the occurrence of conidia on any of the bark samples. However, the concentrations of viable conidia were lower than LC_{50} estimates, suggesting that *A. glabripennis* adults may not become infected based on environmental contamination resulting from natural conidial dispersal from fungal bands.

10.3 Development of Entomopathogens for Control of *Agrilus planipennis*

Agrilus planipennis, the emerald ash borer, is a periodic pest of ash trees (*Fraxinus* spp.) in northeast Asia (Yu 1992). In 2002, this buprestid was identified as the causal agent of ash tree mortality in southern Michigan and Ontario (Haack et al. 2002). *A. planipennis* was likely introduced from China during the 1990s to Lower Michigan in infested wooden packing materials or manufactured goods and became established in the abundant ash resources throughout urban, forested, and riparian ecosystems (Poland & McCullough 2006). Infestations of *A. planipennis* have since been detected in the Upper Peninsula of Michigan, Ohio, Indiana, Illinois, Maryland, Pennsylvania, and West Virginia.

Agrilus planipennis threatens the 16 species of *Fraxinus* native to North America (USDA NRCS 2008), which include abundant ash species used for lumber and wood products. Forest inventories report almost 8 billion ash trees on US timberlands at a compensatory value of $US 282.25 billion (USDA FS 2008). *Fraxinus* species are also the most common trees used to replace landscape plantings of American elms (*Ulmus americana*), decimated by Dutch elm disease in much of North America. The costs for removal and replacement of ash trees killed by *A. planipennis* to communities and smaller landholders are also high, e.g. the expense for ash removal and replacement in six infested southeastern Michigan Counties was estimated at $US 11.7 billion (US Federal Register 2003).

All *Fraxinus* species endemic to the northeastern states and provinces of North America are susceptible to mortality from *A. planipennis*. These include white ash (*F. americana*), green ash (*F. pennsylvanica*), and black ash (*F. nigra*) trees, major components of forests, and the less common blue ash (*F. quadrangulata*) and pumpkin ash (*F. profunda*). Each *Fraxinus* species is adapted to slightly different habitats within forest ecosystems. Several species are tolerant of poorly-drained sites and wet soils, protecting environmentally-sensitive riparian areas, e.g. pure stands of black ash grow in bogs and swamps in northern areas where they provide browse for various wildlife species, thermal cover and protection for ungulates such as deer and moose. In agricultural and shelterbelt areas, ash trees are one of the more prevalent tree species, and protect fragile riparian zones, prevent erosion, and provide shelter for livestock. In forested areas, the bark of young ash trees is a favored food of mammals including beaver, rabbit, and porcupines, whereas older trees provide habitat for cavity-nesting birds such as wood ducks, woodpeckers, chickadees and nuthatches, and seeds are consumed by ducks, song and game birds, small mammals and insects.

In an effort to contain the spread of *A. planipennis* in North America, US and Canadian regulatory agencies imposed quarantines and developed eradication programs. In the US, the eradication program involved *A. planipennis* survey and detection, followed by cutting and chipping of all ash trees in a 0.8 km ($^1/_2$-mile) zone around known infestations. In Canada, the effort to contain *A. planipennis* included cutting a 10×30 km ash-free zone across the Windsor Peninsula, i.e. from Lake Ontario to Lake Erie (CFIA 2008). These efforts were largely unsuccessful due to limited knowledge about *A. planipennis* biology and dispersal potential, lack of detection and control methods, difficulties with quarantine compliance and enforcement, the prevalence of ash, and the sheer size of the *A. planipennis* infestation, which was first discovered ca 10 years after initial introduction. Moreover, regulatory agencies soon learned that humans are responsible for the long-range spread of *A. planipennis* through illegal transport of infested ash nursery stock, firewood, manufactured goods, and timber. In both the US and Canada, eradication strategies are being replaced by management approaches. Researchers are optimistic that improved detection will result in earlier discovery of *A. planipennis* infestations, and various management tools will suppress *A. planipennis* populations below lethal thresholds for North American *Fraxinus* species.

Although the biology of *A. planipennis* was virtually unknown at the time of its discovery in North America, we now have a better understanding of its life cycle (Liu *et al.* 2003, Cappaert *et al.* 2005, Poland & McCullough 2006). *A. planipennis* completes its life cycle in one or two years, depending on the age of the infestation, tree health, and other biotic and abiotic factors. In Michigan, emergence of adults begins in mid to late May and peaks during June. Adults chew and emerge through D-shaped exit holes in the tree bark, begin maturation feeding on ash foliage followed by mating, and after about three weeks, females start to oviposit in bark crevices and between bark layers. Although egg-laying peaks in July, eggs are laid throughout the summer and into early fall due to asynchronous adult emergence and long-lived adults. After egg hatch, neonates tunnel through the tree bark until reaching the phloem where they continue feeding through four larval stages. If mature by fall, larvae chew pupation cells in the outer sapwood or bark, overwinter as mature larvae and pupation occurs during the spring or summer. Early in an infestation, *A. planipennis* oviposit in the upper crown of large ash trees, and as populations increase, the trees become weaker. Tree mortality is caused by larval girdling of the main trunk, when *A. planipennis* populations reach lethal density thresholds for ash. This occurs over a period of several years depending on initial tree health, species, site, rainfall and other factors.

Research on control of *A. planipennis* has focused on the use of protective cover sprays and systemic insecticides, mainly neonicotinoids such as imidacloprid (Poland & McCullough 2006). Efficacy of these products, however, is variable and dependent on infestation level and tree condition when insecticide treatments are initiated, timing, frequency, and method of application, product concentration and formulation, weather, etc. (USDA FS FHTET 2008). *A. planipennis* regulatory activities are limited to quarantine and tree removal. In addition, most communities and homeowners have opted to remove their infested ash trees rather than apply annual insecticide treatments due to the expense, environmental and health risks, and uncertain efficacy.

Research on the development of a microbial control strategy for *A. planipennis* using entomopathogenic fungi was initiated following research on its natural enemies in Michigan field populations from 2002 to 2004 (Bauer *et al*. 2004b, 2005). About 2% of larvae removed from infested ash trees and cultured for entomopathogenic fungi, were infected with strains of *Beauveria bassiana, Isaria farinosa, Isaria fumosorosea, Lecanicillium lecanii,* or *Metarhizium anisopliae*. The successful use of entomopathogenic fungi for insect management (Feng *et al*. 1994, Higuchi *et al*. 1997, Jaronski & Goettel 1997), including trunk sprays of *M. anisopliae* for control of *A. auriventris* (Coleoptera: Buprestidae) on citrus trees in China (Fan *et al*. 1990), led to expanded laboratory, greenhouse and field studies for possible development of entomopathogenic fungi as a management tool for *A. planipennis* on ash trees in North America.

10.3.1 Laboratory and Field Investigations with Entomopathogenic Fungi

10.3.1.1 Laboratory Trials

Preliminary laboratory screening of *A. planipennis* with five isolates of two fungal species (*B. bassiana* and *M. anisopliae*) resulted in higher virulence against adults compared with larvae. All subsequent laboratory studies, therefore, focused on the effects of fungal isolates on *A. planipennis* adults (Liu & Bauer 2006). Adult *A. planipennis*, reared from infested ash trees felled in Michigan, were inoculated by direct immersion in conidial suspensions at two concentrations to determine the time-mortality responses for each fungal isolate. The majority of adult mortality occurred within 4–6 days for all isolates, and for most isolates, higher conidial concentrations resulted in shorter days to death. The cumulative percent mortality 6 days after fungal exposures ranged from 80–97.5 and 97.5–100% for 10^6 and 10^7 conidia per ml, respectively. Within the same time period, only 12.5% of adults died in the control groups. At both concentrations, *B. bassiana* strain GHA treatments resulted in faster mortality for *A. planipennis* adults than the other isolates.

At the time of these studies, two species of entomopathogenic fungi were registered as bioinsecticides in the United States: (1) *B. bassiana* strain GHA, registered in 1995, formulated as BotaniGard ES (petroleum formulation) and Mycotrol O (organic vegetable oil-based) and (2) *M. anisopliae* strain F 52, registered in 2005, formulated as TAE-001 Granular. To compare the concentration-mortality responses of *A. planipennis* adults exposed to each product, a swinging boom spray cabinet with a flat-fan nozzle was used to apply serial dilutions of conidial suspensions to the upper surfaces of leaf rectangles (2×4 cm) cut from fresh, greenhouse-grown ash leaves. Adult beetles were exposed to the treated leaf rectangles for 24 hours, then placed on fresh leaves and daily mortality was determined for 10 days. Each bioassay was replicated twice. At the lowest concentration, adult mortality ranged from 0–35% and at the highest concentration mortality ranged from 95–100%, while average control mortality ranged from 10–20%. The median lethal concentrations (LC_{50}s) were similar for BotaniGard, Mycotrol, and *M. anisopliae* F 52, ranging from 114.5–309.6, 18.4–797.3, and 345.3–362.0 conidia/cm^2, respectively. Subsequent greenhouse and field trials, designed to evaluate the efficacy of cover sprays for *A. planipennis* control, focused on *B. bassiana* GHA-based products because these are (1) registered for use against insects including borers in the US, (2) formulated for aerial application, and (3) a history of use in the US has provided data for a body of literature on environmental persistence and risks to non-target organisms. Fungal bands, containing live cultures of *B. bassiana* strain GHA, were also field tested against *A. planipennis* due to the promise of this deployment method for control of cerambycids (Hajek & Bauer 2007).

10.3.1.2 Greenhouse Pre-Emergent Spray Trials

The efficacy of *B. bassiana* GHA, formulated as BotaniGard, was evaluated as a pre-emergent trunk spray on ash logs infested with *A. planipennis* (Liu & Bauer 2008). It was presumed that ash bark, with its irregular surface, might provide both a large surface area for entrapping sprayed conidia and protection from UV degradation. Moreover, *A. planipennis* adults must chew through the bark of ash trees to emerge from larval phloem-feeding sites; thus, the chance of adults becoming infected was presumably higher following trunk sprays with *B. bassiana* GHA. The latter hypothesis was tested by felling *A. planipennis*-infested green ash trees in southeastern Michigan in March 2003. The trees were cut into 60 cm long logs, refrigerated until July, and then placed in an incubator for 4 weeks at 24°C, 16:8 (L:D) h, and 50–60% RH. Groups of infested ash logs were subsequently sprayed at three rates with BotaniGard ES or BotaniGard ES blank (no fungus) formulation (for the control) using a swinging-boom spray cabinet with a flat fan nozzle. The treated logs were placed individually in aluminum cages and maintained in a greenhouse at ambient conditions ranging from 20–26°C, 20–40% RH, and natural lighting. Each cage was provisioned with a potted evergreen ash tree (*Fraxinus uhdei*) to provide food for emerging *A. planipennis* adults and an uninfested green ash log was placed in each cage for ovipositing females. Adult emergence was monitored daily by marking each new adult emergence hole. Dead adults were removed daily and incubated at 24°C individually in a moist chamber consisting of a 60 mm plastic Petri dish lined with moist sterile filter paper. Fungal infection was confirmed by the presence of mycosis on cadavers 7 days after death. At the end of the study, *A. planipennis* eggs were counted on each oviposition log.

Adult *A. planipennis* mortality, resulting from fungal infection, averaged 33% for beetles emerging from the BotaniGard-treated logs (Liu & Bauer 2008). No fungal infection was detected among adults emerging from the control logs. Interestingly, adult longevity and fecundity was reduced by almost half for the *A. planipennis* adults surviving BotaniGard treatments compared to adults emerging from control logs.

10.3.1.3 Caged Field Trials

The promising results from laboratory and greenhouse studies of fungal control of *A. planipennis* led to expanded field trials using caged sections of ash tree trunks in the field (Liu & Bauer in press). In one caged field trial, BotaniGard ES was sprayed on infested ash trunks in the spring, before adult emergence. The trees selected for this study were green ash trees growing in a small nursery. They were moderately infested with *A. planipennis* and averaged 12 cm in diameter and 9 m in height. A section of tree trunk 50 cm above the ground and 180 cm long was delineated on each tree, and all *A. planipennis* adult emergence holes from previous years were marked and counted. A few days before emergence of adult *A. planipennis*, the trunk sections were sprayed with BotaniGard ES at two application rates with a calibrated professional sprayer fitted with a flat fan nozzle; the control trees were left unsprayed. After treatment, the trunk sections were caged to contain the *A. planipennis* emerging as adults. The cages consisted of aluminum screening stapled around the tree trunks to form a cylinder. To provide food for the emerging adults, an ash branch with several healthy leaves was enclosed inside each cage. The cages were disassembled after 6 weeks and dead adults within cages were cultured for fungal infection. In early fall, the trees were felled and treatment and control trunk sections were dissected for quantification of *A. planipennis* larval and adult densities and percent infection. Percent infection for *A. planipennis* adults emerging from the sprayed trunks averaged 59% at lower and 83% at higher concentrations of BotaniGard. Infected adults that died before or during the emergence process were not found until the trees were felled and dissected in the fall. Larval densities in the treated trees were reduced by more than half compared to control trees. In addition, the larvae dissected from fungal-treated trees were consistently younger than those from control trees.

In another caged field trial, 5 cm wide fungal bands, made by the Hajek laboratory from non-woven polyester fiber bands covered with conidia produced by the culture of *B. bassiana* GHA growing within the band (Higuchi *et al.* 1997; Dubois *et al.* 2004a,b), were stapled around 40 uninfested green ash trees in July. As described above, a cylindrical aluminum screen cage (76.2 cm × 110 cm) was constructed around a section of trunk containing a fungal band. A small ash branch with foliage was enclosed in each cage as adult food. Field-collected *A. planipennis* adults were added at the rate of 10 adults per cage. After 4 weeks, *A. planipennis* adult mortality was determined and each cadaver was cultured to detect mycosis. The fungal bands caused 32% mortality due to fungal infection among *A. planipennis* adults compared to 1% for control adults (H Liu & L Bauer unpublished data).

10.3.1.4 Larval Field Trial

Relatively healthy ash trees respond to *A. planipennis* attack by the formation of callous tissue around larval galleries during the summer. By fall, the growing ridge of callous causes longitudinal splits to develop in the bark over many of the larval galleries, exposing larvae to the external environment. To explore the efficacy of *B. bassiana* GHA sprays for *A. planipennis* larval control, infested white ash trees with bark splits, were sprayed with BotaniGard ES during late fall (Liu & Bauer 2008). These were shade trees in a parking lot and were about 15 years old, 7–10 cm diameter, and 5–6 m tall. The lower 180 cm trunk section of 13 trees was sprayed with BotaniGard at a single concentration using a hand atomizer, and the immediate upper 180 cm trunk section was used as the untreated control. The following winter, the trees were felled, cut into 60 cm logs, dissected in the laboratory, and all *A. planipennis* were screened for fungal infection with mycosis confirmed after 14 days under moist conditions.

The infection rate of the *A. planipennis* larvae was 7.9% in the sprayed trunk sections, significantly higher than the 1.6% infection found in the control trunk sections. The prevalence of fungal infection was positively correlated with larval density in the trunk sections. Once again, larval development was delayed in the fungal-treated ash trunks compared to the controls.

10.3.1.5 Cover-Spray Field Trials

Annual insecticide treatments to preserve high value ash trees must begin before symptoms of *A. planipennis* infestation are visible (Rebek & Smitley 2007). In large ash trees, *A. planipennis* initiates attack in the upper canopy, so early infestation is difficult to detect. One sign of early infestation is upper tree limbs with wood pecks, which are made as woodpeckers remove patches of outer bark while searching beneath for larvae. Another early sign is yellow flagging in the crown, which is caused by yellowing leaves on infested branches. Attack by *A. planipennis* on the main trunk results in small branches, or epicormic shoots, sprouting along the trunk, dead branches and limbs, bark splits, adult emergence holes, and general crown dieback.

To slow the rate of *A. planipennis* colonization of ash trees, topical sprays of *B. bassiana* GHA formulated as BotaniGard ES were tested on 6 year old uninfested white ash trees transplanted to the site the previous year. BotaniGard ES was applied to the foliage and trunks of each of 14 white ash trees four times at 2 week intervals between 25 June and 5 August 2004, with a CO_2 backpack sprayer equipped with a flat fan nozzle. Thirteen white ash were left as untreated controls. In October, the trees were felled, cut into logs, dissected in the laboratory, and all *A. planipennis* were placed in saturated conditions to determine the presence of fungal infection. New colonization of fungal-treated ash trees by *A. planipennis*, as determined by the number of young larvae present in each tree, was reduced by 40.7% compared to control trees. During the course of this study, older larvae were also found in these trees, confirming a 2 year life cycle for some *A. planipennis*. These larvae began developing during the previous year from eggs laid late in the season. In the sprayed trees, fewer young larvae were found but fungal infections were not detected among this younger larval cohort. It is hypothesized that many more young larvae had been present but had died of fungal infection and had decomposed during or soon after egg hatch while tunneling through the fungal-contaminated bark to reach the phloem. This would make finding, collecting, and culturing these small larval cadavers for inclusion in the data set virtually impossible. However, 19.6% of older larvae, developing in the trees since the year before, were infected with *B. bassiana*; none of the large larvae from control trees were infected. These findings support the hypothesis that fungal conidia infiltrate bark splits that form over *A. planipennis* galleries, resulting from the growth of callous around *A. planipennis* feeding damage within the cambial region (Liu & Bauer in press).

The efficacy of BotaniGard ES foliar and trunk cover sprays against well-established *A. planipennis* populations was evaluated in a small nursery of infested green ash trees. The plantation was divided into two plots, each with 25 trees. After leaf flush, crown condition was rated for each tree on a scale of one to three, before fungal application in 2004 and in 2005. Pre-treatment larval densities were estimated in each plot by sampling 50 cm logs, cut at 200–250 cm in height from the main trunks of two randomly selected trees. These logs were dissected and the number of *A. planipennis* was determined on the basis of log surface area. In the treatment plot, BotaniGard ES was sprayed on the trunk and crown of each tree individually using a truck-mounted hydraulic sprayer. The trees were sprayed every two weeks from 23 June to 3 August 2004, a total of four times. Trees in the control plot were not treated. During the following winter, all 50 trees were felled, cut into logs, and transported to the laboratory. A portion of each tree was dissected and *A. planipennis* larvae were used to determine the prevalence of fungal infections. The remaining logs were stored at 4°C until the following summer (2005) when logs were incubated in individual cardboard tubes to collect emerging adults (Liu & Bauer 2006). Adults were collected daily and incubated to determine fungal infection prevalence. Overall, fungus-treated trees contained 46.7% fewer larvae and produced 63.3% fewer adults for the next generation when compared to the controls. The treatment trees sustained 41.5% less crown dieback than did the control trees. The prevalence of fungal infection was positively correlated with larval densities, which likely resulted from increasing horizontal transmission of fungal infection due to increasing numbers of overlapping larval galleries as larval densities increased (Liu & Bauer in press).

In 2004, the persistence of *B. bassiana* GHA conidia was evaluated using treated and control leaves harvested 0, 4, 7, and 11 days after fungal spray. After harvest, leaves were sealed individually in plastic zip-lock bags and transported to the laboratory. One- to six-day old *A. planipennis* adults were exposed to leaves in groups of five for 48 hours. The adults were then transferred to fresh ash leaves in clean dishes, mortality was monitored daily for 14 days and cadavers were placed under high humidity to determine the prevalence of fungal infection. Adult mortality ranged from 100% for those exposed to foliage collected directly after fungal application to 78% for adults exposed to foliage collected 11 days after application, with no statistically significant difference in adult mortality among leaves collected on different days post-application. Control mortality ranged from 20–46%, with no significant difference among leaves collected on different days after application. *B. bassiana* was the primary cause of *A. planipennis* adult mortality in both treatment and controls. Due to the high infectivity of *B. bassiana* strain GHA in *A. planipennis* (Liu & Bauer 2006) and the close proximity of ash trees in the stand, fungal infection from the control leaves likely resulted from drift during application. Time to death for *A. planipennis* adults ranged from 4–7 days after exposure to fungus-sprayed leaves. Longevity was significantly less for adults exposed to fungal-treated leaves compared to control leaves (Liu & Bauer in press).

In summary, ground-based foliar and trunk applications of BotaniGard ES reduced the number of *A. planipennis* feeding in and emerging from infested ash trees, reduced crown dieback in infested ash trees, and reduced new infestation in healthy ash trees (Liu & Bauer in press). These findings, and those from previous studies (Liu & Bauer 2006, 2008), support a role for *B. bassiana* GHA in the management of *A. planipennis* in the field. The use of insect pathogenic fungi for controlling destructive wood-boring insects is not without precedent. Many of these fungal pathogens exhibit high infectivity and virulence, and, given the moist habitat inside trees that provides an excellent environment for fungi, these agents have great potential as important biological control agents. Continued research is needed to reduce the application frequency and area sprayed with *B. bassiana* GHA in order to reduce costs and possible non-target effects. BotaniGard should also be tested as a cover spray for ash trees outside the ash-free zones during eradication to eliminate outlier infestations; at present, all studies have been conducted in the generally infested areas (Liu & Bauer in press).

10.3.2 *Investigations of* **Bacillus thuringiensis** *for* **A. planipennis** *Control*

Strains of *Bt* are found naturally in soil, on leaves, and in other places where insects are abundant, and these strains have restricted host ranges (Crickmore *et al.* 1998). During sporulation, *Bt* produces insecticidal crystal proteins, also known as Cry toxins. *Bt* must be ingested by the insect host because the action of these Cry toxins begins in the midgut; if sufficient toxin is consumed, the result is death by septicemia. *Bt*-based microbial insecticides are the primary tools used to manage forest insect pests due to their limited host ranges, good safety records in human health and the environment, public acceptance, and compatibility with other management strategies such as use of insect biological control agents (Glare & O'Callaghan 2000). Insect larvae are the primary target of *Bt* cover sprays, so larvae living in cryptic environments, such as wood borers, are inaccessible to conventional *Bt* application methods. However, adult insects are also susceptible to *Bt*, including some adult coleopterans, and many beetles, including wood borers, feed on foliage at some point in their life cycles. Thus a *Bt* strain with sufficient virulence could provide adequate control when targeting this life stage.

The first *Bt* strain identified as pathogenic to a coleopteran was *Bt tenebrionis*, which is active against some species of Tenebrionidae and Chrysomelidae (Krieg *et al.* 1983). This discovery stimulated worldwide interest in searching for novel *Bt* strains, and thousands of *Bt* strains are now characterized. At least 30 strains are known to be toxic to various coleopteran pests including *Bt japonensis* (Ohba *et al.* 1992) and *Bt galleriae* SDS-502 (Asano *et al.* 2003) with pathogenicity to certain species of Scarabaeidae. Activity of *Bt* has also been discovered for several wood-boring insects including isolates of *Bt tenebrionis* pathogenic to some species of Bostrichidae, Curculionidae, and Scolytinae (Cane *et al.* 1995, Beegle 1996, Weathersbee *et al.* 2002); *Bt darmstadiensis* with activity against a species of Bostrichidae; *Bt israelensis* with activity against Cerambycidae and Scolytinae (Alfazairy 1986, Méndez-López *et al.* 2003); and *Bt thuringiensis, Bt entomocidus* and *Bt morrisoni* with activity against Scolytinae (Jassim *et al.* 1990, de la Rosa *et al.* 2005). The toxicity spectrum of *Bt* continues to expand as more species are screened, including *Bt* 866 with toxicity against two cerambycids: the Asian longhorned beetle (*A. glabripennis*) and the mulberry longicorn beetle (*Apriona germari*) (Chen *et al.* 2005). For most wood-boring beetles, however, the most effective deployment of *Bt* may require expression of their *cry* toxin genes in transgenic trees.

With international trade, movement of wood-boring beetles that attack and kill live trees has escalated, and there is increasing need to develop environmentally-sensitive control strategies for managing these beetles in natural ecosystems such as forests and riparian areas. Other chapters in this book illustrate how *Bt*-based aerial sprays are used worldwide to suppress or eradicate populations of invasive and native forest insects. Although *Bt* cover sprays have targeted defoliating insect larvae, adult insects can also be susceptible to *Bt*. Moreover, *A. planipennis* adults are defoliators of ash leaves throughout their lives, thus aerial control is feasible. In addition, *A. planipennis* females require a 3 week maturation-feeding period before beginning to lay eggs, providing land managers a window of opportunity for initiating aerial sprays before oviposition begins. Aerial application of a *Bt* strain with high toxicity against *A. planipennis* adults has the potential to reduce the high adult populations, thereby reducing numbers of larvae below a lethal density threshold for North American ash species.

10.3.2.1 Bt *Adulticide Laboratory Bioassays*

Following the discovery of *A. planipennis* in North America in 2002, regulatory agencies and land managers were interested in a registered bioinsecticide for use in the eradication program. Four *Bt*-based products were sprayed on ash leaves in a spray tower and bioassayed against *A. planipennis* adults reared from infested ash logs in the laboratory. The active ingredients of these bioinsecticides are *Bt* strains toxic to certain species of Coleoptera (Novodor®), Lepidoptera (Foray 48B®, Xentari®), or both (Raven®). The products showed some activity against *A. planipennis*, but 4–12 times maximum labeled rates were needed to achieve 66–98% mortality after 6 days of exposure (Bauer et al. 2004a). Further bioassays demonstrated *A. planipennis* adults were not susceptible to the Cry toxins from the *Bt* strains used in these products, although a crude extract of zwittermicin A, another compound produced by *Bt* during fermentation, was toxic (Bauer et al. 2006).

After reviewing the literature and searching patent databases, 18 narrow host-spectrum coleopteran-active *Bt* strains were acquired from culture collections, grown in liquid shake culture, and crystal/spore mixtures were purified. The Cry toxin concentration was estimated for each strain by measuring the intensity of the protein band from SDS-PAGE gels using a densitometer, and comparing the reading to a standard curve prepared from known concentrations of BSA. After standardizing the amount of Cry toxin to a similar concentration for each strain, *A. planipennis* adults were inoculated with a *Bt* crystal/spore mixture using a droplet imbibement bioassay method in which the beetles readily ingest a 0.5 µL droplet containing a known amount of Cry toxin. Using this method, one of the scarab-active strains, *Bt galleriae* (*Btg*) SDS-502 and its Cry8Da toxin (Asano et al. 2003), demonstrated high toxicity against *A. planipennis* adults (LS Bauer unpublished data). The median lethal dose of the *Btg* SDS-502 crystal/spore mixture ranged from 0.16–0.35 µg Cry8Da toxin per beetle and time-to-death ranged from 24–96 hours. Toxicities were similar for solubilized Cry8Da protoxin (130 kDa) and activated toxin (65 kDa); however, the toxicity of the crystal/spore mixture was about 10-fold lower, suggesting somewhat reduced crystal solubilization in the midgut of *A. planipennis*. To evaluate the potential efficacy of *Btg* SDS-502 for aerial application, ten 0.02 µL droplets of crystal/spore mixture, suspended in a 10% sucrose solution, were dispensed onto 1 cm^2 pieces of ash leaf. Each droplet contained ca 0.2 µg Cry toxin; the sucrose served to help the toxin droplets adhere to the ash leaves and to overcome feeding inhibition, which occurs during intoxication of insects by *Bt*. One-week-old adult *A. planipennis* were exposed individually to a treated or control (sucrose only) leaf. After 72 hours, 90% of adults feeding on *Bt*-treated leaves died vs 10% control mortality. In the future, we plan to use the leaf droplet bioassay to optimize formulation ingredients. Once *Btg* SDS-502 is formulated, droplet analyses will be conducted using a rotary atomizer to apply *Btg* and achieve droplet sizes and densities similar to those that would be delivered onto leaves during aerial application in the field. If successful, a new microbial insecticide will be registered for control of *A. planipennis* in ash forests and along riparian areas, thus conserving ecological resources in North America.

10.4 Prospects for Use of Entomopathogens for Control of Invasive Wood Boring Beetles in North America

Methods for microbial control of *A. glabripennis* and *A. planipennis* are under development. In both instances, strategies are constrained by the characteristic difficulties of controlling wood-boring beetles. However, in both cases, the levels of control provided by pathogens (often not 100% immediate mortality) would be appropriate because it takes many beetles to kill individual trees, often over numerous years. Programs for controlling these hosts are quite different due to the different host biologies. Only adults of *A. glabripennis* are directly targeted by fungal bands, although reproduction by females also decreases once infected and conidia can be horizontally transmitted to eggs and larvae, so there are also indirect effects. However, the fungal band strategy still requires that adults contact conidia by walking across a band. The next step in development of fungal bands is to add an attractant so that wandering beetles will be attracted to bands (unfortunately, to date identification of an efficient attractant has been elusive although there now appears to be some progress (S Teale & A Zhang personal communication). Once an attractant can be placed in trees along with fungal bands, the issue of how many fungal bands should be applied to one tree or an area will still need to be addressed but certainly fewer bands will be needed than without attractants. At present, *A. glabripennis* only occurs in urban/suburban areas in North America and its slow spread has helped with a successful eradication strategy. Fungal bands would be highly appropriate to use for protection of high value urban/suburban trees, in conjunction with the eradication program, which relies in part on conventional systemic insecticides. While the applicability of the fungal band approach in forests is not necessary to address at present, fungal bands plus attractants are being very successfully used on a large scale in China against another cerambycid, *Monochamus alternatus* (Li *et al.* 2007).

In contrast, microbial control strategies being investigated against *A. planipennis* have targeted both adults and larvae. Adult *A. planipennis* prefer to navigate a tree by wing rather than by walking, and considerable time is spent feeding on leaves in the canopy, reducing the efficacy of fungal bands for this beetle. However, fungal cover sprays applied to trunks and upper limbs before and after adult emergence, have allowed for conidial contact and infection of *A. planipennis* adults as they (1) chew through the bark to emerge from the tree and (2) during the oviposition period, when adult females search for oviposition sites between layers of bark and in bark crevices. The complex structure and surface area of tree bark serves to conserve fungal inoculum after cover sprays by providing niches where spores are protected from UV. In addition, splits that form in the bark of relatively healthy ash trees directly over *A. planipennis* larval galleries due to callous formation, allow fungal cover sprays to enter, contact, and infect *A. planipennis* larvae in their galleries. Aerial application of a *Bt* cover spray would target adults during the maturation feeding period, and a second application would target late-emerging adults that result from asynchronous development.

For both of these systems, invasive tree-killing pests are of great concern; while eradication may be possible for *A. glabripennis*, eradication programs for widespread and fast-moving *A. planipennis* have been abandoned by regulatory agencies. Without microbial and biological controls (Bauer *et al*. 2004a, 2005, Liu *et al*. 2007), that are generally accepted for use in forested ecosystems, native ash trees will be extirpated from North America. Microbial controls and use of parasitoids can provide the management tools needed by forest managers to suppress *A. planipennis* population densities below the tolerance threshold for ash trees in forests and riparian areas. These methods are also more acceptable to the public and reduced pest populations in forested areas may facilitate survival of ash trees planted throughout urban/suburban landscapes of North America.

Acknowledgments We thank Z. Li and associates, without whose help studies of *A. glabripennis* in China would not have been possible. Laboratory studies with *A. glabripennis* in the US have been possible due to our *A. glabripennis* colony funded by the Alphawood and Milstein/Litwin Foundations. We also thank D. Miller (USDA Forest Service) and H. Liu (Michigan State University) for their tremendous dedication and hard work on *A. planipennis* research since its discovery in Michigan in 2002. We thank A. M. Liebhold for helpful comments on this chapter.

References

Alfazairy AA (1986) The pathogenicity of *Bacillus thuringiensis*, isolated from the *Casuarina* stem borer *Stromatium fulvum* (Coleoptera: Cerambycidae), for larvae of 12 species of mosquitoes. Insect Sci Appl 7:633–636

Asano S, Yamashita C, Iizuka T, Takeuchi K, Yamanaka S, Cerf D, Yamamoto T (2003) A strain of *Bacillus thuringiensis* subsp. *galleriae* SDS-502 containing a *cry8Da* gene highly toxic to *Anomala cuprea* (Coleoptera: Scarabaeidae). Biol Control 28:191–196

Bauer LS, Liu HP, Miller DL (2004a) Microbial control of emerald ash borer. In: Mastro V, Reardon R (eds) Proc Emerald Ash Borer Research and Technology Meeting, Port Huron, MI. US Dep Agric FS FHTET-2004-02. pp 31–32 http://www.fs.fed.us/foresthealth/technology/pdfs/2003EAB.pdf [accessed March 2008]

Bauer LS, Liu HP, Haack RA, Petrice TR, Miller DL (2004b) Natural enemies of emerald ash borer in southeastern Michigan. In: Mastro V, Reardon R (eds) Proc Emerald Ash Borer Research and Technology Meeting, Port Huron, MI. US Dep Agric FS FHTET-2004-02. pp 33–34 http://www.fs.fed.us/foresthealth/technology/pdfs/2003EAB.pdf [accessed March 2008]

Bauer LS, Liu HP, Haack RA, Gao RT, Zhao TH, Miller DL, Petrice TR (2005) Update on emerald ash borer natural enemies in Michigan and China. In: Mastro V, Reardon R (eds) Proc Emerald Ash Borer Research and Technology Meeting, Romulus, MI. US Dep Agric FS FHTET-2004-15. pp 71–72 http://www.fs.fed.us/foresthealth/technology/pdfs/2005EAB.pdf [accessed March 2008]

Bauer LS, Dean D, Handelsman J (2006) *Bacillus thuringiensis*: Potential for management of emerald ash borer. In: Mastro V, Reardon R (eds) Proc Emerald Ash Borer Research and Technology Meeting, Romulus, MI. US Dep Agric FS FHTET-2005-16. pp 40–41 http://www.fs.fed.us/foresthealth/technology/pdfs/2006EAB.pdf [accessed March 2008]

Beegle CC (1996) Efficacy of *Bacillus thuringiensis* against lesser grain borer, *Rhyzopertha dominica* (Coleoptera: Bostrichidae). Biocontr Sci Technol 6:15–21

Cane JH, Cox HE, Moar WJ (1995) Susceptibility of *Ips calligraphus* (Germar) and *Dendroctonus frontalis* Zimmerman (Coleoptera: Scolytidae) to coleopteran active *Bacillus thuringiensis* and Avermectin B. Can Entomol 127:831–837

Cappaert D, McCullough DG, Poland TM, Siegert NW (2005) Emerald ash borer in North America: a research and regulatory challenge. Amer Entomol 51:152–165

CFIA (Canadian Food Inspection Agency) (2008) Emerald ash borer – Ash-free zone backgrounder. http://www.inspection.gc.ca/english/plaveg/pestrava/agrpla/zonee.shtml [accessed January 2008]

Chen J, Dai LY, Want XP, Tian YC, Lu MZ (2005) The *cry3Aa* gene of *Bacillus thuringiensis* encodes a toxin against long-horned beetles. Appl Microbiol Biotechnol 76:351–356

Crickmore N, Zeigler DR, Feitelson J, Schnepf E, Van Rie J, Lereclus D, Baum J, Dean DH (1998) Revision of the nomenclature for the *Bacillus thuringiensis* pesticidal crystal proteins. Microbiol Mol Biol Rev 62:807–813

de la Rosa W, Figueroa M, Ibarra JE (2005) Selection of *Bacillus thuringiensis* strains native to Mexico active against the coffee berry borer *Hypothenemus hampei* (Ferrari) (Coleoptera: Curculionidae: Scolytinae). Vedalia 12:3–9

Dubois TLM (2003) Biological control of the Asian longhorned beetle, *Anoplophora glabripennis*, with entomopathogenic fungi. Ph.D. Dissertation, Cornell University, Ithaca, New York. 200 pp

Dubois T, Hajek AE, Jiafu H, Li Z (2004a) Evaluating the efficiency of entomopathogenic fungi against the Asian longhorned beetle, *Anoplophora glabripennis* (Coleoptera: Cerambycidae), using cages in the field. Environ Entomol 33:62–74

Dubois T, Li Z, Jiafu H, Hajek AE (2004b) Efficacy of fiber bands impregnated with *Beauveria brongniartii* cultures against Asian longhorned beetle, *Anoplophora glabripennis* (Coleoptera: Cerambycidae). Biol Control 31:320–328

Dubois T, Lund J, Bauer LS, Hajek AE (2007) Virulence of entomopathogenic hypocrealean fungi infecting *Anoplophora glabripennis*. BioControl 53:517–528

Fan MZ, Li LC, Guo C, Wu XM, Sun ZB (1990) Control of *Agrilus auriventris* by the strain Ma83 of *Metarhizium anisopliae*. Disinsectional Microorganism 3:278–279

Faria MR de, Wraight SP (2007) Mycoinsecticides and mycoacaricides: A comprehensive list with worldwide coverage and international classification of formulation types. Biol Control 43:237–256

Feng MG, Poprawski TJ, Khachatourians GG (1994) Production, formulation, and application of the entomopathogenic fungus *Beauveria bassiana* for insect control: Current status. Biocontr Sci Technol 4:3–34

Glare TR, O'Callaghan M (2000) *Bacillus thuringiensis*: Biology, ecology and safety. John Wiley and Sons, New York. 350 pp

Haack RA (2006) Exotic bark- and wood-boring Coleoptera in the United States: Recent establishments and interceptions. Can J For Res 36:269–288

Haack RA, Law KR, Mastro VC, Ossenbruggen HS, Raimo, BJ (1997) New York's battle with the Asian long-horned beetle. J For 95:11–15

Haack RA, Jendek E, Liu HP, Marchant KR, Petrice TR, Poland TM, Ye H (2002) The emerald ash borer: A new exotic pest in North America. Newsl Mich Entomol Soc 47(3&4):1–5

Hajek AE (2007) Asian longhorned beetle: Ecology and control. In: Pimentel D (ed) Encyclopedia of pest management, Vol. II. CRC Press, Taylor & Francis Group, Boca Raton, FL. pp 21–24

Hajek AE, Bauer, LS (2007) Microbial control of bark- and wood-boring insects attacking forest and shade trees. In: Lacey LA, Kaya HK (eds) Field manual of techniques in invertebrate pathology, 2nd edition. Springer, Dordrecht, Netherlands. pp 505–525

Hajek AE, Huang B, Dubois T, Smith MT, Li Z (2006) Field studies of control of *Anoplophora* (Coleoptera: Cerambycidae) using fiber bands containing the entomopathogenic fungi *Metarhizium anisopliae* and *Beauveria brongniartii*. Biocontr Sci Technol 16: 329–343

Hajek AE, Lund J, Smith MT (2008) Reduction in fitness of female Asian longhorned beetle (*Anoplophora glabripennis*) infected with *Metarhizium anisopliae*. J Invertebr Pathol 98: 198–205

Higuchi T, Saika T, Senda S, Mizobata T, Kawata Y, Nagai J (1997) Development of biorational pest control formation against longicorn beetles using a fungus, *Beauveria brongniartii* (Sacc.) Petch. J Ferment Bioengin 84:236–243

Jaronski ST, Goettel MS (1997) Development of *Beauveria bassiana* for control of grasshoppers and locusts. In: Goettel MS, Johnson DL (eds) Microbial control of grasshoppers and locusts. Mem Entomol Soc Can 171: 225–237

Jassim HK, Foster HA, Fairhurst CP (1990) Biological control of Dutch elm disease: *Bacillus thuringiensis* as a potential control agent for *Scolytus scolytus* and *S. multistriatus*. J Appl Bacteriol 69:563–568

Krieg VA, Huger AM, Langenbrunch GA, Schnetter W (1983) *Bacillus thuringiensis* var. *tenebrionis*, a new pathotype effective against larvae of Coleoptera. Z Angew Entomol 96:500–508

Li Z, Fan M, Wang S, Ding D (2007) *Beauveria bassiana* plus chemical attractant: A new approach against pine sawyer? Proc Ann Mtg Soc Invertebr Pathol, Quebec City, Quebec Canada. pp 113–114

Lingafelter SW, Hoebeke ER (2002) Revision of the genus *Anoplophora* (Coleoptera: Cerambycidae). Entomol Soc Wash, Washington, DC

Liu HP, Bauer LS (2006) Susceptibility of *Agrilus planipennis* (Coleoptera: Buprestidae) to *Beauveria bassiana* and *Metarhizium anisopliae*. J Econ Entomol 99:1096–1103

Liu HP, Bauer LS (2008) Microbial control of emerald ash borer, *Agrilus planipennis* (Coleoptera: Buprestidae), with *Beauveria bassiana* strain GHA: Greenhouse and field trials. Biol Control 54:124–132

Liu HP, Bauer LS Microbial control of emerald ash borer, *Agrilus planipennis* (Coleoptera: Buprestidae), with *Beauveria bassiana* strain GHA: Field applications. Biocontr Sci Technol (in press)

Liu HP, Bauer LS, Gao RT, Zhao TH, Petrice TR, Haack RA (2003) Exploratory survey for the emerald ash borer, *Agrilus planipennis* (Coleoptera: Buprestidae), and its natural enemies in China. Grt Lks Entomol 36:191–204

Liu HP, Bauer LS, Zhao TH, Gao RT, Song LW, Luan QS, Jin RZ, Gao CQ (2007) Seasonal abundance of *Agrilus planipennis* (Coleoptera: Buprestidae) and its natural enemies *Oobius agrili* (Hymenoptera: Encyrtidae) and *Tetrastichus planipennisi* (Hymenoptera: Eulophidae) in China. Biol Control 42:61–71

Méndez-López I, Basurto-Ríos R, Ibarra J (2003) *Bacillus thuringiensis* serovar *israelensis* is highly toxic to the coffee berry borer, *Hypothenemus hampei* Ferr. (Coleoptera: Scolytidae). FEMS Microbiol Lett 11131:1–5

Ohba M, Iwahana H, Asano S, Sato R, Suzuki N, Hori H (1992) A unique isolate of *Bacillus thuringiensis* serovar *japonensis* with high larvicidal activity specific to scarabaeid beetles. Lett Appl Microbiol 14:54–57

Poland TM, McCullough DG (2006) Emerald ash borer: Invasion of the urban forest and the threat to North America's ash resource. J For 104:118–124

Poland TM, Haack RA, Petrice TR, Miller DL, Bauer LS, Gao R (2006a) Field evaluations of systemic insecticides for control of *Anoplophora glabripennis* (Coleoptera: Cerambycidae) in China. J Econ Entomol 99:383–392.

Poland TM, Haack RA, Petrice TR, Miller DL, Bauer LS (2006b) Laboratory evaluation of the toxicity of systemic insecticides for control of *Anoplophora glabripennis* and *Plectrodera scalator* (Coleoptera: Cerambycidae). J Econ Entomol 99:85–93

Rebek KA, Smitley DR (2007) Homeowner guide to emerald ash borer treatments. Mich State Univ Ext Bull E-2955. http://emeraldashborer.info/files/E2955.pdf [accessed March 2008]

Shanley RP (2007) Using non-woven bands impregnated with the entomopathogenic fungus *Metarhizium anisopliae* (Metchnikoff) for biological control of the Asian longhorned beetle, *Anoplophora glabripennis* (Motschulsky). M.S. Dissertation, Cornell University, Ithaca, New York. 55 pp

Shanley RP, Hajek AE (2008) Environmental contamination with *Metarhizium anisopliae* from fungal bands for control of the Asian longhorned beetle, *Anoplophora glabripennis* (Coleoptera: Cerambycidae). Biocontr Sci Technol 18:109–120

Tsutsumi T (1998) Microbial control of longicorn beeetles using non-woven fabric sheet containing *Beauveria brongniartii* in the orchard. Proc Soc Invertebr Pathol, Sapporo, Japan. pp 203–209

USDA FS FHTET (USDA Forest Service Forest Health Technology Enterprise Team) (2008) Emerald ash borer and Asian longhorned beetle 2006 research and development meeting. http://www.fs.fed.us/foresthealth/technology/pdfs/EAB_ALB_2006.pdf [accessed March 2008]

USDA FS (Forest Service) (2008) Effects of urban forests and their management on human health and environmental quality: Emerald ash borer. http://www.fs.fed.us/ne/syracuse/Data/Nation/data_list_eab.htm [accessed March 2008]

USDA NRCS (Natural Resources Conservation Service) (2008) PLANTS profile. *Fraxinus* L. http://plants.usda.gov/java/profile?symbol=FRAXI [accessed March 2008]

US EPA (Environmental Protection Agency) (2003) *Metarhizium anisopliae* strain F52 (029056) biopesticide fact sheet. http://www.epa.gov/pesticides/biopesticides/ingredients/factsheets/factsheet_029056.htm [accessed 6 March 2008]

US Federal Register (2003) Emerald ash borer: Quarantine and Regulations. 7 CFR Part 301 [Docket No. 02-125-1]

US GAO (Government Accountability Office) (2006) Invasive forest pests: Lessons learned from three recent infestations may aid in managing future efforts. US GAO-06-353

Weathersbee III AA, Tang YQ, Doostdar H, Mayer RT (2002) Susceptibility of *Diaprepes abbreviatus* (Coleoptera: Curculionidae) to a commercial preparation of *Bacillus thuringiensis* subsp. *tenebrionis*. Fla Entomol 85:330–335

Yu CM (1992) *Agrilus marcopoli* Obenberger. In: Xiao GR (ed) Forest insects of China, 2nd edn. China Forestry Publishing House, Beijing, China. pp 400–401

Chapter 11
Control of Gypsy Moth, *Lymantria dispar*, in North America since 1878

Leellen F. Solter and Ann E. Hajek

Abstract Gypsy moth is an outbreak species that was introduced to North America from Europe in 1869, with disastrous consequences. This species is a devastating defoliator in northeastern hardwood forests and continues to spread to the west and south. Four different types of pathogens are of interest for gypsy moth control, making this the invasive arthropod with the greatest diversity of pathogens being utilized for control. *Bacillus thuringiensis kurstaki* HD-1 is commercially available and is usually applied for control instead of synthetic chemical insecticides. *Btk* can provide excellent control of outbreak populations and also gives outstanding results in eradication campaigns when gypsy moth is introduced into new areas. The baculovirus *Ld*MNPV, which is highly specific to gypsy moth, is also mass produced but because of its limited availability is only applied in environmentally sensitive areas. While *Btk* does not cause epizootics in natural gypsy moth populations, *Ld*MNPV has a history of epizootics that have caused crashes in defoliating (high density) populations since the accidental introduction of the virus some time before 1907. The fungal pathogen *Entomophaga maimaiga*, originating from Japan, first reported in North America in 1989 and probably accidentally introduced, also causes dramatic epizootics in both low and high density gypsy moth populations; activity of this fungus is determined, at least in part, by environmental conditions. Several species of microsporidia are known from the native range of gypsy moth and programs are in place to introduce these microsporidia to North American gypsy moth populations to augment the natural enemies already present.

11.1 Biological Control in Forest Systems

Unlike most agricultural systems for which intensive management is required and little insect damage is tolerated, most forest systems are ecologically diverse and tolerate some naturally occurring damage from fire and wind, as well as damage

L.F. Solter
Division of Biodiversity and Ecological Entomology, Illinois Natural History Survey,
1816 S. Oak Street, Champaign, Illinois 61801, USA
e-mail: lsolter@uiuc.edu

Fig. 11.1 Late instar gypsy moth, *Lymantria dispar*, larva. Photo by L. F. Solter

from plant pathogens and herbivorous animals. Such damage may open the canopy and/or stimulate foliage production and are, in fact, natural occurrences in a healthy forest ecosystem (Seastedt *et al*. 1983, DellaSala *et al*. 1995). Even when forest systems are managed for timber production, relatively high economic thresholds allow for less intensive methods of herbivore control, opening up possibilities for methods to control invasive pests that would not be economically feasible in many food crop systems. In addition, broad-scale chemical control methods, which are rarely specific to a single pest or pest complex, are problematic for environmental stewardship of complex ecosystems and use should be minimized where possible. Biological control methods, the "cornerstones" of well-designed integrated pest management programs (Cate & Hinkle 1994), are highly appropriate for management of systems with relatively high economic thresholds and may be the only economically feasible choice in many situations.

Although forest systems are complex and more resistant to disturbance than most agriculturally managed systems, outbreak insect species sometimes cause damage that should be mitigated, particularly if the pest is an invasive species that is not under the control of a full suite of natural enemies and is encroaching on an environment that it can exploit. Such is the situation with the gypsy moth, *Lymantria dispar*, a member of the lepidopteran family Lymantriidae that was imported from France in 1869 (Fig. 11.1). Within 10 years of its accidental release in the Boston area, the gypsy moth was well established and on its way to becoming a major outbreak species in North America, causing millions of dollars in damage to eastern North American oak forests and incurring significant annual costs for control efforts (Leuschner *et al*. 1996).

11.2 The Invasive Gypsy Moth in North American Forests

The larvae of the gypsy moth are voracious foliage feeders with a large host plant range that includes more than 300 plants (Schweitzer 2004). During gypsy moth population outbreaks, the preferred host species, primarily oak, aspen and willow, are extensively defoliated. The major economic loss in North America is to oak forests, which comprise a large percentage of forest habitat in the eastern half of the continent. A large portion of the cost of control, however, is borne by homeowners in treating nuisance problems (Leuschner et al. 1996). Outbreaks occur in periodic cycles, estimated to be approximately every 10 years for major defoliation events, with less intense occurrences mid-cycle, about every 5 years (Johnson et al. 2006).

Pathogens play major roles in the population dynamics of gypsy moth in North America. Fungal and viral pathogens that have been released in North America can naturally prevent outbreaks or, if an outbreak population develops, cause high levels of mortality that result in population crashes. For gypsy moth populations that grow to outbreak densities with few natural controls, bacterial and viral pathogens are the first choice for application from the ground or air. The US federal government no longer supports spraying chemical insecticides for gypsy moth control.

Gypsy moth is consistently spreading to the south and west and a different control approach is practiced on the edge of gypsy moth spread; the Slow the Spread program for control of gypsy moth, *Lymantria dispar*, in the United States is probably one of the most complete large-scale integrated pest management (IPM) programs ever established and maintained (see Chapter 5). The program is a partnership among various agencies including the United States Department of Agriculture Animal & Plant Health Inspection Service (USDA-APHIS), USDA Forest Service (USDA FS), state governments and a number of universities and emphasizes the use of and pheromones for mating disruption and biological control agents.

In this chapter, we summarize the development of a bacterium, *Bacillus thuringiensis kurstaki* as a microbial insecticide; development and use of a baculovirus, the naturally occurring *Lymantria dispar* multinucleocapsid nucleopolyhedrovirus (or nuclear polyhedrosis virus) (*Ld*MNPV) as a microbial pesticide; a naturally cycling and virulent fungus, *Entomophaga maimaiga*; and studies toward release of three species of entomopathogenic microsporidia that occur naturally in European populations of gypsy moth but are not currently found in North American populations. Each major group of microbials in use or under development for use (i.e. bacteria, virus, fungi, and microsporidia) interacts differently with the gypsy moth host, thus resulting in diverse impacts on host populations. This variability also translates to a diversity of strategies for use of the different types of pathogens as well as different expectations for control.

11.3 *Bacillus thuringiensis kurstaki (Btk)*

Bacillus thuringiensis, a Gram positive spore-forming bacterium, was first reported from the silk moth, *Bombyx mori*, in Japan by Ishiwata in 1901, and was formally described 10 years later by Berliner in Germany. The German type strain, originally isolated from the Mediterranean flour moth, *Ephestia kuehniella*, was lost and later re-isolated from the same host in 1927. The first trials with *Bt* for insect control were conducted in the 1920s and 1930s (Federici 2005) and the pathogenicity of *Bt* against gypsy moth was first reported in 1929 (Dubois 1981). Since use of *Bt* for pest control became established and researchers understood that different strains can be very specific, the search for new isolates has been extensive; today, there are many *Bt* culture collections, some with thousands of isolates (Martin & Blackburn 2007). Work toward commercialization of *Bt* for pest control began in earnest in the 1950s and 1960s (Federici 2005). In 1962, a lepidopteran-specific strain of *Bt* was once again isolated from *E. kuehniella* by Kurstak in France (Lord 2005). Dulmage (1970) isolated a similar, virulent strain of *Bt kurstaki* from the pink bollworm, *Pectinophora gossypiella*, and named the strain HD-1. This isolate was found to be highly virulent against Lepidoptera and is the strain currently used in *Btk* formulations for gypsy moth control.

11.3.1 Btk *Biology and Development as a Microbial Pesticide*

Bt infection in insects is typically opportunistic. Epizootics are rare except when and where insects occur in enclosed environments such as grain bins, bee hives or production colonies (Aronson *et al.* 1986). In the insect host, vegetative cells give rise to sporangia, which produce environmentally resistant, infective spores and protein crystals (or "parasporal bodies") composed of protoxins that, when ingested, are broken down by proteolytic enzymes in the insect gut to form delta-endotoxins (Tanada & Kaya 1993). The delta-endotoxins produced by different *Bt* strains are harmful to different insect host orders and to varying degrees; they are frequently lethal to specific hosts. A voluminous literature exists on new toxins that have been isolated from the different *Bt* strains. More than 200 members have been identified for 50 subgroups of *Bt* Cry (= crystal) toxins (Bravo *et al.* 2007). The toxin molecular structure, mode of action and specificity of many strains have been described and the efficacy of naturally occurring and commercially produced strains have been tested (see Glare & O'Callaghan 2000, Charles *et al.* 2000). Here, we briefly address the lepidopteran active HD-1 strain and its activity against the gypsy moth.

Several delta (δ)-endotoxins were isolated from the HD-1 strain and are found in the products currently used in gypsy moth control programs. These include CryIA(a), CryIA(b), CryIA(c), and CryIIA (Valent Biosciences 2001), all of which are toxic to young gypsy moth larvae (Liang & Dean 1994, Liang *et al.* 1995). The specificity and relative toxicity of the toxins is related to the affinity with which they bind to specific receptors in the brush border membranes of host midgut epithelial cells (Jenkins *et al.* 2000, Valaitis *et al.* 2001). Researchers are using molecular techniques to isolate the binding domains in the host cells and evaluate the strength of binding of specific toxins (e.g. Jenkins *et al.* 2000). Genetic engineering of toxins is being carried out to eliminate non-toxic moieties (Audtho *et al.* 1999) and improve binding efficiency (Rajamohan *et al.* 1996). Binding affinities are related to toxicity and to host specificity (Liang *et al.* 1995, Rajamohan *et al.* 1996).

In general, *Bt* toxins begin the process of killing hosts by creating pores in the membrane of the midgut epithelial cells, destroying the selective permeability of the cells and causing ion leakage. The gut cells lyse and die (Bravo *et al.* 2007).

Perforation of the gut membranes paralyzes the gut and allows leakage of bacteria and other gut contents into the hemolymph of the host. Gut paralysis causes cessation of feeding fairly quickly after ingestion of the toxins (Reardon *et al.* 1994). The prevailing model suggests that insects die due to septicemia after bacteria in the gut invade the hemocoel (see Broderick *et al.* 2006). Gypsy moth larvae have long been known to be less susceptible to many lepidopteran-active *Bt* toxins than other target hosts (DH Dean personal communication). Dubois and Dean (1995) showed synergism of *Btk* with a wide variety of naturally occurring epiphytic bacteria in gypsy moth larvae. Recent research has continued these studies to show that opportunistic bacteria present in the gypsy moth midgut are required for gypsy moth larvae to die after eating *Btk* (Broderick *et al.* 2006).

Bt can sporulate and reproduce in the soil; thus, it is ideally adapted to mass production in media that is not too complex or expensive. *Bt* is grown in large fermentation tanks. Ingredients of the "fermentation beer" vary but include sources of carbon and nitrogen that must be certified as acceptable for residue in the final product. The final product is 6–8% solids with 1–3% being spores and crystals. The "fermentation beer" is centrifuged and/or filtered and resuspended in the formulation materials (Couch 2000), which may be oil-based or aqueous for HD-1 products (Valent Biosciences 2001).

11.3.2 Efficacy Testing and Enhancers

Years of formulation studies and recommendations for aerial and ground sprays are summarized in the Valent BioSciences Forestry Technical Manual (2001) and by Reardon *et al.* (1994). Early developers identified the importance of factors such as spray volume to ensure coverage of foliage; use of multiple applications to overcome the diversity of larval ages present at one time, especially since early instars are more susceptible; and the use of stickers so that materials remained on the foliage (Dubois 1981). Since then, parameters such as spray type (van Frankenhuyzen *et al.* 1991), size and density of droplets (Dubois *et al.* 1993), distribution of spray in the canopy (Yendol *et al.* 1990), spray nozzle types and sizes (Dubois *et al.* 1994), concentration of active ingredients in the environment (Dubois *et al.* 1993, Thorpe *et al.* 1998), number of applications (Dubois *et al.* 1988), persistence of *Btk* activity (Sundaram *et al.* 1994), effect of host density (Podgwaite *et al.* 1993, Ridgway *et al.* 1994) and other factors for successful coverage and efficacy have been addressed in many studies. In addition, efforts have been made to enhance the efficacy and persistence of the spores and toxins by adding phagostimulants (Farrar & Ridgway 1995), other bacteria (Dubois & Dean 1995), optical brighteners (Martin 2004), zwittermycin A (isolated from *B. cereus*) (Broderick *et al.* 2003), and varying mixtures of Cry toxins (Lee *et al.* 1996). Additional studies showed that the addition of ß-exotoxin, also produced by *Btk*, synergized the effects of HD-1 preparations (Dubois 1986), but this toxin is not permitted in commercial products due to deleterious effects on vertebrates. In laboratory tests, oak tannins plus *Btk* inhibited feeding by larvae (Falchieri *et al.* 1995), while both oak tannins and aspen phenolic glycosides enhanced the efficacy of *Btk* (Hwang *et al.* 1995). These and other studies suggested that formulations should be adjusted depending on the tree species and forest stand quality (Appel & Schultz 1994). Studies showed that overall *Btk* efficacy is somewhat less than that of Dimilin® (diflubenzuron) (Ridgway *et al.* 1994, Liebhold *et al.* 1996) but *Btk* had less environmental impact (Reardon *et al.* 1994).

11.3.3 Environmental Safety of Btk

Although commercial formulations of *Btk* meet stringent US Environmental Protection Agency safety requirements, extensive monitoring studies have been conducted to evaluate safety of *Btk* to vertebrate animals including humans, beneficial invertebrates and aquatic systems. *Btk* is primarily toxic to Lepidoptera. In laboratory studies of a broad range of native Lepidoptera, mortality rates were high for 64% of treated species (n = 42) (Peacock *et al.* 1998). Laboratory testing of the endangered Karner blue butterfly, *Lycaeides melissa samuelis*, indicated susceptibility similar to that of gypsy moth (Herms *et al.* 1997), and late instars of the cinnabar moth, *Tyria jacobaeae*, a biological control agent of tansy ragwort, were also susceptible (James *et al.* 1993). Field studies have been more inconclusive; they have shown effects of annual treatments of *Btk* but site-specific results were common and timing of treatment may be a factor in whether a susceptible host is at risk (Herms *et al.* 1997). In a 4-year field study in Virginia, some macro-lepidopteran species were more common during treatment years, compared with pre-treatment years, and some were less common (Rastall *et al.* 2003). Other studies found greater effects: species richness and abundance were reduced in treated plots on Vancouver Island, BC, Canada, the year of treatment (1999) (Boulton 2004) and the year after treatment (Boulton *et al.* 2007), but there was no significant variation in lepidopteran species diversity after treatment. Collections made 5 years after the sprays found that although all of the 11 species that had been reduced in numbers in 1999 had increased 5 years later, densities of 4 species remained significantly reduced in treated areas compared with controls (Boulton *et al.* 2007). Impacts of treatments were evaluated in two 3-year studies in West Virginia: the first study found significant reductions in species richness and in total numbers of Lepidoptera after *Btk* treatment (Miller 1990), and the second found similar effects but noted that declines were also observed at untreated defoliation sites (Sample *et al.* 1996). *Btk* drift was evaluated in a study conducted when gypsy moth was being eradicated in the western state of Utah. *Btk* was detected more than 3,000 meters from the spray boundaries (Barry *et al.* 1993) and mortality was significant for populations of two selected lepidopteran species, *Callophrys* (= *Incisalia*) *fotis* and *Callophrys sheridanii* (Whaley *et al.* 1998). Ant populations were not affected in 3-year treatment plots (Wang *et al.* 2000) and predatory carabid populations were also not significantly impacted (Cameron & Reeves 1990).

Some concerns have been raised about indirect effects, for example, the impacts of reduced caterpillar numbers on vertebrate animals. Sopuck *et al.* (2002) and

Marshall *et al.* (2002) found no significant effects on birds, and there were also no changes in salamander diets in treated areas (Raimondo *et al.* 2003).

Human health concerns have been addressed in several before-and-after spray evaluations (see Chapter 16). While the presence of *Btk* in nasal passages increased after sprays in Victoria BC, Canada, well over half the subjects tested positive for *Btk* before the sprays were applied (de Amorim *et al.* 2001). In Oregon, 95% of human patients who tested positive for *Btk* were probably contaminated by sprays but were not adversely affected because no clinical illness was associated with the *Btk*. It appears that the most serious safety concerns regarding *Btk* applications are for Lepidoptera, and particularly in areas where there is little untreated space for refugia to harbor sensitive species to repopulate treated areas. An excellent review of additional safety issues and reference list can be found in Glare and O'Callaghan (2000).

11.3.4 Application in Gypsy Moth Management

The gypsy moth was the first pest for which *Btk* was used in broad scale application (Reardon *et al.* 1994). Field trials were conducted from the early 1930s through the 1960s with isolates that were less virulent than HD-1 and using methods that did not achieve levels of mortality close to those of chemical controls (Dubois 1981). The subsequent basic research, development and evaluation of the HD-1 isolate has been extensively covered by others (Glare & O'Callaghan 2000, Reardon *et al.* 1994) and will not be reiterated here; the following is a brief update on current practices.

The calculations for toxicity of the HD-1-S-1980 standard reference used currently are based on the original standard for *B. thuringiensis thuringiensis* of 1,000 international toxin units of insecticidal activity (IUs) per mg of powder. The potency of the HD-1-S-1980 *Btk* isolate is calculated as 16,000 IU/mg powder (Reardon *et al.* 1994).

Btk is used successfully against gypsy moth populations of varying densities, from increasing populations to outbreak populations to newly colonizing populations. *Btk* is currently the pesticide of choice for gypsy moth eradication treatments far from areas where gypsy moth is established, as well as for eradication of newly-forming populations at the leading edge of the gypsy moth invasion in the Slow the Spread Program (see Chapter 5). When established gypsy moth populations increase to outbreak densities, *Btk* is the primary pesticide used during suppression programs in the US. In the years 2005–2007, over 600,000 acres (242,820 ha) were treated with *Btk* in state programs. In contrast, the most commonly used synthetic chemical insecticide, Dimilin, was used by only four states (USDA Forest Service 2008); most other states and the federal government have excluded use of Dimilin due to concerns about nontarget impacts on arthropods. Recommendations for the most commonly used *Btk* formulations in state suppression programs, Foray® (produced by Valent BioSciences Corporation), vary from single or double applications of 24–36 billion international units (BIU)/acre (= 60–89 BIU/ha) (Valent Biosciences 2001) (see also Chapter 5). Applications are made by air or ground when early instars are present on the foliage. For homeowners, over-the-counter *Bt*-based products such as Thuricide® are available.

11.4 *Lymantria dispar* Nucleopolyhedrovirus (*Ld*MNPV)

The *Lymantria dispar* nucleopolyhedrovirus, *Ld*MNPV, was first reported as being present in North American gypsy moth populations in 1907 (Hajek *et al.* 2005) and was suggested as a microbial control agent early in the 20th century (Reiff 1911). It is not known how the virus was introduced to North America, but it was possibly present in the accidentally introduced host, or introduced with parasitoids. *Ld*MNPV is found worldwide in gypsy moth populations, including those of the Asian gypsy moth (Lewis 1981, Pemberton *et al.* 1993). Efforts to develop the virus as a microbial control agent for gypsy moth began in earnest in the 1960s.

11.4.1 Basic Biology and Epizootiology

*Ld*MNPV is a member of the family Baculoviridae, double-stranded DNA viruses unique to arthropods and relatively common in Lepidoptera. This species is one of a group of nucleopolyhedrosis viruses in which multiple nucleocapsids containing viral DNA are packaged within one envelope; thus, the "M" in *Ld*MNPV designates that virions are multiply enveloped. These enveloped nucleocapsids (virions) are occluded by a matrix of the protein polyhedrin. The resulting occlusion bodies (OBs) may be from 1 to 10 μm in size, often between 2 and 5 μm, and contain a varying number of virions depending on the size of the occlusion body (Blissard & Rohrmann 1990).

The *Ld*MNPV infection cycle is nearly identical to that of the *Autographa californica* MNPV described by Blissard & Rohrmann (1990), except that the cycle is longer, probably an evolutionary response to the longer larval period of the gypsy moth (Riegel and Slavicek 1997). When gypsy moth larvae ingest polyhedra, alkaline conditions in the insect gut dissolve the polyhedrin protein matrix, releasing virions that then enter the midgut epithelial cells. The virions are transported to the nucleus where they replicate and form "budded virus", virions that are not occluded in a polyhedra matrix. These non-enveloped virions exit the host cell nucleus and bud through the cytoplasmic membrane to infect other cells. In addition, some nucleocapsids that enter midgut cells bypass the nucleus and proceed directly to budding out of the cell. At approximately 48 hours after inoculation of the host, "occluded virions" (OBs, as described above) begin forming in host cell nuclei. Host death usually occurs 10–14 days post inoculation (Grove & Hoover 2006), depending on the size of larva, virus dose and ambient temperature (Reardon *et al.* 1996). Gypsy moth larvae tend to die hanging from twigs or tree bark by the first pair of prolegs in an inverted "V" configuration. Infection destroys the cells, including dermal cells, resulting in a fragile host cadaver that breaks open or shatters at the slightest disturbance (Reardon *et al.* 1996). Once the cuticle of the cadaver is no longer intact, the polyhedra are easily scattered in the environment where they are available to be ingested by susceptible larvae.

Transmission of *Ld*MNPV within a host population has been difficult to evaluate. Occlusion bodies protect the virions to a certain extent but the virions within the OBs are still extremely sensitive to UV radiation. If protected from UV in leaf litter, under tree bark and in the soil, the virus particles are thought to survive for up to 1 year, long enough to infect the next host generation (Podgwaite *et al.* 1979). There is evidence that egg masses may be contaminated by the ovipositing female moth and larvae become infected when they hatch and ingest the polyhedra. Various studies suggested the following possibilities: 1. female adults may harbor a latent infection that is passed transovarially (Il'inykh 2007), however, Murray *et al.* (1991) found no evidence for this; 2. females become contaminated in the environment and shed polyhedra on the egg mass when they oviposit (Evans 1986); and 3. the egg masses become contaminated by polyhedra in the environment (Murray and Elkinton 1989). It is also possible that dispersing early instar larvae contact inoculum remaining from the previous year on the host trees (Woods *et al.* 1989) or in the leaf litter. Early instar larvae, which are highly susceptible to *Ld*MNPV, may transmit virus after ballooning to disperse (Dwyer & Elkinton 1995) and early instars appear to initiate the first epizootic wave when host densities are relatively high (Woods *et al.* 1991). A second wave occurs as older uninfected larvae eat virus encountered in the environment.

In addition to host-to-host transmission, birds and small mammals may have a role in transmitting *Ld*MNPV. In sites where *Ld*MNPV was naturally prevalent, several mammals and birds were shown to have viable polyhedra in their alimentary tracts (Lautenschlager *et al.* 1980, JR Reilly personal communication), and in laboratory tests, virus eliminated from two mammals infected 95% of larvae fed fecal samples (Lautenschlager & Podgwaite 1977). Parasitoids may also contribute to increased transmission of the virus (Raimo *et al.* 1977) and/or viral movement into naïve populations.

*Ld*MNPV is considered to be a strongly host density dependent pathogen and, indeed, it often produces high mortality rates in outbreak gypsy moth populations (Leonard 1974, 1981). Reilly and Hajek (2008) found that resistance of gypsy moth larvae to the virus was lower when gypsy moth population densities were high. Nevertheless, many low density gypsy moth populations have been reported to suffer high virus mortality, and high density populations may, at least initially, have a relatively low prevalence of the virus. Dwyer *et al.* (1997) showed via models and transmission experiments that *Ld*MNPV transmission based on host density is not linear. Various factors may account for this finding, such as seasonality in host reproduction, delays in host infection and death, and heterogeneity in susceptibility among hosts (Dwyer *et al.* 1997), as well as clumping of pathogens within host cadavers in the environment (D'Amico *et al.* 2005), and variability in susceptibility to viral infection at different times within each larval instar (Hoover *et al.* 2002, Grove & Hoover 2006). Although selected food plants (e.g. oaks, *Quercus* spp., which are highly preferred by gypsy moth larvae) were shown in the laboratory to increase susceptibility of gypsy moth larvae to *Ld*MNV (Keating *et al.* 1990), field experiments showed only weak effects on transmission (Dwyer *et al.* 2005) and defoliation-induced tannin production in oaks was "too little too late" to affect virus transmission (D'Amico *et al.* 1998). Until the arrival of *E. maimaiga* in North America (next section), however, *Ld*MNPV was the major factor in gypsy moth population declines in North America (Doane1970), and this pathogen still plays a significant role in natural control of the pest. Although few studies have quantified this, before the presence of *E. maimaiga* in North America (Section 11.5), epizootics caused by *Ld*MNPV commonly caused "crashes" in outbreak gypsy moth populations (Leonard 1974, 1981, Elkinton & Liebhold 1990), but the crashes often occurred after gypsy moth populations were sufficiently dense to cause defoliation.

11.4.2 Environmental Safety of **Ld*MNPV***

*Ld*MNPV is highly specific to the gypsy moth. In laboratory tests, 46 species of non-target Lepidoptera were refractory to *Ld*MNPV (Barber *et al.* 1993). In subsequent field testing, resident mammal species, both caged and free-living (Lautenschlager *et al.* 1978), and caged and free-living birds (Lautenschlager *et al.* 1979), were found to be unaffected by aerial sprays of the formulated *Ld*MNPV, Gypchek®. To our knowledge, it is the most host specific of the natural control agents of gypsy moth, rivaled only by *E. maimaiga*. Interestingly, an *Ld*MNPV "host specificity" gene, *hrf-1*, enables production of *Autographa californica* MNPV, a species with a relatively broad host range within the Noctuidae, in a nonpermissive *L. dispar* cell line (Thiem *et al.* 1996). Toxicological testing was also negative except that the Gypchek formulation was an eye irritant, probably due to insect parts in the mixture (Reardon *et al.* 1996).

11.4.3 Development of **Ld*MNPV* *as a Microbial Control Agent***

The development of *Ld*MNPV as a microbial pesticide by the USDA Forest Service and USDA APHIS among others, has been summarized in several excellent publications (see multiple authors in Ch. 6.3 and 6.5 of Doane & McManus 1981, Reardon *et al.* 1996). Recognized for its natural effects on host populations and efficacy in the laboratory, persistence in the environment, and host specificity (Lewis 1981, Lewis *et al.* 1981), the major drawback to use of *Ld*MNPV as a microbial pesticide is that the virus must be produced in living cells. Production is labor-intensive and costly and, thus, availability is low. Nevertheless, its positive characteristics justified evaluation, and *Ld*MNPV now has a 40-year history of product development; in 1978, it became the third baculovirus registered with the US Environmental Protection Agency for pest control use (Reardon & Podgwaite 1994).

Formulated *Ld*MNPV has been produced by several entities in five countries and products have included Virin-ENSh® (Russia), Biolavirus-LD® (Czech Republic), Lymantrin® (France), Disparvirus® (Canada) and Gypchek (US). None of these products are commercially available and registrations appear to have lapsed on Lymantrin, Biolavirus-LD and Disparvirus. The major issue appears to be the lack of interest on the part of non-governmental companies to transfer production of a relatively costly and labor-intensive product to the private sector (JD Podgwaite personal communication). Here, we briefly discuss the development of Gypchek, still in production and use in the US.

*Ld*MNPV is produced *in vivo* at the USDA APHIS, PPQ Otis Pest Survey, Detection & Exclusion Laboratory and processed into the Gypchek product at the USDA Forest Service Laboratory in Ansonia, Connecticut. Larvae are infected with the "Hamden" wildtype *Ld*MNPV and 14 days after inoculation cadavers are blended with water to release the virus; the mix is then filtered and centrifuged to remove insect parts. Solids are lyophilized and finely ground to make a powder with 15% virus occlusion bodies (Reardon *et al.* 1996). A range of storage methods and formulations have been tested, including a molasses-based formulation that worked well for preserving the virus and stimulating feeding but was problematic in aerial spray equipment. The current product is formulated in "Carrier 038-A", originally produced by Novo Nordisk and now supplied by OMNOVA Solutions (JD Podgwaite personal communication).

Synergistic materials have been tested for tank mixes of virus, including extensive work on optical brighteners, which were found to synergize *Ld*MNPV in the laboratory (Shapiro & Robertson 1992, Sheppard & Shapiro 1994) and in field tests (Cunningham *et al.* 1997), but which have not yet been included in the product. In addition, field experiments with the neem extract azadirachtin (Cook *et al.* 1997), the phenolic glycoside salacin from aspen (Cook *et al.* 2003) and granulovirus (Webb *et al.* 2001) have been conducted with varying degrees of success. Without commercial interest, though, the more effective enhancers have not been incorporated into Gypchek.

A major reason that the present methods for mass production of *Ld*MNPV are not more acceptable to industry is that production is costly. The virus is produced in living gypsy moth larvae; maintaining and infecting gypsy moth larvae, which have an extended obligate egg-stage diapause, makes virus production relatively complex. At present, the USDA Forest Service laboratory at Delaware, Ohio, has been investigating larger scale production of *Ld*MNPV in cell culture. Recent genetic studies of *Ld*MNPV have aimed at identifying strains that are potent and stable in cell culture. While *in vitro* production would produce a cleaner product without the necessity of rearing gypsy moths, problems with genetic mutants from the wildtype virus have occurred that lowered the production of occluded virus (Bishoff and Slavicek 1996, Slavicek *et al.* 1998). A genetic variant (A21-MPV) that is stable in cell culture has been isolated from the genetically variable wildtype virus by Slavicek *et al.* (1996). *Ld*MNPV was the first nucleopolyhedrovirus for which enhancin genes, previously reported from granuloviruses, were found (Bischoff & Slavicek 1997). Two genes have been identified and both contribute to potency of the virus (Popham *et al.* 2001).

11.4.4 Applications for Gypsy Moth Control

Various formulations of *Ld*MNPV have been tested extensively for efficacy in the field and compared to Dimilin® and *Btk* products. Reardon *et al*. (1996) summarized the aerial and ground spray testing during the years 1989–1993 in which the best formulations reduced egg mass numbers by 88–98%, comparing favorably with Dimilin and *Btk*, which are less host specific. The current recommendation for eradication using Gypchek is two or more aerial treatments 3 days apart at 5×10^{11} OB in 1 gal 038-A carrier mix/acre (9.34 L/ha) ($= 1.24 \times 10^{12}$ OB/ha). For suppression, the recommendation is either two applications 3 days apart of 2×10^{11} OB in 1 gal/acre (4.94×10^{11} OB/ha), or one application of 4×10^{11} OB in 1 gal/acre (9.88×10^{11} OB/ha). Some land managers prefer to use 0.5 gal/acre (4.67 L/ha) to reduce costs. The recommendation for ground application is 1×10^{12} OB in 100 gal (379 L) aqueous formulation (with 2% Bond sticker)/acre ($= 2.47 \times 10^{12}$ OB/ha). Ground applications are not commonly carried out (JD Podgwaite personal communication).

In the past 14 years, areas sprayed with Gypchek averaged approximately 7,800 acres (3.157 ha), with lows of less than 2,000 acres (810 ha) per year in 1995 and 1996, and over 15,000 acres (6,073 ha) in 2003–2005 (USDA Forest Service data, JD Podgwaite). Of all the microbial pesticides available for gypsy moth control, *Ld*MNPV has the least impacts (virtually none) on the environment and is strongly recommended for environmentally sensitive areas (Schweitzer 2004). It is hoped that this product will have a permanent future in gypsy moth control programs, but efforts to produce Gypchek and other *Ld*MNPV products remain stymied by the cost and labor intensive nature of the present production methodologies.

11.5 Fungal Pathogen: *Entomophaga maimaiga*

Several species of fungal entomopathogens native to North America have been reported infecting gypsy moth larvae (see Hajek *et al*. 1997). All are anamorphs of Hypocreales and none have been known to cause epizootics in North American *L. dispar* populations; maximum infection levels recorded among native fungal pathogens was < 13% for *Isaria* (= *Paecilomyces*) *farinosa*. Since 1989, by far the most abundant fungal pathogen in North America has been *Entomophaga maimaiga*, a species that originated in Asia.

11.5.1 How Did E. maimaiga *Reach North America?*

In 1909, G. P. Clinton traveled to Japan where he worked near Tokyo to collect specimens of an entomophthoralean pathogen infecting gypsy moth larvae. He returned to Cambridge, Massachusetts with two diseased larvae. Scientists at Harvard University amplified this fungal isolate in gypsy moth larvae and then released infected larvae at six sites near Boston in 1910–1911 (Speare & Colley 1912). Speare and

Colley reported in 1912 that they detected no transmission of the "gypsy fungus" in the field, but that the resting spores could survive the northeastern winter. They suggested that virulence was not optimal in the strain of the fungus they were studying. Subsequent surveys of gypsy moth pathogens from 1912 to 1989 never detected entomophthoralean fungi infecting northeastern gypsy moths (Hajek et al. 1995). In 1984, R.S. Soper and M. Shimazu isolated an entomophthoralean pathogen of gypsy moth from western Honshu, Japan, and named the species *Entomophaga maimaiga* (Soper et al. 1988). This isolate was released in Allegany State Park, New York, in 1985 and in Shenandoah National Park, Virginia, in 1986 but no infections were found at release sites and the isolate evidently failed to establish (Hajek 1999).

In 1989, gypsy moth populations increased in parts of the northeastern U.S. and many dead late instar larvae were found attached to tree trunks. Although it is typical to find cadavers of late instar larvae hanging on tree trunks during and after LdMNPV epizootics, an LdMNPV epizootic had not been expected. Gypsy moth populations were not very dense at that time and LdMNPV epizootics characteristically occur only when gypsy moth population density is relatively high. Upon dissection of cadavers, resting spores of an entomophthoralean fungus were discovered. An epizootic occurred and the fungal pathogen was detected in seven northeastern states during the exceptionally rainy spring in 1989. During the following years it became evident that epizootics also occurred during springs when rainfall levels were normal for the Northeast, and *E. maimaiga* spread naturally and with some purposeful introductions across the contiguous northeastern gypsy moth distribution (Hajek 1999).

Molecular studies confirmed that the fungus causing epizootics in North American gypsy moth since 1989 is *E. maimaiga*, which is native to Japan, Far Eastern Russia and northeastern China (Nielsen et al. 2005). A recurring question has been whether or not the fungus released early in the century actually became established and, if it did, why it was not detected until 1989? Amplified fragment length polymorphism (AFLP) assays clearly show that the fungus now established in North America came from Japan. It is not most closely related to the isolates of *E. maimaiga* from the Tokyo area, so the suggestion that *E. maimaiga* in North America today was initially introduced in 1910–1911 and remained at low densities or quiescent for decades is not supported. Molecular results also do not support the suggestion that *E. maimaiga* became established in North America from 1985 to 1986 releases; North American isolates are in a clade distinct from the isolate released in 1985–1986 and the amount of variability in AFLP haplotypes per locality over time is relatively small. In the AFLP study, isolates from only five locations in Japan were available and it seems most likely that an isolate from the source location for the North American introduction was not included. Although we do not know how *E. maimaiga* was introduced to North America, we have some ideas about when and where it was introduced. Models of rainfall and gypsy moth density suggest that *E. maimaiga* would have been observed in 1945 and 1971 if it had been present, suggesting that this fungus became established in North America more recently than 1971 (Weseloh 1998). Also, Connecticut was the center of the 1989 epizootic, suggesting that this was the region of introduction.

11.5.2 Basic Biology and Epizootiology

Entomophaga maimaiga is an obligate pathogen that infects the larval stage of gypsy moth. After an infective spore lands on a larva, the fungus penetrates through the larval cuticle to infect; time from infection to death depends on instar, dose, temperature and fungal isolate but often ranges between 4 and 7 days at 20°C (Hajek 1999). *E. maimaiga* produces two types of spores, relatively short-lived conidia that are actively ejected from cadavers, and long-lived azygospores (= resting spores) that are produced within cadavers. Early instars killed by *E. maimaiga* generally die within the tree canopy holding onto twigs and branches and conidia are ejected from these cadavers. Later instars killed by *E. maimaiga* generally die attached to tree trunks, and azygospores produced within these cadavers are eventually deposited in the soil when cadavers decompose after falling to the ground. Azygospores in the soil germinate only in spring when moisture is available. They produce infective germ conidia but only some of the azygospores germinate each year, creating a reservoir of resting spores in the soil.

E. maimaiga is well known for its ability to cause epizootics that can result in control of gypsy moth populations (e.g. Hajek 1997, Webb *et al.* 1999), even acting to prevent incipient outbreaks (Elkinton 2003). Titers of resting spores in the soil at bases of trees can be very high, especially after epizootics, and these resting spores begin germinating at the approximate time that first instar larvae are dispersing. Second and third instar larvae remain in the canopy feeding while healthy late stage instars walk down trunks to rest in the leaf litter during daylight hours, climbing to the tree canopy at dusk and back down at dawn; it is postulated that resting spore infections predominantly occur when larvae are on or near the soil. Infections initiated by resting spores produce conidia after larvae die and these conidia then infect other larvae. Simulation models of the *E. maimaiga*/gypsy moth system have estimated that 4–9 cycles of infection (from infection of one larva to infection of the next) occur in one year (Hajek *et al.* 1993). In the field, we find that few early instars are infected but levels of infection increase with increasing larval instar, and the majority of infections are seen in late instars, the stage that would normally cause the majority of defoliation.

During individual years or at individual sites, high levels of infection have been recorded from both low (Hajek *et al.* 1990) and high density gypsy moth populations (Hajek 1997). At different times researchers have found that levels of infection by *E. maimaiga* were associated with gypsy moth density (Webb *et al.* 2004) or not (Elkinton *et al.* 1991, Webb *et al.* 1999). Weseloh (2004) found that larval density was more important to modeling infection levels than was resting spore load in the forest. We hypothesize that the amount of horizontal transmission of *E. maimaiga* in gypsy moth populations is influenced by the amount of inoculum present in the area (e.g. density of airborne conidia and/or density of resting spores in the soil) as well as the amount of moisture and the temperatures in the gypsy moth/*E. maimaiga* microhabitats. Numerous studies have clearly demonstrated that moisture is critically important for infection to occur (Hajek 1999).

Aerial sampling revealed that episodes of airborne conidia occur during epizootics (Hajek *et al.* 1999) and some conidia remain in local areas (e.g. Hajek *et al.* 1996b). However, between 1989 and 1992, *E. maimaiga* spread rapidly across the contiguous northeastern distribution of gypsy moth, at >100 km/year. It has been assumed that the majority of this spread was due to long distance dispersal by conidia during favorable conditions for conidial production and survival. Models confirm that mechanisms for short and long distance conidial dispersal differ (e.g. Weseloh 2003b).

E. maimaiga now plays a pivotal role in population dynamics of gypsy moth in North America. Because this pathogen has prevented outbreaks and caused crashes of outbreak populations, the public and land managers would like to be able to predict epizootics in order to prevent costly insecticide applications. However, once present in an area, the activity of *E. maimaiga* is dependent to a large extent on rainfall and temperature and it is difficult to predict the weather far enough ahead of the gypsy moth season to impact management decisions. N. Siegert (personal communication) worked with climate models to predict the frequency of occurrence of *E. maimaiga* epizootics in the North Central states where gypsy moth populations are spreading. Long-term weather data for the North Central states were matched with the weather conditions during 11 epizootics from 1989 to 1999. Results from these studies suggest that epizootics may not occur as frequently in Minnesota, Wisconsin and Michigan as in more southern states of this region.

11.5.3 *Environmental Safety of* E. maimaiga

Numerous studies have been conducted to investigate the host specificity of *E. maimaiga*. Initial laboratory bioassays established that this fungus can only infect lepidopteran larvae (Hajek *et al.* 2003). During laboratory bioassays using high dosages and optimal conditions, *E. maimaiga* infected a diversity of lepidopteran species at low levels; the only insect family with consistently high levels of infection was the Lymantriidae, the family that includes gypsy moth. Rearing lepidopteran larvae collected from foliage in the forest revealed almost no infection in non-target lepidopteran larvae while high percentages of gypsy moth larvae were infected (Hajek *et al.* 1996a). Because *E. maimaiga* resting spores reside in the upper layers of the soil, often at high titers, we also sampled and reared lepidopteran larvae in the leaf litter. Results were similar to those from foliage-feeding non-target lepidopteran larvae; almost no infection was found in any species except gypsy moth. Although laboratory assays had demonstrated that lymantriids were particularly at risk, few lymantriids were present during studies of lepidopteran larvae on foliage or in the litter. A 5-year study followed (1997–2001), specifically comparing infection levels of native lymantriid larvae with gypsy moth larvae in sites in Virginia and West Virginia (Hajek *et al.* 2004). While native lymantriid populations were always low, of the seven species of endemic lymantriids collected, *E. maimaiga* infection was only found in three species and was always <50% per species per year. Our studies clearly demonstrated that the physiological host range documented in the laboratory was not similar to infection levels found in the field (the ecological host range), even for native North American lymantriids, the most susceptible species in the laboratory. Field-cage studies demonstrated that gypsy moth larvae were most at risk of infection when caged on the soil at bases of trees (Hajek 2001). We hypothesize that *E. maimaiga* infection levels are low in non-target species because few lepidopteran larvae, particularly native lymantriids, spend much time at the soil surface, although this is a common habitat for late instar gypsy moth larvae.

11.5.4 Applications for Gypsy Moth Control

Since 1989, the public and land managers have been interested in obtaining *E. maimaiga* for introduction. The stage that is most practical for production, distribution and release is the resting spore. However, production of resting spores has some hurdles that have not been overcome for any species of entomophthoralean fungi. *E. maimaiga* resting spores can be produced in the laboratory *in vivo* and, for only some isolates, *in vitro* (Kogan & Hajek 2000). Neither process presently allows efficient production of massive quantities of resting spores. After maturation, *E. maimaiga* resting spores enter constitutive dormancy that lasts from at least 10 months to over 10 years.

The principal use of *E. maimaiga* resting spores has been to release them in areas newly colonized by gypsy moth where this fungus does not yet occur, because gypsy moth populations are continuing to spread to the west and south. Initially, the major source of *E. maimaiga* resting spores for microbial control was soil at the bases of trees after epizootics but, more recently, late instar cadavers have been collected during epizootics for redistribution of resting spores. Appropriate permits should be obtained whenever introductions to new areas are planned. Cadavers or soil containing *E. maimaiga* resting spores have been released in at least 10 states as well as in Bulgaria, Russia and possibly Romania (Hajek *et al.* 2005, AE Hajek unpublished data). Releases in the US have usually resulted in establishment of *E. maimaiga*, even though relatively small numbers of resting spores were released (e.g. Smitley *et al.* 1995, Hajek *et al.* 1996b). Studies are presently under way to investigate the time period needed for *E. maimaiga* to spread on its own into areas newly colonized by gypsy moth.

Limited studies have also been conducted to evaluate augmentative releases of *E. maimaiga* in areas where this fungus already occurs. The goal of these releases was to add to the titer of *E. maimaiga* inoculum in the area so that infection levels would build up earlier in the season. A study by Hajek and Webb (1999) demonstrated a trend toward higher levels of infection in resting spore-amended plots, but results were variable across plots and the gypsy moth population throughout the area then declined. Webb *et al.* (2004) also reported a trend toward increased infection when watering resident resting spores in the soil. The augmentative approach has not been investigated further, because gypsy moth populations throughout the northeast have generally been relatively low since 1989 but also because *E. maimaiga* resting spores are not generally available. It is predicted that with the presence of

E. maimaiga, gypsy moth outbreaks will still occur but that outbreak population densities will be lower than in the past (Weseloh 2003a).

11.6 Microsporidia

Several species of microsporidia, single-celled eukaryotic pathogens, are common in gypsy moth populations throughout Central and Eastern Europe but have never been found in North American populations (Podgwaite 1981, Solter & Maddox 1998). These organisms possess a unique morphology and biology, and have a decidedly convoluted taxonomic history. Considered to be protozoans until the late 1990s, analyses of several genes have clearly shown that microsporidia are not closely related to other protozoan taxa (Adl *et al.* 2005) and they are currently considered to be a basal group in or near the Kingdom Fungi (Gill & Fast 2006). Over 1,200 species have been described and the majority, over 700 species, were isolated from insect hosts (Wittner 1999, Becnel & Andreadis 1999).

Although microsporidia are a diverse group of organisms with life cycles that vary from relatively simple to quite complex, all species have some characteristics in common that limit their use as pesticidal microbial control agents but enhance their potential as natural enemies of pest insects. Microsporidia are obligate pathogens; they only reproduce and mature within living cells. The mechanisms of spread within the host are still not well understood, although the unusual method of host cell invasion is the major characteristic that defines the taxon. Nearly all species possess a polar filament that is coiled within the mature, infective spore. When the infective spore is in the appropriate medium, usually the gut lumen of the host following ingestion, the spore germinates by everting the filament. The contents of the spore are "injected" through the filament into the cytoplasm of the host cell (reviewed by Keohane & Weiss 1999).

A simple life cycle of a terrestrial microsporidium involves germination into the host cell and intracellular vegetative reproduction with one or more divisions before sporulation. After sporulation occurs, the infective "environmental" spores are released into the environment in the feces, or possibly silk, of living infected hosts and/or from decomposing cadavers of hosts that have died from infections. Spores may survive in the environment for some period of time until ingestion by a susceptible host.

Microsporidia typically cause chronic disease in a host (Gaugler & Brooks 1975), but some species are relatively virulent, causing death of most hosts during the larval stages. Species that are transmitted from the female host to offspring in the egg or embryo frequently cause high mortality in neonate host larvae. Other genera may be more benign; for example, some species infect only the midgut tissues and are readily disseminated into the environment via the feces. Larvae infected as early instars by the more chronic microsporidian species may die (Siegel *et al.* 1986), but larvae infected as mid- to late instars may survive to pupate, eclose and reproduce. The more virulent species, frequently infecting the important metabolic fat body tissues, appear to be transmitted primarily via decomposing host cadavers, which are completely filled with infective spores released into the environment.

11.6.1 History and Description of Gypsy Moth Microsporidia

The first published record of microsporidia from the gypsy moth was a description of *Pleistophora schubergi* by Zwölfer (1927). This species was eventually re-described and placed in the genus *Endoreticulatus* (Cali & El Garhy 1991). Subsequently, several *Endoreticulatus* isolates were recovered from gypsy moth, including a cryptic species from Portugal that may have been an adventitious or non-target infection in the gypsy moth (Solter *et al.* 1997), and an isolate from Asenovgrad, Bulgaria, that appears to be a true gypsy moth pathogen (Solter *et al.* 2000).

Several other microsporidian isolates were identified by European scientists in the 1950s, all of them in the genera *Nasema* and *Vairimorpha* (Table 11.1). The first attempts to describe what would become known as the gypsy moth *Nosema/Vairimorpha* species complex were complicated by understandable confusion over the multiple spore types produced by these very closely related species (David *et al.* 1989, David & Novotny 1990, J Weiser & A Linde personal communication). This species complex should probably be assigned to one genus based on their close phylogenetic relationship (Baker *et al.* 1994). *Nosema lymantriae*, *Nosema portugal* and, presumably, *Nosema serbica*, a species that is only available on stained slides, all produce two spore types (Maddox *et al.* 1999). The first is an internally infective spore that develops early in the infection process and probably germinates in the tissues of the host to infect adjacent cells. The second spore type is the mature and infective, "environmental" spore. *Vairimorpha disparis* produces an additional spore type that is haploid and monokaryotic, eight of which are enclosed in a membrane formed by the parasite (Vavra *et al.* 2006). These forms are called octospores and are probably a meiospore or sexual form (Ironside 2007).

Table 11.1 Microsporidia described from *Lymantria dispar*[a]

Originally described species name	Currently recognized species name
Thelohania disparis Timofejeva 1956	*Vairimorpha disparis* (Timofejeva 1956)
Thelohania similis Weiser 1957	(now *V. disparis*)
Nosema muscularis Weiser 1957	(spore form of *V. disparis* & *N. lymantriae*)
Nosema lymantriae Weiser 1957	*Nosema lymantriae* Weiser 1957
Nosema serbica Weiser 1964	*Nosema serbica* Weiser 1964
Nosema portugal Maddox & Vavra 1999	*Nosema portugal* Maddox & Vavra 1999
Pleistophora schubergi Zwolfer 1927	now *Endoreticulatus schubergi* (Zwolfer 1927)

[a]Modified from Maddox *et al.* (1999) and McManus and Solter (2003).

11.6.2 Field Observations and Exploration

Prevalence of microsporidia in field populations of the gypsy moth and effects on the host populations have been reported by researchers in several areas of Europe. Prevalence in natural populations fluctuates widely, but is often maintained at low enzootic levels (Pilarska *et al.* 1998) and appears to be strongly host density dependent, similar to *Ld*MNPV. European researchers have reported very high prevalence of various microsporidian species, and also high host mortality (Romanyk 1966, Sidor 1976, Purrini & Skatulla 1978, Zelinskaya 1980, 1981, Glowacka-Pilot 1983, Sierpinska 2000). In the only published long-term study of microsporidian prevalence in field populations of the gypsy moth, Pilarska *et al.* (1998) recorded the prevalence of three species of microsporidia, *Endoreticulatus* sp. (now known to be *E. schubergi*) (Wang et al. 2005), *Nosema* sp. (= *lymantriae*) and *Vairimorpha* sp. (= *disparis*), each of which occurred alone in a relatively isolated host population in Bulgaria. Data for *V. disparis* spanned 15 years and provided insights into the ability of the microsporidium to be maintained at a very low prevalence in the host population between outbreaks.

Between 1985 and 2005, collaborators in several European countries and the US participated in an exploration program for European and Asian gypsy moth microsporidia (Jeffords *et al.* 1986, McManus & Solter 2003). No microsporidia were found in Asian gypsy moth populations (Siberia) but more than 25 isolates were recovered from European populations in Austria, Bulgaria, Czech Republic, Germany, Hungary, Poland, Portugal, Romania, and Slovakia. The collected isolates have provided material for extensive field and laboratory studies on the taxonomy, biology, host specificity, genetic relationships, and transmission of gypsy moth microsporidia. The microsporidian isolates from the gypsy moth that have been manipulated experimentally in the laboratory and in the field include *V. disparis, N. portugal, N. lymantriae*, and *E. schubergi*. These isolates were produced in gypsy moth larvae in the laboratory, the rDNA has been sequenced, and they are stored in liquid nitrogen.

11.6.3 Field Studies in Europe and the United States

Weiser and Novotny (1987) reported the first attempts to manipulate microsporidia in field populations of the gypsy moth. The authors pointed out the difficulty of measuring the effects of a chronic pathogen (in this case, *Nosema lymantriae*) in the field, particularly when other disease organisms and parasites were present. The effect of microsporidian treatment was, however, calculated at nearly a 79% increase in mortality by the time of adult eclosion. Progeny produced in the treated plot were reduced to 10% of the numbers produced in the control plot. David and Novotny (1990) produced the first evidence from a field study that *Nosema lymantriae* is quite host specific, even in a confined space (sleeve cages) saturated by infective spores.

In the US, Jeffords *et al.* (1989) evaluated persistence of microsporidia in a gypsy moth host population in Maryland. *Vavraia* sp. (= *Endoreticulatus* sp.,) and *Nosema* sp. (= *N. portugal*) were released via contaminated egg masses into natural gypsy moth populations in isolated woodlots, one isolate per site. No infected larvae were recovered the following year from the site where *Endoreticulatus* sp. was released; *Nosema portugal*, however, was recovered from 8% of early stage larvae, 13% of late stage larvae, and 10% of adults the year following release.

11.6.4 Microsporidia for Classical Biological Control of the Gypsy Moth

Extensive laboratory and field studies have been conducted since the mid-1990s to elucidate the biology of the gypsy moth microsporidia. These studies were intended to provide the basic information needed to obtain permission to release these pathogens for classical biological control of gypsy moth in North America. In addition, these studies were conducted to determine the effects of these pathogens on gypsy moth populations in Europe, where they occur naturally and might be used in augmentative release programs. More stringent rules for inoculative release for classical biological control are now in effect in the US, and permission would not have been granted based on the 1986 release in Maryland. Research cooperators in the US and Europe agreed to share isolates, standardize pathogen dosages when possible, use gypsy moth larvae from the colony at the United States Department of Agriculture, Animal & Plant Health Inspection Service for laboratory and field studies, and use the same laboratory rearing regime.

11.6.4.1 Transmission of Gypsy Moth Microsporidia

The mechanisms of microsporidian transmission in host populations can usually be predicted based on the host tissues that the pathogen utilizes for development; for example, when gut tissues are infected, infective spores are usually found in the host feces and horizontal transmission among larvae by ingestion of infective spores is assumed to occur. Likewise, when the gonads of larval and adult female hosts are infected, transovarial transmission (vertical transmission from infected female to the offspring) usually occurs. *N. lymantriae* targets the fat body, Malpighian tubules, and gonads of the larval host and is transmitted both horizontally and vertically (Goertz *et al.* 2007, Goertz & Hoch 2008). *V. disparis*, primarily a fat body pathogen, is transmitted mainly by release of spores from decomposing cadavers. Although *V. disparis* also infects the gonads, there is no indication that transovarial infection occurs, even in the rare instances when an infected female survives to pupate and eclose (Goertz & Hoch 2008). It is difficult to discern from the literature on persistence and prevalence of *N. lymantriae* and *V. disparis* whether vertical transmission or horizontal transmission via cadavers is more successful. In Europe, both species appear to persist at low enzootic levels and increase to significant prevalence when host densities increase (Sidor 1976, Zelinskaya 1980, Sierpinska 2000, Pilarska *et al.* 1998).

11.6.4.2 Virulence Studies

The most virulent microsporidium infecting gypsy moth is *V. disparis* (Solter *et al.* 1997), and no differences in virulence or rDNA sequences have been noted among *V. disparis* isolates. Different isolates of *N. lymantriae*, however, vary slightly both in virulence (Goertz *et al.* 2004, 2007) and in rDNA sequence.

11.6.4.3 Environmental Safety

The host specificity of *V. disparis* and *N. lymantriae* was evaluated in a series of laboratory and field experiments. Laboratory studies involved testing 50 lepidopteran species native to eastern North American forests for susceptibility to *V. disparis, N. portugal*, and two *N. lymantriae* isolates (Solter *et al.* 1997). Although all four isolates could initiate infections in some of the non-target hosts, the infections, particularly those of the *Nosema* species, were noted to be atypical compared with infections observed in the gypsy moth host. Further studies using the gypsy moth as a model non-target host challenged with microsporidia from native forest Lepidoptera from the northeastern US showed that these atypical infections are not transmitted to conspecific hosts (Solter & Maddox 1998). Solter *et al.* (2000) then surveyed non-target Lepidoptera feeding with gypsy moth populations in Bulgaria. Microsporidia were recovered from non-target species but the three gypsy moth pathogens were only found in gypsy moth. It appears that other lepidopteran species do not serve as reservoir hosts. Recently, studies were conducted in Slovakia to determine if release of *V. disparis* and *N. lymantriae* into naïve gypsy moth populations would result in non-target infections (LF Solter unpublished data). Spores were sprayed onto the foliage of low-growing oak trees in concentrations calculated to infect 100% of gypsy moth larvae in the laboratory. *V. disparis* infected several non-target individuals but in low numbers, and follow-up monitoring for 2 years revealed no further infections in the non-target population. *N. lymantriae* did not infect non-target species, even when spore concentrations were quadrupled.

11.6.4.4 Release and Persistence Monitoring

Inoculative releases of *V. disparis, N. portugal* and *N. lymantriae* with the goal of permanent establishment are planned for May 2008 under an USDA APHIS PPQ Permit (L Solter & M McManus). The pathogens will be released in living infected third instar larvae to facilitate spread of infective spores in the environment via feces and cadavers. This method also reduces exposure of native non-target insects while protecting the microsporidian spores from ultraviolet radiation, which quickly kills them. A release of *V. disparis* and *N. lymantriae* is also planned for south central Bulgaria (D Pilarska & A Linde personal communication), where only *E. schubergi* has been recovered. The populations in both countries will be monitored for persistence of the microsporidia and for susceptibility of non-target species. These releases are intended to augment the natural enemy complex of the gypsy moth.

11.7 Pathogen and Parasitoid Interactions in Gypsy Moth Populations

Competitive displacement of natural enemies by additional introduced or augmented natural enemies is an important factor to consider in biological control programs. Because *Btk* is primarily used as a biopesticide and persistence of the toxins is relatively low (Sundaram *et al.* 1994), interaction with most gypsy moth natural enemies is probably indirect, but *Btk* sprays can alter the progression of *E. maimaiga* and *Ld*MNPV epizootics by reducing the density of the host (Mott & Smitley 2000). Laboratory studies have demonstrated that sublethal doses of *Btk* reduced percentage parasitism by the tachinid *Compsilura concinnata*, as well as reducing the size of emerging parasitoids (Erb *et al.* 2001). The effect of one hymenopteran parasitoid, *Cotesia melanoscelus*, however, was synergized by *Btk* sprays. This parasitoid only parasitizes smaller larvae that develop quite rapidly, so *C. melanoscelus* normally has only a narrow window for parasitism. Apparently the slower development of larvae due to sublethal *Btk* dosages provides a longer window for *C. melanoscelus* parasitism (Weseloh *et al.* 1983).

Studies on interactions of *Ld*MNPV with other natural enemies have shown that parasitoids may be affected by virus-infected hosts to varying degrees; for example, aerial sprays of Gypchek resulted in reduced parasitism by *C. melanoscelus* (Webb *et al.* 1989). However, Reardon & Podgwaite (1976) found a positive correlation between *Ld*MNPV infection and levels of parasitism by *C. melanoscelus* and the tachinid *Parasetigena silvestris*. Tachinid parasites were particularly tolerant of *Ld*MNPV-infected gypsy moth larvae; however more parasitoids emerged from gypsy moth larvae collected 19 days after ground sprays of Gypchek than from larvae collected after 28 days, when host larvae were older and many were dying of *Ld*MNPV infections (White & Webb 1994).

The only study of interactions between *E. maimaiga* and parasitoids indicated that relative timing of infection and parasitism affects the outcome. Parasitism by the tachinid *C. concinnata* was higher for control larvae than for fourth instars infected with *E. maimaiga* 1 day prior to exposure to parasites, although parasitism of controls was less than larvae infected with *E. maimaiga* 4 days prior to parasitoid exposure. The authors hypothesized that differential larval movement at different stages of *E. maimaiga* infection could help to explain results (Smith 1999).

E. maimaiga generally kills larvae more quickly than *Ld*MNPV and, while both pathogens can co-infect and reproduce in the same cadaver, the relative success of each depends on the sequence and relative timing of infection. Malakar *et al.* (1999) found that the first *Ld*MNPV epizootic wave in young gypsy moth larvae was not affected by *E. maimaiga*, which is more prevalent in older larvae, but virus infection was reduced by activity of *E. maimaiga* in the second wave.

Bauer *et al.* (1998) determined that, unlike the antagonistic effects reported for other NPV-microsporidia interactions (reviewed in Solter & Becnel 2003), *Ld*MNPV infections were synergized by prior infection of gypsy moth larvae with *Nosema portugal*. Although more rapid host mortality negatively affected reproduction of one or both pathogens depending on the sequence of inoculation, the results correspond with older reports in Eastern Europe of gypsy moth populations that collapsed, with these two pathogens playing major roles in the same populations (Zelinskaya 1980, 1981).

Competitive interactions of gypsy moth microsporidia versus parasitoids were studied, as well as competition among different species of microsporidia infecting gypsy moth larvae. In laboratory studies, *Vairimorpha disparis* (*Vairimorpha* sp.)

negatively affected larvae of the parasitoid *Glyptapanteles liparidis* (Hoch *et al.* 2000, Hoch *et al.* 2002), but results from field studies suggested that, at the prevalence of *V. disparis* typically found in Europe, there was little effect on the parasitoid complex (Hoch *et al.* 2001, G Hoch personal communication). Solter *et al.* (2002) and Pilarska *et al.* (2006) found competition between the microsporidia *V. disparis* and *N. lymantriae*, both fat body pathogens, but not between either of these two microsporidia and *E. schubergi*, a gut pathogen. A review of the literature (Solter *et al.* 2002) and evidence from the field (Pilarska *et al.* 1998, Solter *et al.* 2000) suggest that *V. disparis* and *N. lymantriae* rarely occur in the same host population, but each may be found occurring on occasion with *E. schubergi*. The authors suggest that *V. disparis* and *N. lymantriae* not be released in the same sites (Solter *et al.* 2002, Pilarska *et al.* 2006).

11.8 Conclusions

The gypsy moth is not a typical invading pest; even in undisturbed native habitat, it is an outbreak species that periodically causes serious defoliation of its preferred hosts. Despite a rich complex of natural enemies, the gypsy moth manages to reproduce to levels that result in one or more years of defoliation before populations are controlled. In North America, the gypsy moth is particularly successful because of the quantity of oak forest habitat available and the lack of a full complex of predators, parasites and pathogens. The use of *Btk* and *Ld*MNPV as microbial insecticides, and the accidental introductions of *Entomophaga maimaiga* and *Ld*MNPV, have provided excellent tools for preventing some outbreaks, controlling outbreaks, and slowing the spread of this well-established pest. Perhaps microsporidian species, introduced as classical biological control agents, will provide one more type of natural enemy to suppress population outbreaks. Nearly two-thirds of the susceptible forests in the US remain free of gypsy moth attack. It is hoped that a combination of resources, including pathogens, will prevent tree mortality and changes in forest composition due to gypsy moth as well as the nuisance caused by outbreak populations.

References

Adl SM, Simpson AGB, Farmer MA, Andersen RA, Anderson OR, Barta JR, Bowser SS, Brugerolle G, Fensome RA, Fredericq S, James TY, Karpov S, Kugrens P, Krug J, Lane CE, Lewis LA, Lodge J, Lynn DH, Mann DG, McCourt RM, Mendoza L, Moestrup Ø, Mozley-Standridge SE, Nerad TA, Shearer CA, Smirnov AY, Spiegel FW, Taylor MFJR (2005) The new higher level classification of eukaryotes with emphasis on the taxonomy of protists. J Euk Microbiol 52:399–451

Appel HM, Schultz JC (1994) Oak tannins reduce effectiveness of Thuricide (*Bacillus thuringiensis*) in the gypsy moth (Lepidoptera: Lymantriidae). J Econ Entomol 87:1736–1742

Aronson AI, Beckman W, Dunn P (1986) *Bacillus thuringiensis* and related insect pathogens. Microbiol Rev 50:1–24

Audtho M, Valaitis AP, Alzate O, Dean DH (1999) Production of chymotrypsin-resistant *Bacillus thuringiensis* Cry2Aa1 delta-endotoxin by protein engineering. Appl Environ Microbiol 65:4601–4605

Baker MD, Vossbrinck CR, Maddox JV, Undeen AH (1994) Phylogenetic relationships among *Vairimorpha* and *Nosema* species (Microspora) based on ribosomal RNA sequence data. J Invertebr Pathol 64:100–106

Barber KN, Kaupp WJ, Holmes SB (1993) Specificity testing of the nuclear polyhedrosis virus of the gypsy moth, *Lymantria dispar* (L.) (Lepidoptera: Lymantriidae). Can Entomol 125: 1055–1066

Barry JW, Skyler PJ, Teske ME, Rafferty JA, Grimm BS (1993) Predicting and measuring drift of *Bacillus thuringiensis* sprays. Environ Toxicol Chem 12:1977–1989

Bauer LS, Miller DL, Maddox JV, McManus ML (1998) Interactions between a *Nosema* sp. (Microspora: Nosematidae) and nuclear polyhedrosis virus infecting the gypsy moth, *Lymantria dispar* (Lepidoptera: Lymantriidae). J Invertebr Pathol 74:147–153

Becnel JJ, Andreadis TG (1999) Microsporidia in insects. In: Wittner M, Weiss LM (eds) The microsporidia and microsporidiosis. Amer Soc Microbiol Press, Washington, D.C. pp 447–501

Bishoff DS, Slavicek JM (1996) Characterization of the *Lymantria dispar* nucleopolyhedrovirus *25K FP* gene. J Gen Virol 77:1913–1923

Bishoff DS, Slavicek JM (1997) Molecular analysis of an *enhancin* gene in the *Lymantria dispar* nuclear polyhedrosis virus. J Virol 71:8133–8140

Blissard GW, Rohrmann GF (1990) Baculovirus diversity and molecular biology. Ann Rev Entomol 35:127–155

Boulton TJ (2004) Responses of nontarget Lepidoptera to Foray 48B®, *Bacillus thuringiensis* var. *kurstaki* on Vancouver Island, British Columbia, Canada. Environ Toxicol Chem 23:1297–1304

Boulton TJ, Otvos IS, Halwas KL, Rohifs DA (2007) Recovery of nontarget Lepidoptera on Vancouver Island, Canada: one and four years after a gypsy moth eradication program. Environ Toxicol Chem 26:738–748

Bravo A, Gill S, Soberón M (2007) Mode of action of *Bacillus thuringiensis* Cry and Cyt toxins and their potential for insect control. Toxicon 49:423–435

Broderick NA, Goodman RM, Handelsman J, Raffa K (2003) Effect of host diet and insect source on synergy of gypsy moth (Lepidoptera: Lymantriidae) mortality to *Bacillus thuringiensis* subsp. *kurstaki* by zwittermicin A. Environ Entomol 32:387–391

Broderick, NA, Raffa KF, Handelsman J (2006) Midgut bacteria required for *Bacullus thuringiensis* insecticidal activity. Proc Natl Acad Sci USA 103:15196–15199

Cali A, El Garhy M (1991) Ultrastructural study of the development of *Pleistophora schubergi* Zwölfer, 1927 (Protozoa, Microsporida) in larvae of the spruce budworm, *Choristoneura fumiferana* and its subsequent taxonomic change to the genus *Endoreticulatus*. J Protozool 38:271–278

Cameron EA, Reeves RM (1990) Carabidae (Coleoptera) associated with gypsy moth, *Lymantria dispar* L. (Lepidoptera: Lymantriidae) populations subjected to *Bacillus thuringiensis* Berliner treatments in Pennsylvania USA. Can Entomol 122:123–130

Cate JR, Hinkle MK (1994) Integrated pest management: The path of a paradigm. Natl Audubon Soc Special Rep. 43 pp

Charles J-F, Delécluse A, Nielsen-Le Roux C (eds) (2000) Entomopathogenic bacteria: From laboratory to field application. Kluwer Acad Publ, Dordrecht, Netherlands

Cook SP, Webb RE, Thorpe KW, Podgwaite JD, White GB (1997) Field examination of the influence of azadirachtin on gypsy moth (Lepidoptera: Lymantriidae) nuclear polyhedrosis virus. J Econ Entomol 90:1267–1272

Cook SP, Webb RE, Podgwaite JD, Reardon RC (2003) Increased mortality of gypsy moth *Lymantria dispar* (L.) (Lepidoptera: Lymantriidae) exposed to gypsy moth nuclear polyhedrosis virus in combination with the phenolic glycoside salicin. J Econ Entomol 96:1662–1667

Couch TL (2000) Industrial fermentation and formulation of entomopathogenic bacteria. In: Charles JF, Delécluse A, Nielsen-LeRoux C (eds) Entomopathogenic bacteria: From laboratory to field application. Kluwer Academic Publishers, Dordrecht, Netherlands. pp 297–316

Cunningham JC, Brown KW, Payne NJ, Mickle RE, Grant GG, Fleming RA, Robinson A, Curry RD, Langevin D, Burns T (1997) Aerial spray trials in 1992 and 1993 against gypsy moth, *Lymantria dispar* (Lepidoptera: Lymantriidae), using nuclear polyhedrosis virus with and without an optical brightener compared to *Bacillus thuringiensis*. Crop Prot 16:15–23

D'Amico VD, Elkinton JS, Dwyer G, Willis RB, Montgomery ME (1998) Foliage damage does not affect within-season transmission of an insect virus. Ecology 79:1104–1110

D'Amico VD, Elkinton JS, Podgwaite JD, Buonaccorsi JP, Dwyer G (2005) Pathogen clumping: an explanation for non-linear transmission of an insect virus. Ecol Entomol 30:383–390

David L, Mirchev P, Pilarska D (1989) Application of the microsporidian *Nosema* sp. to biological protection of oak forests from the gypsy moth, *Lymantria dispar* L. Acta Entomol Bohemoslov 86:269–274

David L, Novotny J (1990) Laboratory activity and field effect of *Nosema lymantriae* Weiser (Microsporidia) on larvae on the gypsy moth, *Lymantria dispar* L. Biologia 45:3–79

de Amorim GV, Whittome B, Shore B, Levin DB (2001) Identification of *Bacillus thuringiensis* subsp. *kurstaki* strain HD-1-like bacteria from environmental and human samples after aerial spraying of Victoria, British Columbia, Canada, with Foray 48B. Appl Environ Microbiol 67:1035–1043

DellaSala DA, Olson DM, Barth SE, Crane SL, Primm SA (1995) Forest health: Moving beyond rhetoric to restore healthy landscapes in the inland Northwest. Wildlife Soc Bull 23:346–356

Doane CC (1970) Primary pathogens and their role in the development of an epizootic in the gypsy moth. J Invertebr Pathol 15, 21–33

Doane CC, McManus ML (eds) (1981) The gypsy moth: Research toward integrated pest management. USDA Forest Service Tech Bull 1584

Dubois NR (1981) *Bacillus thuringiensis*. In: Doane CC, McManus ML (eds) The gypsy moth: Research toward integrated pest management. USDA Forest Service Tech Bull 1584 pp 455–461

Dubois NR (1986) Synergism between beta-exotoxin and *Bacillus thuringiensis* ssp. *kurstaki* HD-1 in gypsy moth, *Lymantria dispar*, larvae. J Invertebr Pathol 48:146–151

Dubois NR, Dean DH (1995) Synergism between CryIA insecticidal crystal proteins and spores of *Bacillus thuringiensis*, other bacterial spores, and vegetative cells against *Lymantria dispar* (Lepidoptera: Lymantriidae) larvae. Environ Entomol 24:1741–1747

Dubois NR, Reardon RC, Kolodny-Hirsch DM (1988) Field efficacy of the NRD-12 strain of *Bacillus thuringiensis* against gypsy moth (Lepidoptera: Lymantriidae). J Econ Entomol 81:1672–1677

Dubois NR, Reardon RC, Mierzejewski K (1993) Field efficacy and deposit analysis of *Bacillus thuringiensis*, Foray 48B, against gypsy moth (Lepidoptera: Lymantriidae). J Econ Entomol 86:26–33

Dubois NR, Mierzejewski K, Reardon RC, McLane W, Witcosky JJ (1994) *Bacillus thuringiensis* field applications: effect of nozzle type, drop size, and application timing on efficacy against gypsy moth. J Environ Sci Heal B 29:679–695

Dulmage HT (1970) Insecticidal activity of HD-1, a new isolate of *Bacillus thuringiensis* var. *alesti*. J Invertebr Pathol 15:232–239

Dwyer G, Elkinton JS (1995) Host dispersal and the spatial spread of insect pathogens. Ecol 76:1262–1275

Dwyer G, Elkinton JS, Buonaccorsi JP (1997) Host heterogeneity in susceptibility and disease dynamics: Tests of a mathematical model. Am Nat 150:685–707

Dwyer G, Firestone J, Stevens TE (2005) Should models of disease dynamics in herbivorous insects include the effects of variability in host-plant foliage quality? Am Nat 165:16–31

Elkinton JS (2003) Gypsy moth. In Resh VH, Cardé RT (eds) Encyclopedia of insects, Academic Press, Amsterdam. pp 493–497

Elkinton JS, Liebhold AS (1990) Population dynamics of gypsy moth in North America. Annu Rev Entomol 35:571–596

Elkinton JS, Hajek AE, Boettner GH, Simons EE (1991) Distribution and apparent spread of *Entomophaga maimaiga* (Zygomycetes: Entomophthorales) in gypsy moth (Lepidoptera: Lymantriidae) populations in North America. Environ Entomol 20:1601–1605

Erb SL, Bourchier RS, van Frankenhuyzen K, Smith SM (2001) Sublethal effects of *Bacillus thuringiensis* Berliner subsp. *kurstaki* on *Lymantria dispar* (Lepidoptera: Lymantriidae) and the tachinid parasitoid *Compsilura conncinata* (Diptera: Tachinidae). Environ Entomol 30:1174–1181

Evans HF (1986) Ecology and epizootiology of baculoviruses. In: Federici BA, Granados RR (eds) The biology of baculoviruses. Vol 2, CRC Press, Boca Raton, FL. pp 89–132

Falchieri D, Mierzejewski K, Maczuga S (1995) Effects of droplet density and concentration on the efficacy of *Bacillus thuringiensis* and carbaryl against gypsy moth larvae (*Lymantria dispar* L.). J Environ Sci Heal B 30:535–548

Farrar RR, Ridgway RL (1995) Enhancement of activity of *Bacillus thuringiensis* Berliner against four lepidopterous insect pests by nutrient-based phagostimulants. J Entomol Sci 30:29–42

Federici B (2005) Insecticidal bacteria: An overwhelming success for invertebrate pathology. J Invertebr Pathol 89:30–38

Gaugler RR, Brooks WM (1975) Sublethal effects of infection by *Nosema heliothidis* in the corn earworm, *Heliothis zea*. J Invertebr Pathol 26:57–63

Gill EE, Fast NM (2006) Assessing the microsporidia-fungal relationship: Combined phylogenetics analysis of eight genes. Gene 375:103–109

Glare T, O'Callaghan M (2000) *Bacillus thuringiensis*: Biology ecology and safety. J Wiley & Sons, NY

Glowacka-Pilot B (1983) Rola patogenow w przebiegu gradacji brudnicy neiparki (*Lymantria dispar* L.) na Bagnach Beibrzanskich w latach 1976–1978. Prace Instytutu Badawczego Lesnictwa Poland, no 609

Goertz D, Hoch G (2008) Vertical transmission and overwintering of microsporidia in the gypsy moth, *Lymantriae dispar*. J Invertebr Pathol 99:43–48

Goertz D, Pilarska D, Kereselidze M, Solter L, Linde A (2004) Studies on the impact of two *Nosema* isolates from Bulgaria on the gypsy moth (*Lymantria dispar* L.). J Invertebr Pathol 87:105–113

Goertz D, Solter LF, Linde A (2007) Horizontal and vertical transmission of a *Nosema* sp. (Microsporidia) from *Lymantria dispar* (L.) (Lepidoptera: Lymantriidae). J Invertebr Pathol 95:9–16

Grove MJ, Hoover K (2006) Intrastadial developmental resistance of third instar gypsy moths (*Lymantria dispar* L.) to *L. dispar* nucleopolyhedrovirus. Biol Control 40:355–361

Hajek AE (1997) Fungal and viral epizootics in gypsy moth (Lepidoptera: Lymantriidae) populations in central New York. Biol Control 10:58–68

Hajek AE (1999) Pathology and epizootiology of the Lepidoptera-specific mycopathogen *Entomophaga maimaiga*. Microbiol Molecul Biol Rev 63:814–835

Hajek AE (2001) Larval behavior in *Lymantria dispar* increases risk of fungal infection. Oecologia 126:285–291

Hajek AE, Webb RE (1999) Inoculative augmentation of the fungal entomopathogen *Entomophaga maimaiga* as a homeowner tactic to control gypsy moth (Lepidoptera: Lymantriidae). Biol Control 14:11–18

Hajek AE, Humber RA, Elkinton JS, May B, Walsh SRA, Silver JC (1990) Allozyme and RFLP analyses confirm *Entomophaga maimaiga* responsible for 1989 epizootics in North American gypsy moth populations. Proc Natl Acad Sci 87:6979–6982

Hajek AE, Larkin TS, Carruthers RI, Soper RS (1993) Modeling the dynamics of *Entomophaga maimaiga* (Zygomycetes: Entomophthorales) epizootics in gypsy moth (Lepidoptera: Lymantriidae) populations. Environ Entomol 22:1172–1187

Hajek AE, Humber RA, Elkinton JS (1995) The mysterious origin of *Entomophaga maimaiga* in North America. Am Entomol 41:31–42

Hajek AE, Butler L, Walsh SRA, Silver JC, Hain FP, Hastings FL, Odell TM, Smitley DR (1996a) Host range of the gypsy moth (Lepidoptera: Lymantriidae) pathogen *Entomophaga maimaiga* (Zygomycetes: Entomophthorales) in the field versus laboratory. Environ Entomol 25:709–721

Hajek AE, Elkinton JS, Witcosky JJ (1996b) Introduction and spread of the fungal pathogen *Entomophaga maimaiga* along the leading edge of gypsy moth spread. Environ Entomol 25:1235–1247

Hajek AE, Elkinton JS, Humber RA (1997) Entomopathogenic hyphomycetes associated with gypsy moth. Mycologia 89:825–829

Hajek AE, Olsen C, Elkinton JS (1999) Dynamics of airborne conidia of the gypsy moth (Lepidoptera: Lymantriidae) fungal pathogen *Entomophaga maimaiga* (Zygomycetes: Entomophthorales). Biol Control 16:111–117

Hajek AE, Delalibera Jr I, Butler L (2003) Entomopathogenic fungi as classical biological control agents. In: Hokkanen HMT, Hajek AE (eds) Environmental impacts of microbial insecticides. Kluwer Academic Publishers, Dordrecht, NL. pp 15–34

Hajek AE, Strazanac JS, Wheeler MM, Vermeylen F, Butler L (2004) Persistence of the fungal pathogen *Entomophaga maimaiga* and its impact on native Lymantriidae. Biol Control 30:466–471

Hajek AE, McManus ML, Delalibera Jr I (2005) Catalogue of introductions of pathogens and nematodes for classical biological control of insects and mites. USDA Forest Service FHTET-2005-05. 59 pp http://www.fs.fed.us/foresthealth/technology/pdfs/catalogue.pdf [accessed March 2008]

Herms CP, McCullough DG, Bauer LS, Haack RA, Miller DL, Dubois NR (1997) Susceptibity of the endangered Karner blue butterfly (Lepidoptera: Lycaenidae) to *Bacillus thuringiensis* var. *kurstaki* used for gypsy moth suppression in Michigan. Grt Lks Entomol 30:125–141

Hoch G, Schopf A, Maddox JV (2000) Interactions between an entomopathogenic microsporidium and the endoparasitoid *Glyptapanteles liparidis* within their host, the gypsy moth larva. J Invertebr Pathol 75:59–68

Hoch G, Zubrik M, Novotny J, Schopf A (2001) The natural enemy complex of the gypsy moth, *Lymantria dispar* (Lep., Lymantriidae), in different phases of its population dynamics in eastern Austria and Slovakia: A comparative study. J Appl Entomol 125:217–227

Hoch G, Schafellner C, Henn MW, Schopf A (2002) Alterations in carbohydrate and fatty acid levels of *Lymantria dispar* larvae caused by a microsporidian infection and potential adverse effects on a co-occurring endoparasitoid, *Glyptapanteles liparidis*. Arch Insect Biochem Physiol 50:109–120

Hoover K, Grove MJ, Su S (2002) Systemic component to intrastadial developmental resistance in *Lymantria dispar* to its baculovirus. Biol Control 25:92–98

Hwang SY, Lindroth RL, Montgomery ME, Shields KS (1995) Aspen leaf quality affects gypsy moth (Lepidoptera: Lymantriidae) susceptibility to *Bacillus thuringiensis*. J Econ Entomol 88:278–282

Il'inykh AV (2007) Epizootiology of baculoviruses. Biol Bull 34:434–441

Ironside JE (2007) Multiple losses of sex within a single genus of Microsporidia. BMC Evol Biol 7, Art No 48

James RR, Miller JC, Lighthart B (1993) *Bacillus thuringiensis* var. *kurstaki* affects a beneficial insect, the cinnabar moth (Lepidoptera: Arctiidae). J Econ Entomol 86:334–339

Jeffords MR, Maddox JV, O'Hayer KW (1986) Microsporidian spores in gypsy moth larval silk: a possible route of horizontal transmission. J Invertebr Pathol 49:332–333

Jeffords MR, Maddox JV, McManus ML, Webb RE, Wieber A (1989) Evaluation of the overwintering success of two European microsporidia inoculatively released into gypsy moth populations in Maryland, USA. J Invertebr Pathol 53:235–240

Jenkins JL, Lee MK, Valaitis AP, Curtiss A, Dean DH (2000) Bivalent sequential binding model of a *Bacillus thuringiensis* toxin to gypsy moth aminopeptidase N receptor. J Biol Chem 275:14423–14431

Johnson DM, Liebhold AM, Bjornstad ON (2006) Geographical variation in the periodicity of gypsy moth outbreaks. Ecography 29:367–374

Keohane EM, Weiss LM (1999) The structure, function, and composition of the microsporidian polar tube. In: Wittner M, Weiss LM (eds) The microsporidia and microsporidiosis. Amer Soc Microbiol Press, Washington, D.C. pp 196–224

Keating ST, Schultz JC, Yendol WG (1990) The effect of diet on gypsy moth (*Lymantria dispar*) larval midgut pH, and its relationship with larval susceptibility to a baculovirus. J Invertebr Pathol 56:317–326

Kogan PH, Hajek AE (2000) Formation of azygospores by the insect pathogenic fungus *Entomophaga maimaiga* in cell culture. J Invertebr Pathol 75:193–201

Lautenschlager RA, Podgwaite JD (1977) Passage of infectious nuclear polyhedrosis virus through the alimentary tracts of two small mammal predators of the gypsy moth. Environ Entomol 6:737–738

Lautenschlager RA, Rothenbacher H, Podgwaite JD (1978) Response of small mammals to aerial application of the nucleopolyhedrosis virus of the gypsy moth, *Lymantria dispar*. Environ Entomol 7:676–684

Lautenschlager RA, Rothenbacher H, Podgwaite JD (1979) Response of birds to aerial application of the nucleopolyhedrosis virus of the gypsy moth, *Lymantria dispar*. Environ Entomol 8:760–764

Lautenschlager RA, Podgwaite JD, Watson DE (1980) Natural occurrence of the nucleopolyhedrosis virus of the gypsy moth *Lymantria dispar* (Lepidoptera: Lymantriidae) in wild birds and mammals. Entomophaga 25:261–268

Lee MK, Curtiss A, Alcantara E, Dean DH (1996) Synergistic effect of the *Bacillus thuringiensis* toxins CryIAa and CryIAc on the gypsy moth, *Lymantria dispar*. Appl Environ Microbiol 62:583–586

Leonard D (1974) Recent developments in ecology and control of the gypsy moth. Annu Rev Entomol 19:197–229

Leonard D (1981) Bioecology of the gypsy moth. In: Doane CC, McManus ML (eds) The gypsy moth: Research toward integrated pest management. USDA Forest Service Tech Bull 1584. pp 9–29

Leuschner WA, Young JA, Ravlin FW (1996) Potential benefits of slowing the gypsy moth's spread. S J Appl For 120:65–73

Lewis FB (1981) Gypsy moth nucleopolyhedrosis virus. In: Doane CC, McManus ML (eds) The gypsy moth: Research toward integrated pest management. USDA Forest Service Tech Bull 1584. pp 454–455

Lewis FB, Rollinson WD, Yendol WG (1981) Laboratory evaluations. In: Doane CC, McManus ML (eds) The gypsy moth: Research toward integrated pest management. USDA Forest Service Tech Bull 1584. pp. 455–461

Liang Y, Patel SS, Dean DH (1995) Irreversible binding kinetics of *Bacillus thuringiensis* CryIA delta-endotoxins to gypsy moth brush border membrane vesicles is directly related to toxicity. J Biol Chem 270:24719–24724

Liang Y, Dean DH (1994) Location of a lepidopteran specificity region in insecticidal crystal protein CryIIA from *Bacillus thuringiensis*. Mol Microbiol 13:569–575

Liang Y, Patel SS, Dean DH (1995) Irreversible binding kinetics of *Bacillus thuringiensis* CryIA delta-endotoxins to gypsy moth brush border membrane vesicles is directly correlated to toxicity. J Biol Chem 270:24719–24724

Liebhold A, Luzader E, Reardon R, Bullard A, Roberts A, Ravlin W, Delost S, Spears B (1996) Use of a geographic information system to evaluate regional treatment effects in a gypsy moth (Lepidoptera: Lymantriidae) management program. J Econ Entomol 89:1192–1203

Lord JC (2005) From Metchnikoff to Monsanto and beyond: The path of microbial control. J Invertebr Pathol 89:19–29

Maddox JV, Baker M, Jeffords MR, Kuras M, Linde A, McManus M, Solter L, Vavra J, Vossbrinck C (1999) *Nosema portugal* n.sp., isolated from gypsy moths (*Lymantria dispar* L.) collected in Portugal. J Invertebr Pathol 73:1–14

Malakar R, Elkinton JS, Carroll SD, D'Amico VD (1999) Interactions between two gypsy moth (Lepidoptera: Lymantriidae) pathogens: Nucleopolyhedrovirus and *Entomophaga maimaiga*

(Zygomycetes: Entomophthorales): Field studies and a simulation model. Biol Control 16:189–198

Marshall MR, Cooper RJ, Dececco JA, Strazanac J, Butler L (2002) Effects of experimentally reduced prey abundance on the breeding ecology of the red-eyed vireo. Ecol Applic 12:261–280

Martin PA (2004) A stilbene optical brightener can enhance bacterial pathogenicity to gypsy moth (Lepidoptera: Lymantriida) and Colorado potato beetle (Coleoptera: Chrysomelidae). Biocontr Sci Technol 14:375–383

Martin PAW, Blackburn MB (2007) Using combinatorics to screen *Bacillus thuringiensis* isolates for toxicity against *Manduca sexta* and *Plutella xylostella*. Biol Control 42:226–232

McManus ML, Solter L (2003) Microsporidian pathogens in European gypsy moth populations. Proc: ecology, survey, and management of forest insects. USDA Forest Service, NE Res Stn Gen Tech Rep NE-311. pp 44–51

Miller JC (1990) Field assessment of the effects of a microbial pest control agent on nontarget Lepidoptera. Am Entomol 36:135–139

Mott M, Smitley D (2000) Impact of *Bacillus thuringiensis* application on *Entomophaga maimaiga* (Entomophthorales: Entomophthoraceae) and *Ld*MNPV-induced mortality of gypsy moth (Lepidoptera: Lymantriiidae). Environ Entomol 29:1312–1322

Murray KD, Elkinton JS (1989) Environmental contamination of egg masses as a major component of transgenerational transmission of gypsy moth nuclear polyhedrosis virus (LdMNPV). J Invertebr Pathol 53:324–334

Murray KD, Shields KS, Burand JP, Elkinton JS (1991) The effect of gypsy moth metamorphosis on the development of nuclear polyhedrosis virus infection. J Invertebr Pathol 57:352–361

Nielsen C, Milgroom MG, Hajek AE (2005) Genetic diversity in the gypsy moth fungal pathogen *Entomophaga maimaiga* from founder populations in North America and source populations in Asia. Mycol Res 109:941–950

Peacock JW, Schweitzer DF, Carter JL, Dubois NR (1998) Laboratory assessment of the effects of *Bacillus thuringiensis* on native Lepidoptera. Environ Entomol 27:450–457

Pemberton RW, Lee JH, Reed DK, Carlson RW, Han HY (1993) Natural enemies of the Asian gypsy moth (Lepidoptera: Lymantriidae) in South Korea. Ann Entomol Soc Amer 86:423–440

Pilarska DK, Solter LF, Maddox JV, McManus ML (1998) Microsporidia from gypsy moth (*Lymantria dispar* L.) populations in Central and Western Bulgaria. Acta Zool Bulgarica 50:109–113

Pilarska DK, Solter LF, Kereselidze M, Linde A, Hoch G (2006) Microsporidian infections in *Lymantria dispar* larvae: Interactions and effects of multiple species infections on pathogen horizontal transmission. J Invertebr Pathol 93:105–113

Podgwaite JD (1981) Natural disease within dense gypsy moth populations. In: Doane CC, McManus ML (eds) The gypsy moth: Research toward integrated pest management. US Dept. of Agric Tech Bull 1584. pp 125–134

Podgwaite J, Shields K, Zerillo R, Bruen R (1979) Environmental persistence of the nucleopolyhedrosis virus of the gypsy moth. Environ Entomol 8:523–536

Podgwaite JD, Dubois NR, Reardon RC, Witcosky J (1993) Retarding outbreak of low-density gypsy moth (Lepidoptera: Lymantriidae) populations with aerial applications of Gypchek and *Bacillus thuringiensis*. J Econ Entomol 86:730–734

Popham HJR, Bishoff DS, Slavicek JM (2001) Both *Lymantria dispar* nucleopolyhedrovirus enhancin genes contribute to viral potency. J Virol 75:8639–8648

Purrini VK, Skatulla U (1978) Über die natürlichen Krankheiten des Schwammspinners, *Lymantria dispar* L. (Lep., Lymantriidae) in Sardinien, Italien. Anz Schadlingskd PFL 51:9–11

Raimo B, Reardon RC, Podgwaite JD (1977) Vectoring gypsy moth nuclear polyhedrosis virus by *Apanteles melanoscelus* (Hymenoptera:Braconidae). Entomophaga 22:207–216

Raimondo S, Pauley TK, Butler L (2003) Potential impacts of *Bacillus thuringiensis* var. *kurstaki* on five salamandar species in West Virginia. NE Naturalist 10:25–38

Rajamohan F, Alzate O, Cotrill JA, Curtiss A, Dean DH (1996) Protein engineering of *Bacillus thuringiensis* delta-endotoxin: Mutations at domain II of CryIAb enhance receptor affinity and toxicity toward gypsy moth larvae. Proc Natl Acad Sci USA 93:14338–14343

Rastall K, Kondo V, Strazanac JS, Butler L (2003) Lethal effects of biological insecticide application on nontarget lepidopterans in two Appalachian forests. Environ Entomol 32:1364–1369

Reardon RC, Podgwaite JD (1976) Disease-parasitoid relationships in natural populations of *Lymantria dispar* in the northeastern United States. Entomophaga 21:333–341

Reardon RC, Podgwaite JD (1994) Summary of efficacy evaluation using aerially applied Gypchek against gypsy moth in the U.S.A. J Environ Sci Heal B 29:739–756

Reardon RC, Dubois N, McLane W (1994) *Bacillus thuringiensis* for managing gypsy moth: A review. USDA Forest Service Tech Transfer FHM-NC-01-94

Reardon RC, Podgwaite J, Zerillo R (1996) Gypchek-The gypsy moth nucleopolyhedrosis virus product. USDA Forest Service Tech Transfer FHTET-96-16

Reiff W (1911) The wilt disease or flacherie of the gypsy moth. Contrib Entomol Lab, Bussey Inst, Harvard Univ. p 36

Riegel CI, Slavicek JM (1997) Characterization of the replication cycle of the *Lymantria dispar* nuclear polyhedrosis virus. Virus Res 51:9–17

Reilly JR, Hajek AE (2008) Density-dependent resistance of the gypsy moth *Lymantria dispar* to its nucleopolyhedrovirus, and the consequences for population dynamics. Oecologia 154:691–701

Ridgway RL, Thorpe KW, Webb RE, Venables L (1994) Gypsy moth management in suburban parks: Program evaluation. J Entomol Sci 29:557–569

Romanyk N (1966) Natural enemies of *Lymantria dispar* in Spain. Bol Serv Plagas For 9:157–163

Sample BE, Butler L, Zivkovich C, Whitmore RC, Reardon R (1996) Effects of *Bacillus thuringiensis* Berliner var. *kurstaki* and defoliation by the gypsy moth (*Lymantria dispar* L.) (Lepidoptera: Lymmantriidae) on native arthropods in West Virginia. Can Entomol 128:573–592

Schweitzer D (2004) Gypsy moth (*Lymantria dispar*): Impacts and options for biodiversity-oriented land managers. NatureServe, Arlington, VA. 59 pp

Seastedt TR, Crossley DA, Hargrove WW (1983) The effects of low-level consumption by canopy arthropods on the growth and nutrient dynamics of black locust and red maple trees in the southern Appalachians. Ecology 64:1040–1048

Shapiro M, Robertson JL (1992) Enhancement of gypsy moth (Lepidoptera: Lymantriidae) baculovirus activity by optical brighteners. J Econ Entomol 85:1120–1124

Sheppard CA, Shapiro M (1994) Physiological and nutritional effects of a fluorescent brightener on nuclear polyhedrosis virus-infected *Lymantria dispar* (L.) larvae (Lepidoptera: Lymantriidae). Biol Control 4:404–411

Sidor C (1976) Oboljenja Izazvana microorganizmima kod nekih Lymantriidae u Jugoslavifi I Njihov Znacaj za entomofaunu. (Diseases provoked with microorganisms by some Limantriidae in Yugoslavia and their importance for entomofauna.) Arh Biol Nauka Beograd 28:127–137

Siegel JP, Maddox JV, Ruesink WG (1986) Lethal and sublethal effects of *Nosema pyrausta* on the European corn borer, *Ostrinia nubilalis* in Central Illinois USA. J Invertebr Pathol 48:167–173

Sierpinska A (2000) Preliminary results on the occurrence of microsporidia of the gypsy moth (*Lymantria dispar* L.) from different habitats of Poland. IOBC WPRS Bull 23:291–295

Slavicek JM, Mercer MJ, Kelly ME, Hayes-Plazolles N (1996) Isolation of a baculovirus variant that exhibits enhanced polyhedra production stability during serial passage in cell culture. J Invertebr Pathol 67:153–160

Slavicek JM, Mercer MJ, Pohlman D, Kelly ME, Bishoff DS (1998) Identification of a novel *Lymantria dispar* nucleopolyhedrovirus mutant that exhibits abnormal polyhedron formation and virion occlusion. J Invertebr Pathol 72:28–37

Smitley DR, Bauer LS, Hajek AE, Sapio FJ, Humber RA (1995) Introduction and establishment of *Entomophaga maimaiga*, a fungal pathogen of gypsy moth (Lepidoptera: Lymantriidae) in Michigan. Environ Entomol 24:1685–1645

Smith P (1999) Interactions between a tachinid parasitoid and a fungal pathogen of gypsy moth. Proc Ann Mtg Soc Invertebr Pathol, Irvine, CA 22–27 August. p 70

Solter LF, Becnel JJ (2003) Environmental safety of microsporidia. In: Hokkanen HMT, Hajek AE (eds) Environmental impacts of microbial insecticides: Need and methods for risk assessment. Kluwer Academic Publishers, Netherlands. pp 93–118

Solter LF, Maddox JV (1998) Physiological host specificity of microsporidia as an indicator of ecological host specificity. J Invertebr Pathol 71:207–216

Solter LF, Maddox JV, McManus ML (1997) Host specificity of microsporidia (Protista: Microspora) from European populations of *Lymantria dispar* (Lepidoptera: Lymantriidae) to indigenous North American Lepidoptera. J Invertebr Pathol 69:135–150

Solter LF, Pilarska DK, Vossbrinck CF (2000) Host specificity of microsporidia pathogenic to forest Lepidoptera. Biol Control 19:48–56

Solter LF, Siegel JP, Pilarska DK, Higgs MC (2002) The impact of mixed infection of three species of microsporidia isolated from the gypsy moth, *Lymantria dispar* L. (Lepidoptera: Lymantriidae). J Invertebr Pathol 81:103–113

Soper RS, Shimazu M, Humber RA, Ramos ME, Hajek AE (1988) Isolation and characterization of *Entomophaga maimaiga* sp. nov., a fungal pathogen of gypsy moth, *Lymantria dispar*, from Japan. J Invertebr Pathol 51:229–241

Sopuck L, Ovaska K, Whittington B (2002) Responses of songbirds to aerial spraying of the microbial insectide *Bacillus thuringiensis* var. *kurstaki* (Foray 48B®) on Vancouver Island, British Columbia, Canada. Environ Toxicol Chem 21:1664–1672

Speare AT, Colley R (1912) The artificial use of the brown-tail fungus in Massachusetts, with practical suggestions for private experiment, and a brief note on a fungous disease of the gypsy caterpillar. Wright & Potter, Boston

Sundaram KMS, Sundaram A, Hammock BD (1994) Persistence of *Bacillus thuringiensis* deposits in a hardwood forest after aerial application of a commercial formulation at two dosage rates. J Environ Sci Heal B 29:999–1052

Tanada Y, Kaya HK (1993) Insect pathology. Academic Press, San Diego, CA

Thiem SM, Du X, Quentin ME, Berner MM (1996) Identification of a baculovirus gene that promotes *Autographa californica* nuclear polyhedrosis virus replication in a nonpermissive insect cell line. J Virol 70:2221–2229

Thorpe KW, Podgwaite JD, Slavicek JM, Webb RE (1998) Gypsy moth (Lepidoptera: Lymantriidae) control with ground-based hydraulic applications of Gypchek, in vitro-produced virus, and *Bacillus thuringiensis*. J Econ Entomol 91:875–880

USDA Forest Service (2008) Gypsy Moth Digest. http://na.fs.fed.us/fhp/gm [accessed 3 March 2008]

Valaitis AP, Jenkins JL, Lee MK, Dean DH, Garner KJ (2001) Isolation and partial characterization of gypsy moth BTR-270, an anionic brush border membrane glycoconjugate that binds *Bacillus thuringiensis* CryIA toxins with high affinity. Arch Insect Biochem Physiol 46:186–200

Valent Biosciences (2001) Protecting our forests-protecting our future. Forestry technical manual. Valent Biosciences, USA

van Frankenhuyzen K, Wiesner CJ, Riley CM, Nystrom C, Howard CA, Howse GM (1991) Distribution and activity of spray deposits in an oak canopy following aerial application of diluted and undiluted formulations of *Bacillus thuringiensis* Berliner against the gypsy moth, *Lymantria dispar* L. (Lepidoptera: Lymantriidae). Pesticide Sci 33:159–168

Vavra J, Hylis M, Vossbrinck CR, Pilarska DK, Linde A, Weiser J, McManus ML, Hoch G, Solter LF (2006) *Vairimorpha disparis* n. comb. (Microsporidia: Burenellidae): A redescription of the *Lymantria dispar* (L.) (Lepidoptera: Lymantriidae) microsporidium, *Thelohania disparis* Timofejeva 1956. J Euk Microbiol 53:292–304

Wang C, Strazanac J, Butler L (2000) Abundance, diversity, and activity of ants (Hymenoptera: Formicidae) in oak-dominated mixed Appalachian forests treated with microbial pesticides. Environ Entomol 29:579–586

Wang CY, Solter LF, T'sui WH, Wang CH (2005) An *Endoreticulatus* species from *Ocinara lida* (Lepidoptera: Bombycidae) in Taiwan. J Invertebr Pathol 89:123–135

Webb RE, Shapiro M, Podgwaite JD, Reardon RC, Tatman KM, Venables L, Kolodny-Hirsch DM (1989) Effect of aerial spraying with Dimilin, DiPel or Gypchek on two natural enemies of the gypsy moth (Lepidoptera: Lymantriidae). J Econ Entomol 82:1695–1701

Webb RE, White GB, Thorpe KW, Talley SE (1999) Quantitative analysis of a pathogen-induced premature collapse of a "leading edge" gypsy moth (Lepidoptera: Lymantriidae) population in Virginia. J Entomol Sci 34:84–100

Webb RE, Shapiro M, Thorpe KW, Peiffer RA, Fuester RW, Valenti MA, White GB, Podgwaite JD (2001) Potentiation by a granulosis virus of Gypchek, the gypsy moth (Lepidoptera: Lymantriidae) nuclear polyhedrosis virus product. J Entomol Sci 36:169–176

Webb RE, Bair MW, White GB, Thorpe KW (2004) Expression of *Entomophaga maimaiga* at several gypsy moth (Lepidoptera: Lymantriidae) population densities and the effect of supplemental watering. J Entomol Sci 39:223–234

Weiser J, Novotny J (1987) Field application of *Nosema lymantriae* against the gypsy moth, *Lymantria dispar* L. J Appl Ent 104:58–62

Weseloh RM (1998) Possibility for recent origin of the gypsy moth (Lepidoptera: Lymantriidae) fungal pathogen *Entomophaga maimaiga* (Zygomycetes: Entomophthorales) in North America. Environ Entomol 27:171–177

Weseloh RM (2003a) People and the gypsy moth: A story of human interactions with an invasive species. Am Entomol 49(3):180–190

Weseloh RM (2003b) Short and long range dispersal in the gypsy moth (Lepidoptera: Lymantriidae) fungal pathogen, *Entomophaga maimaiga* (Zygomycetes: Entomophthorales). Environ Entomol 32:111–122

Weseloh RM (2004) Effect of conidial dispersal of the fungal pathogen *Entomophaga maimaiga* (Zygomycetes: Entomophthorales) on survival of its gypsy moth (Lepidoptera: Lymantriidae) host. Biol Control 29:138–144

Weseloh RM, Andreadis TG, Moore REB, Anderson JF, Dubois NR, Lewis FB (1983) Field confirmation of a mechanism causing synergism between *Bacillus thuringiensis* and the gypsy moth parasitoid *Apanteles melanoscelus*. J Invertebr Pathol 41:99–103

Whaley WH, Anhold J, Schaalje GB (1998) Canyon drift and dispersion of *Bacillus thuringiensis* and its effects on select nontarget lepidopterans in Utah. Environ Entomol 27:539–548

White GB, Webb RE (1994) Survival of dipteran parasitoids (Diptera: Tachinidae) during a virus-induced gypsy moth population collapse. Proc Entomol Soc Wash 96:27–30

Wittner M (1999) Historic prospective on the microsporidia: Expanding horizons. In: Wittner M, Weiss LM (eds) The microsporidia and microsporidiosis. Amer Soc Microbiol Press, Washington, D.C. pp 447–501

Woods SA, Elkinton JS, Podgwaite JD (1989) Acquisition of nuclear polyhedrosis virus from tree stems by newly emerged gypsy moth (Lepidoptera: Lymantriaidae) larvae. Environ Entomol 18:298–301

Woods SA, Elkinton JS, Murray KD, Liebhold AM, Gould JR, Podgwaite JD (1991) Transmission dynamics of a nuclear polyhedrosis virus and predicting mortality in gypsy moth (Lepidoptera: Lymantriidae) populations. J Econ Entomol 84:423–430

Yendol WG, Bryant JE, McManus ML (1990) Penetration of oak canopies by a commercial preparation of *Bacillus thuringiensis* applied by air. J Econ Entomol 83:173–179

Zelinskaya LM (1980) Role of microsporidia in the abundance dynamics of the gypsy moth (*Porthetria dispar*) in forest plantings along the lower Dnepr River (Ukrainian Republic, USSR). Vestnik zoology (Zoology Bulletin) 1:57–62

Zelinskaya LM (1981) Using the index of imago infection by spores of microsporidia for predicting the reproduction of *Lymantria dispar*. Lesnoye Khozyaistvo (Forestry) 4:58–60

Zwölfer W (1927) Die pebrine des schwammspinners (*Porthetria dispar* L.) and goldafters (*Nygmia phaeorrhoea* Don. = *Euproctis chrysorrhoea* L.), eine neue wirtschaftlich bedeutungsvolle infektionskrankheit. Verh Dtsch Ges Angew Entomol 6:98–109

Chapter 12
Controlling the Pine-Killing Woodwasp, *Sirex noctilio*, with Nematodes

Robin A. Bedding

Abstract The pine-killing woodwasp *Sirex noctilio*, a native to Eurasia/Morocco, was accidentally introduced into various Southern Hemisphere countries during the last century and has recently (2005) been detected in north-eastern North America. The parasitic nematode *Beddingia siricidicola* is by far the most important control agent of sirex and has been introduced into each Southern Hemisphere country soon after sirex became established. The nematode has a complex life cycle with morphologically very different forms. One form feeds on the tree-pathogenic, sirex-symbiotic fungus (*Amylostereum areolatum*) as this fungus grows throughout the tree, while the other form grows in and then sterilises adult female *S. noctilio*. The fungal-feeding form of *B. siricidicola* is used to mass-produce the nematode. Methods are described for liberating nematodes in pine plantations. The nematode has caused major crashes in *S. noctilio* populations so that sirex-infested trees can no longer be found in many plantations. A problem arose when it was discovered that long-term *in vitro* culture using only the fungal cycle without intervention of parasitic cycles had selected, over many years, for a nematode strain (the "defective strain") that rarely formed the infective stage and was therefore much less effective in the field. Isolation of the "Kamona strain", annual replenishment from liquid nitrogen storage and other procedural changes are enabling strain replacement in the field. While nematode control in most of the Southern Hemisphere has proved to be highly successful, there are problems in the KwaZulu-Natal region of South Africa where warm dry winters cause sirex-infested trees to dry out before the nematode populations can spread throughout the tree. In North America an inferior strain of nematode appears to have been accidentally introduced with sirex. The symbiotic fungus of sirex introduced to North America is a different strain of *A. amylostereum* to that in the Southern Hemisphere and does not permit optimal nematode development.

R.A. Bedding
CSIRO Entomology, P.O. Box 1700, Canberra ACT 2601 Australia
e-mail: robin.bedding@csiro.au

12.1 Introduction

Interest in nematode control of *Sirex noctilio* (sirex) has increased greatly since this severe pest of pine (*Pinus* spp.) trees spread north in South Africa into KwaZulu-Natal in 2002 (Dyer 2007) and was found in northern USA in 2004 (Hoebeke *et al.* 2005) and southern Canada during 2005 (de Groot 2007).

Sirex noctilio is the only siricid out of about 40 siricid species infesting conifers world-wide that can kill relatively healthy pine trees. Endemic to Eurasia/Morocco, where it does little harm to native pines (Spadberry & Kirk 1978), *S. noctilio* has become a major pest in the Southern Hemisphere. It was first detected in New Zealand (1940s), then Tasmania (1952), mainland Australia (1961), Uruguay (1980), Argentina (1985), Brazil (1988), South Africa (1994) and Chile (2001). In some of the worst affected areas sirex has resulted in up to 80% tree death.

In the whole of the Southern Hemisphere there are about 7.6 million hectares of commercial pine plantations currently threatened by *S. noctilio* (Iede *et al.* 2000, Wood *et al.* 2001) with potential damage estimated at US$16 million–US$60 million per annum for Australia's 1 million ha of pine (Bedding & Iede 2005) and US$23.2 million per annum for Brazil's 1.8 million ha (Iede *et al.* 2007).

The recent establishment of *S. noctilio* in North America represents a huge increase in potential damage with 58 million hectares of susceptible forests in the USA (Schneeberger 2007) and over 142 million hectares in Canada (CanFI2001 Database 2007). Using the computer program CLIMEX (Sutherst *et al.* 1999), Carnegie *et al.* (2006) show that sirex could become established even more widely after accidental introduction to additional countries, including in Paraguay, Bolivia, Peru, Ecuador, Columbia and Venezuela in South America, most of Mexico, USA and Canada in North America, and Guatemala, Costa Rica and Panama in Central America. In Africa they showed that in addition to South Africa, pine forests in Zimbabwe, Mozambique, Madagascar, Tanzania, Uganda, Kenya and Ethiopia were climatically suitable for sirex. In China, pine forests from Yunnan Province in south-central China through most provinces to Heilongjiang Province in northeastern China would be climatically suitable for the establishment of sirex.

Forest hygiene and particularly timely thinning has an important impact on sirex populations. However, while several insect parasitoids have been introduced into the Southern Hemisphere, these, perhaps with the exception of the endoparasite, *Ibalia leucospoides*, have limited effect. Of seven species of the nematode genus *Beddingia*[1] (= *Deladenus*) (Bedding 1968, 1975) found parasitising siricids and their parasitoids (Bedding & Akhurst 1978), only *B. siricidicola* was found to be suitable for the control of sirex (Bedding 1984). This nematode has now been released and established in Australia, New Zealand, Brazil, Uruguay, Argentina, Chile and South Africa. The nematode is strongly density dependent, can achieve levels

---

[1] Those species previously included in the genus *Deladenus* having both free-living and parasitic life cycles associated with extreme adult female dimorphism, were assigned to a new genus, *Beddingia* by Blinova and Korenchenko (1986). This nomenclature was adopted by Remillet and Laumond (1991) and by Poinar *et al.* (2002) (who also established a new family, Beddingiidae with *Beddingia* as its only genus).

of parasitism approaching 100% and is generally recognised as the main controlling agent of sirex in the Southern Hemisphere (Iede *et al.* 2000, Carnegie *et al.* 2005, Bedding & Iede 2005).

12.2 Pest Biology

Generally sirex attack only pine plantations that are over 10–12 years old. In the early stages of infestation, female sirex are attracted to suppressed, drought-stressed or damaged pine trees (Madden 1977). As the sirex population builds up over several years and there are increasing numbers of sirex available to attack each tree, more and more vigorous trees can also be killed.

Sirex kills pine trees by injecting a toxic mucous (Coutts 1969a, b) and spores of a tree pathogenic fungus *Amylostereum areolatum* (Gaut 1969) 10–20 mm deep into the wood (see Fig. 12.1). If the sirex female detects that the tree is suitable, she will lay one or more eggs in adjacent drills. Depending on size, a female sirex can oviposit from 30 to 450 eggs (Madden 1974) and may oviposit in several different trees (JL Madden personal communication).

The injected mucous suppresses translocation of sugars, and as a result, formation of polyphenols at the site of oviposition that would normally suppress fungal growth. Depending on the resistance of the tree, the fungus begins to grow significantly a few weeks to several months after oviposition and the local drying out of the wood resulting from the fungal growth causes the sirex eggs to hatch (Madden 1968). At about this stage, the tree dies and subsequently the fungus spreads throughout the tree. The sirex larvae bore through the tree feeding on the symbiotic fungus and after many months may have grown several centimetres in length before pupating near the surface. Sirex adults bore out of the wood to emerge from late December to April in mainland Australia but later in Tasmania, whereas in Brazil emergence begins in September (Iede *et al.* 1998). Usually there is one generation per year but there may be two in Brazil and one generation can take 1–3 years in Tasmania. Adult sirex live for 2–3 weeks.

Fig. 12.1 A female *Sirex noctilio* inserting toxic mucous, fungal spores and eggs into a pine tree

12.3 Nematode Biology

The nematode, *B. siricidicola*, was first discovered infecting *S. noctilio* in New Zealand, by Zondag (1962). Its life cycle and the biology of various strains in various siricid species were described in detail by Bedding (1967, 1972, 1984). The life cycle is an extraordinary one involving two very different forms of adult female, each associated with a different life cycle (Bedding 1967, 1972, 1984, 1993). One form of the nematode parasitises and sterilises female sirex, and the other form feeds on the sirex's symbiotic fungus as the fungus grows throughout the infested tree (Fig. 12.2).

12.3.1 Parasitism of Sirex

There may be 1 to more than 100 adult parasitic nematodes within the haemocoel of a single sirex; usually they are found within the abdominal cavity but may be in the thorax and even within the legs and testes. They are cylindrical in form and vary from a few mm in length up to 25 mm in rare cases; occasionally they are green in colour. Just before a sirex adult emerges from a tree, each adult parasitic nematode will have released many hundreds of juveniles into the sirex haemocoel. Most of these juvenile nematodes migrate into the testes or ovaries of their host. The testes may become greatly hypertrophied and often fused. The ovaries are usually somewhat atrophied and juvenile nematodes penetrate each egg before the shell has hardened (in most strains of sirex and nematodes). The female sirex is thus completely sterile. Parasitised sirex oviposit readily but introduce packets of nematodes into the tree instead of viable eggs. From an evolutionary standpoint this situation can only occur because *S. noctilio* is a communal insect with many sirex (some unparasitised) usually ovipositing on the same tree. In various other species of siricids that are solitary, juvenile nematodes do not penetrate the eggs but are introduced, surrounding the outside of the egg, during oviposition (Bedding 1972).

In the parasitised male sirex, spermatozoa pass from the testes into the vesiculae seminales well before juvenile nematodes invade the testes. No juvenile nematodes pass down the sirex's vas deferens into the vesiculae seminales so that nematodes cannot be transferred from male to female sirex during copulation and the male is thus a dead end for the nematode. (However, the otherwise sterile testis filled with thousands of nematodes makes an easy vehicle to establish aseptic cultures of the nematodes (see Fig. 12.3B)).

12.3.2 The Fungal Feeding Cycle

Bedding (1967) found that when juvenile nematodes were removed from parasitised sirex and placed on cultures of the sirex symbiotic fungus, *A. areolatum*, the nematodes readily fed on the growing fungus, grew into adults quite unlike the parasitic

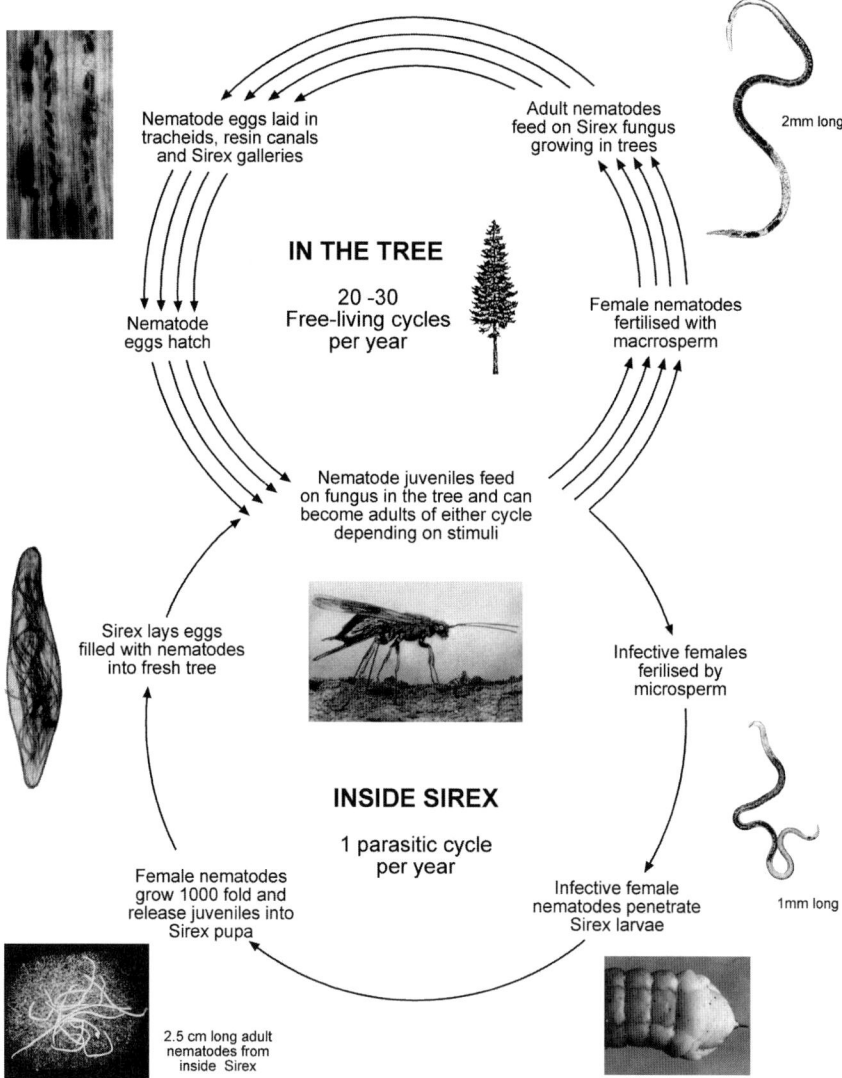

Fig. 12.2 Biology of the nematode parasite of sirex, *Beddingia siricidicola* (after Bedding 1993)

form and the female adult nematodes laid eggs (Fig. 12.3A). When the nematode eggs hatched, the resulting juveniles also fed on symbiotic fungus and grew into adult males and females that laid eggs. This cycle could be repeated indefinitely without intervention of a cycle parasitic in sirex. Similarly, when juvenile nematodes are introduced into trees by nematode-parasitised sirex, these juveniles move through the tracheids of the tree feeding on the fungus, growing into adults and breeding in vast numbers as the fungus grows throughout the tree. Without the

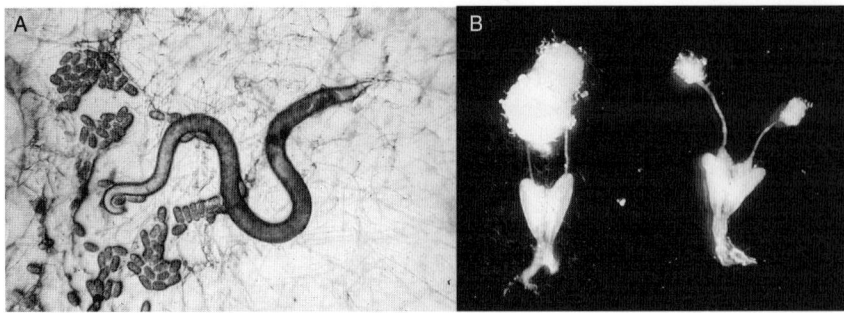

Fig. 12.3 A. Mycetophagous female and eggs of *B. siricidicola* on potato dextrose agar culture of *Amylostereum areolatum*. **B.** Male reproductive organs from parasitised sirex (*right*) and unparasitised sirex (*left*). Infected testes are used to establish monoxenic cultures of *Beddingia siricidicola* on the symbiotic fungus *Amylostereum areolatum*

multiplication of free-living nematodes within a sirex-infested tree, there would be, at the most, a few thousand juveniles in often hundreds of kg of wood comprised of hundreds of millions of tracheids within the tree and the chances of these juvenile nematodes reaching significant numbers of sirex larvae would be minimal. As it is, the free-living nematode population breeds throughout the tree and can reach most of the sirex larvae (progeny of unparasitised sirex) feeding within it (provided moisture levels are adequate).

Until nematode populations reach sirex larvae these free-living cycles continue, with a cycle possible every 12 days or so at 24°C (Akhurst 1975). However, when part of the nematode population is in close proximity to a sirex larva, the local microenvironment (high CO_2 and low pH) has a dramatic impact on nematode eggs and young juvenile nematodes (Bedding 1993). Instead of becoming mycetophagous, egg-laying adult females and associated males, they become infective females and their associated males.

12.3.3 The Infective Stage

The infective female is morphologically very distinct from the mycetophagous female (Bedding 1968, 1972); in fact it is so different that when discovered it would have been placed in a separate family. The extreme dimorphism reflects the differences in function of the two forms. Whereas the stylet of the mycetophagous female is like a fine hypodermic syringe, adapted to pierce and suck fluid from the fungal hyphae, the stylet of the infective female is twice as long and a much stouter spear used to puncture the cuticle of the sirex larva so that the infective female can gain entry. The very different gland structures are no doubt similarly adapted to function. The reproductive system of the infective female consists largely of a long cylindrical oviduct terminated by a few generative cells. Because only the female penetrates sirex larvae, she cannot be fertilised after entry and must take all the spermatozoa that are required to produce thousands of progeny with her. This has required another morphological adaptation: whereas mycetophagous females (that can re-mate when necessary) are fertilised by males having giant amoeboid sperm, another kind of male producing microsperm (almost entirely nucleus) fertilises the infective female. Because of the very small size of the micro spermatozoa, the infective female oviduct can hold many thousands of them.

The infective female nematode penetrates sirex larvae, and occasionally sirex pupae, after repeatedly probing with its spear-like stylet to puncture the larval cuticle. Then it slips into the sirex larval haemocoel usually leaving a scar that remains until the next sirex larval moult (Bedding 1972). It then moves around inside its host for a few days before shedding its cuticle (not at true moult) to reveal its entire body surface covered with microvilli (Riding 1970). The microvilli are able to rapidly absorb food from the sirex larval blood and within a few weeks the infective female nematode (now a parasitic female), initially only about 0.6 mm long, can grow up to 1000 fold in volume and reach 20 mm in length. However, the reproductive system of the infective female remains more or less the same size until the onset of the host's pupation. Then, presumably as a result of the hormonal changes within its host, the few cells at the tip of the nematode's oviduct proliferate profusely and dramatically, eventually developing into many hundreds of eggs that are fertilised by the microsperm filling the nematode's oviduct.

Before the sirex female host emerges from a tree, the nematode eggs hatch and the parent nematode becomes little more than a tube filled with juveniles that now force their way out from all over the parent's surface and into the blood of the sirex host and thence to the reproductive organs. Although these juveniles appear to be identical to those hatching from eggs in the free-living cycle, when they are placed on fungal cultures and exposed to high levels of CO_2 and low pH they develop into only mycetophagous adults and never into infective females (RA Bedding & J Calder unpublished data). Presumably the same is true within the tree: when sirex introduces juvenile nematodes into a tree during oviposition, there must be at least one fungal feeding/breeding cycle before infection of a sirex larva can occur.

12.4 Manipulation of the Nematode for Control of Sirex

B. siricidicola is essentially used as a classical biological control agent; after nematodes have been introduced into a plantation, parasitised sirex females emerge from infested trees and disperse the nematodes to trees freshly attacked by unparasitised sirex. As the sirex population builds up over several years, more and more parasitised sirex attack each susceptible tree and this leads to higher and higher percentage parasitism of sirex larvae parented by unparasitised sirex ovipositing on the same tree. Finally the sirex population crashes because once the most susceptible trees have been killed it takes larger numbers of sirex than are available to kill the healthier trees. However, human intervention is necessary for continual control because nematode-parasitised sirex are usually smaller (they have had to compete with the nematode population for fungal food) and don't fly as far as unparasitised sirex so that it is the latter that usually initiate infestations in newly susceptible plantations. (Plantations/compartments do not usually become susceptible until 10–12 years of age). As a result, nematodes must be introduced as soon as sirex is detected in a new area and until nematodes are confirmed to be established.

All is not quite so simple though since it has been important over many years to isolate the best species and strain of nematode and to develop methods for rearing, storing, formulation, inoculation, distribution and quality control.

The free-living, fungal-feeding cycles of *Beddingia* species not only significantly increase the nematodes' ability to find and parasitise their hosts in the natural state, they have greatly facilitated manipulation of these nematodes for the biological control of sirex (Bedding & Akhurst 1974, Bedding 1979, 1984). Using fungal cultures enabled the storage of a large library of cultures of different species and strains of *Beddingia* over many years and also allowed for the development of a method for mass producing the chosen nematode strain for liberation.

12.4.1 Selection of Best Species and Strain for Original Liberations

Throughout the 1960s and early 1970s, CSIRO personnel, Drs. Philip Spradberry and Alan Kirk, and various consultants conducted a comprehensive search, in hundreds of localities from Europe, USA, Canada, India, Pakistan, Turkey, Morocco and Japan, for coniferous trees infested with various siricid species. Thousands of logs from these trees were then caged in quarantine, mainly at Silwood Park, UK, where all emerging insects were investigated for biological control potential. As a result of these collections, seven species of *Beddingia* were found parasitising 19 siricids (associated with two fungal symbionts) and 12 parasitoids from 31 tree species and 29 countries (Bedding & Akhurst 1978).

Beddingia species were dissected from thousands of insects from hundreds of sources. These nematodes could not have survived for more than a few days had it not been possible to culture them on the symbiotic fungi of their hosts. As it was, hundreds of monoxenic nematode cultures were readily established on fungal cultures on potato dextrose agar (PDA) plates and slants in tubes and placed under refrigeration after initial establishment. Most cultures were sub-cultured every 6 months or so and maintained for several years during evaluation, classification and experimentation.

Out of the hundreds of nematode cultures assessed during 1970–1974 for suitability to control sirex in Australia, all cultures of five species were rejected because they parasitised siricids associated with the symbiotic fungus *Amylostereum chaillettii* and could not feed on the symbiont (*A. areolatum*) of *S. noctilio* (Bedding & Akhurst 1978). Of the two remaining species, *Beddingia wilsoni* was rejected because, although it parasitised sirex and fed on both *A. chaillettii* and *A. areolatum*, it also parasitised the beneficial rhyssine parasitoids. *B. siricidicola* remained the chosen species but it was then a matter of selecting which strain would be most suitable for development for biological control.

Several strains of *B. siricidicola* parasitised sirex but did not enter the eggs of their host (Bedding 1972) and were therefore rejected. The many remaining strains were inoculated into hundreds of randomly selected sirex-infested logs to compare rates of parasitism. Four strains (from Corsica, Thasos in Greece, Sopron in Hungary and New Zealand) were found to give nearly 100% parasitism of emerging sirex (Bedding & Iede 2005). Small numbers of each of these strains were liberated in Victoria (where sirex first established on the mainland of Australia) in the early 1970s. However, it was found that sirex parasitised by the 198 strain from Sopron (which was derived from a single parasitised female *S. juvencus*) were significantly larger than those parasitised by the other strains. The size of parasitised sirex females was important, because flight mill studies established that although nematode parasitism itself did not affect the distance flown, size had a major impact. Thus, small sirex could often fly no further than 2 km on flight mills whereas the largest sirex flew up to 200 km. Large sirex also oviposit more and survive longer and so if parasitised are much better at distributing the nematode. For these reasons all liberations (in Australia, South America and South Africa), subsequent to these findings (post 1971), were of the 198 strain.

12.4.2 Mass Rearing B. siricidicola *for Liberation*

It is neither necessary nor feasible to use the parasitic life cycle to produce the millions of nematodes required for liberation. The nematodes can be cultured readily on the symbiotic fungus, *A. areolatum*. Initial cultures are made on potato dextrose agar plates and these are then used as inoculum for 500 ml erlenmeyer flasks containing 90 g wheat or wheat/rice autoclaved in 150 ml water to produce a matrix of sterile swollen grains with air spaces between each grain (Fig. 12.4A). All procedures are conducted under aseptic conditions.

Aseptic fungus is initially isolated by dipping a living sirex female into 100% ethanol and then igniting it before plunging it under sterile water and dissecting out its ooidal glands (found at the base of the sirex's ovipositor) in a laminar flow cabinet. An ooidial gland is then streaked across a PDA plate prior to subculture. Similarly, the testes full of juvenile nematodes (Fig. 12.3B), are removed from a parasitised male sirex and then placed on a 4 day culture of the fungus on PDA.

Once monoxenic cultures are established these can then be readily subcultured onto fresh PDA plates and there should be no need to re-establish cultures from parasitised sirex again. Because the nematodes feed readily only on the growing front of the fungal culture, experience is necessary to determine when to sub-culture and how much inoculum to use; if there are too many nematodes/fungus, the fungus is unable to grow readily, whereas if there are not enough nematodes/fungus, the fungus grows too rapidly and may smother the nematodes and invade the nematode eggs.

Flasks are incubated for between 5 and 8 weeks and are harvested when most fungus has been consumed (see Fig. 12.4A). Harvesting is achieved simply by adding

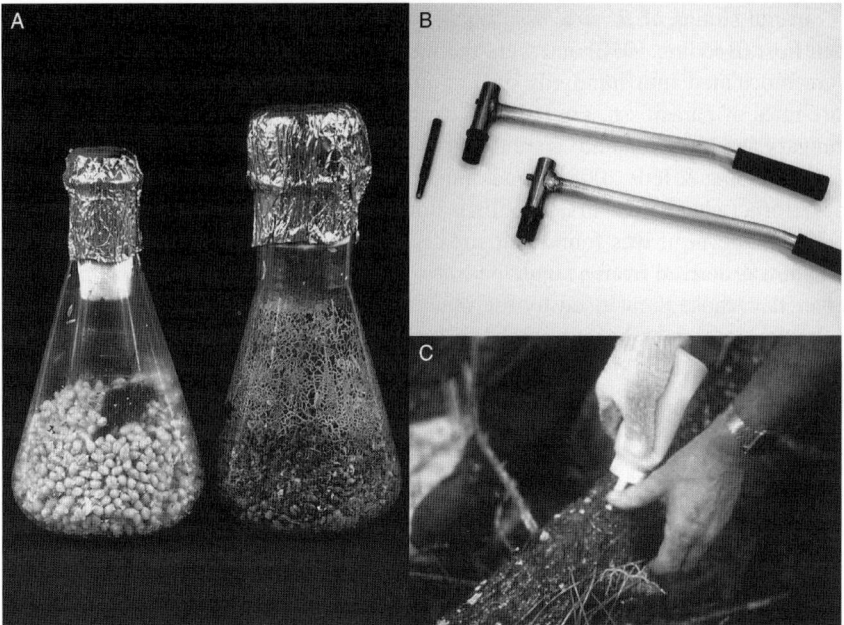

Fig. 12.4 A. Culture flasks just after inoculation with nematode/fungus culture (*left*) and just prior to harvest (*right*). **B.** Wad punch is mounted in a hammer to enable clean cutting of the tree's tracheids to permit nematode entry. **C.** Nematode/gel is squeezed from a sauce bottle into each inoculation hole

tap water to mature flasks to just cover the medium within, leaving for about 15 minutes with occasional gentle swirling and then decanting the resulting nematode suspension through a sieve into bowls to settle for about 20 minutes. This is repeated three times for each flask. After settling, the nematodes are washed by decanting, counted and adjusted to a concentration of 100,000 nematodes per ml. Usually 5 million nematodes are added to each breathable (e.g. Everfresh®) plastic bag; the bags are sealed and then layered between packing foam inside polystyrene boxes together with insulated freezer bricks. Boxes are consigned for next day delivery and packets of finely ground (<600 micron) polyacrylamide gel, allowing 5 g for each million nematodes, are included for final mixing. Detailed standard operating procedures for rearing, harvesting, counting, storing, formulation and quality control have been summarized by Calder & Bedding (2002).

12.4.3 Inoculation of Sirex-Infested Trees

The way in which sirex/*Amylostereum* infected trees are inoculated is particularly important. During initial experiments, holes were drilled into sirex-infested billets and a water suspension of nematodes was added to each hole but this resulted in negligible parasitism in the emerging sirex. It was later found that the drilling resulted in twisted, blocked tracheids (wood tubes) and that the water was rapidly absorbed into the wood leaving the nematodes "high and dry". However, cleanly cutting tracheids to make inoculation holes, using an inoculum of nematodes suspended in gel, inoculating timber of adequate moisture content and using correct inoculum size and spacing resulted in nearly 100% of emerging sirex being parasitized, without the size of these sirex being adversely affected (Bedding & Akhurst 1974).

Currently, inoculation holes are made in felled sirex-infested trees, with frequently sharpened, rebound hammer wad punches (Fig. 12.4B), and the holes are injected with nematodes suspended in 1% < 600 micron, polyacrylamide gel (in Australia and South Africa) or as per Bedding & Akhurst (1974) in foamed, 12% gelatine solution (in South America). Inoculation holes are made about 10 mm deep and every 30 cm. There is one row of holes where the tree diameter is less than 15 cm and two rows of staggered holes where tree diameter is greater than 15 cm. Approximately 2000 nematodes are added per inoculation hole in about 1 ml of gel, which is pressed into the hole using a finger. Usually one operator makes the holes while another dispenses the gel/nematodes (Fig. 12.4C). The spacing of inoculations and numbers of nematodes is important: too many nematodes and/or more closely spaced inoculations result in smaller, less effectively dispersing sirex (nematodes compete with sirex larvae for fungal food) whereas too few nematodes with wider spacing results in lower levels of parasitism. It is possible that parameters may have to be adjusted for different species of host tree and/or climate with, for example, closer inoculations being better where colder climates may reduce the number of generations of mycetophagous cycles per year.

12.4.4 Liberation Strategy

Although nematodes can sometimes be spread from forest to forest by sirex females (Bedding 1979), this is too unreliable and may occur too late in the infestation of a plantation to prevent a serious outbreak. Nematodes are therefore introduced as early as possible after sirex arrives in a plantation by artificially inoculating trees already infested by sirex (as described above) at easily accessible, strategic points within a forest. In Australia and South America, inoculation with *B. siricidicola* is part of a national strategy for sirex control, also comprising quarantine, detection, monitoring and silvicultural control (Haugen *et al*. 1990, Iede *et al*. 2000) and in Australia there are also operations worksheets (National Sirex Co-ordination Committee 2002). The Australian National Strategy and worksheets are currently being updated (N Collett personal communication).

In most Southern Hemisphere countries, sirex has not yet completed its likely geographic spread (current evidence suggests an unassisted spread of about 30 km per annum) and it has only just begun to establish in North America. In addition, since sirex does not usually attack pine trees until they are about 12 years old, there are always plantations or compartments that sirex has only just reached. This situation is determined by aerial and ground surveys and by results from trap tree plots and/or chemical lures. Trap trees (Madden & Irvine 1971) are particularly useful both for monitoring and, if found to be infested by sirex, for inoculation with nematodes. Although trap trees can be more sparsely distributed for monitoring, as a means of introducing nematodes, a trap tree plot of about five trees is established at an easily accessible spot (e.g. road side) for each 20–30 ha plantation compartment. To do this, just enough herbicide (e.g. DiCamba) to almost kill trees is injected at the base of each tree 3 months before expected sirex emergence; this makes the trees highly susceptible to sirex. Trees found to be infested with sirex are then inoculated with nematodes. Ideally this is all that is required to establish nematodes within a compartment. However, where sirex populations are already established in over 1% of trees, it may be necessary to inoculate all infested trees in every fifth row (20% of infested trees) to achieve rapid control. Obviously careful monitoring is therefore particularly important.

12.4.5 Monitoring

Even after nematodes have been liberated in an area it is important to ensure that they are established. To do this, logs collected from sirex-infested trees are caged and emerging sirex are dissected to determine levels of parasitism. Whether nematodes are present in particular sirex-infested trees can also be detected by cutting chips from along the tree and standing these in shallow water for 24 hours. Experience is required to distinguish these nematodes microscopically from other species that may be present and of course chipping cannot determine percentage parasitism of sirex within the tree. Despite these difficulties, chipping has the advantages of being much less labour consuming, it can be conducted on a greater number of trees than caging and it does not deplete the numbers of parasitised sirex emerging and transmitting nematodes to trees newly infested by sirex.

Much work remains to be conducted on exactly what are the most efficient and effective guidelines for introducing nematodes where nematode parasitism is already present but low. Current Australian guidelines are that if no nematode parasitism is detected in a compartment and more than 1% of trees are infested by sirex, 20% of these trees should be inoculated; with 1–5% sirex infected 10% of infested trees should be inoculated, and with 5–10% sirex infected 5% of infested trees should be inoculated. However, where parasitism is higher than 10%, no further inoculation is likely to be worthwhile.

12.5 Success of Nematode Control

Bedding (1993) claimed that nematode parasitism of sirex is density dependent. The density referred to is of the percentage of trees infested by sirex in a given area (but may not apply to the density of sirex larvae within a given tree). Essentially, if nematode parasitism is present in a population, the more sirex there are in a given area, the more oviposition (by both healthy and parasitized female sirex) there will be on each tree. Hopefully this leads to high percentage parasitism and collapse of the sirex population well before a high percentage of trees are killed; this is provided that non-defective nematodes are in place and well distributed when sirex infestation is low.

Because detailed forest assessments are very labour intensive and expensive, the number of case studies has been limited to those reported by Bedding & Iede (2005). Thus, one year after inoculating all infested trees in every tenth row in a 400 ha pine plantation in Mt. Helen, Northern Tasmania, during 1972, parasitism reached 86% and in the following year, extensive ground and aerial surveys revealed no sirex-infested trees. This was despite 5–10% tree death from sirex attack during the several previous years when no nematodes were present.

Nematode treatment was conducted on a much larger scale in the 113,000 ha *Pinus radiata* plantations of the "Green Triangle" area of Victoria/South Australia where, in the absence of nematodes, a total of nearly 5 million trees were killed by sirex between 1987 and 1989 (80% tree death in some areas). Here, in 1987, in an operation costing AU$1.3 million, all infested trees in every fifth row were inoculated with nematodes (Haugen & Underdown 1990). By 1989, nearly 100% of sirex were parasitised, the population crashed and it has been difficult to find any sirex-infested trees there ever since. These results were achieved even though, as was found later, the nematodes used were of the defective strain (see below); presumably it was because of the very high density of sirex that the nematodes were effective nonetheless.

Nematodes were released from 1990 to 1993 in a 12,000 ha plantation in Encruzilhado do Sul, Brazil, where 30% of trees were infested by sirex in some compartments during 1991. This resulted in parasitism of 45% in 1991, 75% in 1992 and 90% in 1994 and it was difficult to find any sirex-infested trees in 1995. Elsewhere in Brazil, parasitism by *B. siricidicola* was evaluated annually in seven localities and ranged from 17, 39, 57 and 65% in four localities and over 92% in three other localities. Such variation also occurs in Australia and seems to be related to the density of sirex-infested trees (Bedding & Iede 2005).

In the Cape Peninsula of South Africa where sirex was first detected in 1994, 296 sirex-infested trees were inoculated in a 90 km arc around Cape Town during 1995–1996. Resulting nematode parasitism increased from 22.6% in 1996 to 54% in 1997 and to 96.1% in 1998, with never more than 3.2% tree death in any forest compartment (Tribe & Cillié 2004). Currently it is difficult to find sirex-infested trees in the Cape region.

12.6 Problems Arising in Nematode Control

12.6.1 Australia

12.6.1.1 Decline in Infectivity

The main problem that has arisen in Australia with nematode control was revealed during the Green Triangle outbreak when, instead of achieving yields of nearly

100% parasitised sirex from inoculated trees as expected from using the methods of Bedding & Akhurst (1974), only about 25% of emerging sirex were parasitised (Bedding & Iede 2005). Although this could have been the result of incorrect procedures having developed during the 15 years that the Victorian Forest Commission had been responsible for inoculations, this was found not to be the case. In fact, it transpired that there had been declining parasitism from inoculated trees over many years and that this was the result of genetic changes in the nematodes used (Bedding 1992). These changes were the result of continual sub culture of the nematodes in the fungal feeding cycle over some 20 years without the intervention of the parasitic cycle.

When fungal cultures of *B. siricidicola* are sub-cultured at an optimal stage and are kept so that there is little accumulation of CO_2 and acidity, there is little tendency for the cultures to produce infective females. However, until the problem of genetic change was discovered, culture plates were often kept for long periods in plastic bags before sub-culturing; this allowed for the accumulation of CO_2 and acidity and this initially stimulated the formation of many infective females on the plates. When these plates were sub-cultured, only those nematodes that had not responded to the CO_2 and acidity levels occurring were able to grow and reproduce, since infective females need to infect sirex to complete their cycle. In other words, there was a continual selection against the tendency to form the infective stage. Gradually this selection produced what Bedding (1993) termed the "defective strain". Unfortunately, decline of the nematodes being liberated occurred over many years and it is not known when the decline started to become important and therefore what areas became "contaminated" by inoculations with the less effective strain as sirex spread across Australia and later South America.

Although it was low levels of parasitism in inoculated logs that drew attention to the problem, of far greater significance was what that meant in terms of the ability of this defective strain of nematodes to control sirex populations once liberated. There is every reason to expect that nematode control with the defective strain may not occur until sirex infestations are severe (perhaps > 10% tree death) whereas the original strain produced high levels of parasitism at very much lower tree death (probably <1%). The defective strain was almost certainly only effective in the "Green Triangle" because of the very high density of sirex infestation (up to 80% tree death) and intensive liberation.

12.6.1.2 Re-Isolation of Original Strain

Once the problem of the defective strain was evident, it was obviously important, but not at all easy, to replace it. To collect and select new strains from overseas would have been a huge task. Even collections from the original location in Sopron might not have yielded a suitable strain since the original strain was from a single parent nematode, was inbred for many generations and was likely to have been subjected to genetic drift before testing; it is possible that this could have led to a strain highly suitable for manipulation but one which, because it is so virulent, would not have survived indefinitely in the field. Most sirex-infested localities within Australia had had recent introductions of defective strain so it was not possible to obtain the uncontaminated original strain except from one area. Dick Bashford from the Tasmanian Forestry Department confirmed that Kamona plantations, near Scottsdale in Northern Tasmania, had not had any nematode introductions since the very first liberations of the 198 Sopron strain in Australia in 1970. In 1991, Bashford was able to find only 9 sirex-infested trees in the whole area but one of these had nematodes. The nematodes were established in monoxenic culture on *A. areolatum*, and a series of test inoculations of sirex-infested logs produced over 95% parasitism compared to 23% with the defective strain (Bedding 1993). The new isolate also produced infective females readily on culture plates exposed to acidity and 10% CO_2. Using randomly amplified polymorphic DNAs (RAPDs), J. Calder found that three primers (OP-A04, OP-X11, OP-FO3), out of over 100 tested, could differentiate between the defective strain and the strain re-isolated from Kamona, even though both strains were derived from the original isolate from Sopron (Bedding & Iede 2005). This new strain, named the Kamona strain (Bedding 1993), has been used for inoculations in Australia since 1991 and was introduced into Brazil and South Africa in 1995.

12.6.1.3 Avoiding Infectivity Decline

Even the Kamona strain had been originally sub-cultured on fungus without intervention of the parasitic phase for several years prior to liberation and was found after re-isolation to lose some infectivity after only 6 months of sub-culture on fungus (Bedding & Iede 2005). To minimise this problem, soon after the Kamona strain was isolated, hundreds of vials of it were stored in liquid nitrogen using the method developed by Bedding (1993). *B. siricidicola* could not be stored successfully in liquid nitrogen using methods developed for *Caenorhabditis elegans* or for entomopathogenic nematodes (Popiel *et al.* 1988, Popiel & Vasquez 1991). To store *B. siricidicola* in liquid nitrogen requires suspending aseptic nematodes in sterile 5% glycerol and then evaporating off water in a laminar flow cabinet over several days to achieve nematodes suspended in 50% glycerol before vials, each containing 300 µl of this suspension, are plunged directly into liquid nitrogen. *B. siricidicola* treated in this way have so far survived for 15 years with over 75% viability. Each year, a single vial of original stock is rapidly thawed under running warm water and the resulting suspension added aseptically to young cultures of the symbiotic fungus, *A. areolatum*, on $1/4$ strength potato dextrose agar; sub-cultures are from young parent cultures kept in paper bags (to reduce CO_2 accumulation) and mass cultures are made before any decline in infectivity. As soon as the first mass cultures are mature, some are harvested and the resulting nematodes stored in vials in liquid nitrogen. This allows for possible (though unlikely) deterioration of the original stocks and ensures that for the foreseeable future there are stocks of the material originally stored, or at least of material having had only a few generations of further sub-culturing. Material stored in liquid nitrogen is maintained in three separate locations.

12.6.1.4 Testing for Infectivity

Laboratory cultures or isolates obtained from the field can be tested for their ability to form infective females (which in turn reflects levels of parasitism of sirex that will be obtained from logs inoculated with these isolates). Nematode eggs, harvested under sterile conditions from the mycetophagous cultures of *B. siricidicola*, are placed on *A. areolatum* growing on 0.2% lactic acid/PDA plates inside desiccators containing 10% CO_2. After 10–12 days, the resulting adults are washed off and counted as mycetophagous or infective females. With the defective strain there will be hardly any infective females, whereas with the Kamona strain most adult females will be infective forms.

12.6.1.5 Replacing Defective Strain *B. siricidicola* in the Field

The defective strain is particularly pernicious. R. A. Bedding and J. Calder found that at least six back crosses between Kamona and defective strains (pure Kamona crossed back to progeny of the previous cross) were required before the final hybrids were fully infective (Bedding & Iede 2005). Therefore, it has been an on-going priority to re-introduce Kamona repeatedly into the forests of Victoria, South Australia and southern New South Wales until it can be shown by RAPDs and infectivity tests that the Kamona strain dominates.

12.6.1.6 Temperature Effects

It appears that sirex is well under control in Australia although it is now close to the border of Queensland and has not yet reached the state of West Australia. Areas of pine plantation in both these states as well as northern NSW can be particularly hot in the summer so that sirex-infested trees could heat up enough to kill or disrupt any nematodes breeding within. Thus, Akhurst (1975) found that even at 27.5°C, there was an abortion rate of eggs of 95.9% although after culture at this temperature for 3 weeks, this rate dropped to 15%. Whether temperature adaptation was physiological or genetic is unknown. 30°C was lethal to all stages after initial rearing at 24°C. Although temperatures are buffered within the tree, it is likely that temperatures exceeding 30°C could occur within a tree although pockets of lower temperature might allow survival at some points. Low temperatures that occur in some areas of Australia are unlikely to significantly impact parasitism since high levels of parasitism have been found in Tasmanian forests which are amongst the coldest. However, Akhurst (1975) found that there was negligible egg hatch at 5°C and, even at 10°C, eggs took 13 days to hatch compared to 3 days at 25°C. Considerably more work is needed on the effects of temperature on the life cycle of *B. siricidicola* and this has commenced in Queensland (M Ramsden personal communication).

12.6.1.7 Other Problems

The use of trap trees is an integral part of monitoring and initial introduction of nematodes into new areas. Increasingly, bark beetles are infesting trap trees and rendering them unsuitable for sirex attack. A solution to this situation is being addressed by a study on the use of bark beetle repellents (A Carnegie personal communication).

Particularly in Queensland, there are a number of plantations of *Pinus carabea* and *Pinus carabea* x *Pinus elliottii* hybrids and because sirex has not reached Queensland yet, it has not yet been possible to determine whether this tree species has any effect on the nematode parasitism of sirex. Aspects such as tracheid and bordered pit diameter, resin content, fungal growth rates and production of toxic metabolites of different tree species might have some effect on nematode development and migration within infested trees.

12.6.2 South America

Before it was appreciated that the 198 Sopron strain of *B. siricidicola* had become defective, defective nematodes were introduced into Brazil from Australia in 1989. It was not until 1995 that the Kamona strain was liberated there and so there are likely to be residual populations of defective strain in Brazil and possibly in other South American countries. Kamona strain is not frozen in liquid nitrogen but cultures (hopefully of pure Kamona strain) are re-isolated periodically from the field (ET Iede personal communication). There may be a problem with this approach because, firstly, such isolates could be contaminated with the defective strain and, secondly, the number of generations that the "wild nematodes" will have been cultured in the laboratory will tend to increase each year, while only one parasitic cycle will have occurred. Nematodes isolated from infected sirex in Brazil in 1995 were sent to Argentina, and later cultures, presumably of the Kamona strain, were sent to Uruguay and Chile (Hurley *et al.* 2007).

12.6.3 South Africa

The situation in South Africa is particularly interesting. After sirex was discovered, in stands of *P. radiata* in the Cape Peninsula in 1994, Kamona strain nematodes were introduced and parasitism reached over 96% within 3 years (Tribe & Cillié 2004). However, by 2002 sirex had spread north to the *Pinus patula* plantations of KwaZulu-Natal and by 2007 had killed over 1.5 million trees in 30,000 hectares (Dyer 2007). When sirex-infested trees were inoculated in this region in 2004 and 2005, less than 10% parasitism was recorded from sirex emerging from those trees (Hurley *et al.* 2007) and trees inoculated on Sappi landholdings yielded only 2.1% parasitism in 2004, 9.3% in 2005 and 8.5% in 2006 (Verleur 2007).

Exactly why there should be this huge discrepancy between the Cape (and everywhere else in the Southern Hemisphere) and KwaZulu-Natal has not yet been fully elucidated although there are a number of probable contributing factors. These include rapid drying of infested trees, *P. patula* as the main tree species, incorrect inoculation procedures, poor quality nematodes, nematodes not adapted to the KwaZulu-Natal strain of *A. areolatum*, death of parasitised sirex prior to emergence, interactions between fungus and *P. patula* unfavourable to the nematodes, and, poor initial spread of fungus in infested trees so that nematodes are isolated from the sirex larvae prior to trees drying.

Unlike the Cape, which has winter rainfall, KwaZulu-Natal has summer rainfall and dry warm winters (frequently over 20°C) and this results in sirex-infested trees drying rapidly. Bedding & Akhurst (1974) demonstrated that *B. siricidicola* could not migrate and therefore infect sirex in wood with a water content of less than 30% ($=$ 50% Australian forestry formula). This would certainly be one factor leading to inoculated trees having low parasitism, particularly if trees were inoculated at the same time as in Australia. (Sirex in KwaZulu-Natal emerges at least 2 months earlier than in Australia and therefore infested trees should be inoculated 2 months earlier). When nematodes are introduced by parasitised female sirex, this would occur when the trees had not yet begun to dry and so resulting parasitism should be higher. This is the case, since Verleur (2007) found that female sirex were 54% parasitised in the basal third (the moistest part) of naturally struck trees. It is of course quite possible that parasitism in naturally struck trees could build up much higher as there is increasing parasitism in the wild population (as it does in other regions). In that case, the low levels of parasitism resulting from inoculations could be no more than an expensive nuisance. Whether drying of timber is the only factor producing low levels of parasitism is doubtful since even where moisture levels are much higher than the minimum, parasitism is nowhere near what is to be expected elsewhere.

It appears unlikely that *P. patula* as the tree species could be an important factor since Zondag (1966) reported the first finding of *B. siricidicola*, as heavily parasitising sirex in this tree species: "A nematode disease of *S. noctilio* which caused 95% infection of the adults was discovered in January 1962 in a *Pinus patula* stand in Rotoehu, S. F." Here, a total of 288 males were dissected and 277 were found parasitised (96.2%), while of 81 females, 75 were parasitised (92.6%). Even if tree species is involved in low parasitism then it might be because of some unknown reaction with the South African strain of *A. areolatum*.

That low parasitism from inoculated trees could be operator error or poor quality nematodes was demonstrated to be very unlikely when the operators concerned achieved high levels of parasitism after inoculating sirex-infested *P. radiata* in the Cape Peninsula with nematodes sent from Australia by the same supplier (M Verleur personal communication).

As described below for the USA, strains of *A. areolatum* can vary considerably in their growth rates and suitability as food for *B. siricidicola*. There seems to be no problem with nematodes on this symbiont in the Cape and DNA analysis indicates that *A. areolatum* from KwaZulu-Natal is the same as that from the Cape. However, that does not definitely preclude there being a difference in suitability as food for

B. siricidicola (B Slippers personal communication). Should the fungal strain prove to be a problem, repeated rearing of the Kamona strain on it should lead to better reproduction of the nematode as has been observed for various strains of *B. siricidicola* on various fungal isolates by the author (RA Bedding unpublished data).

12.6.4 North America

The Kamona strain of *B. siricidicola* has been imported into the USA (Williams et al. 2007) and results from recent inoculation of logs are pending (D Williams personal communication). In the meantime, and not connected to these inoculations, nematodes have been found parasitizing sirex both in New York state and in southern Canada; these nematodes most likely entered North America with sirex, as occurred in New Zealand. Yu Qing of the Canadian Forest Service has found 30% of several hundred sirex to be parasitised by a *Beddingia* species (Q Yu personal communication), probably *B. siricidicola*, which can be distinguished from the Kamona strain using ITS sequencing (I Leal personal communication). A problem is that the nematodes did not appear to have entered the sirex's eggs (Q Yu personal communication). This means that this particular nematode would have little effect in controlling sirex but it could also have much more serious implications. While failure to enter eggs can be dependent on the strain of *B. siricidicola* concerned, as found in one isolate from New Zealand (Zondag 1975), it can also be a result of the strain of sirex, as found in Belgium (Bedding 1972). In the former case it is possible, although perhaps a little difficult, to replace the nematode with Kamona strain. If the latter, it would make the control of sirex with nematodes unlikely even after extensive searching for new strains. Whether juvenile nematodes enter the eggs of their host is dependent on when the nematodes are released by the parent nematodes into the sirex haemocoel in relation to when the sirex hardens its egg shells (see earlier). The Kamona strain is so "potent" that few eggs are produced by Australian sirex, these are much smaller than normal and juvenile nematodes fill the eggs well before shell hardening; the chances of finding an even more "potent" nematode are slim. However, if required it would be well worth testing the easily obtainable *B. siricidicola* strain from New Zealand, which gave high levels of parasitism in the inoculation trials mentioned above and has a similar effect on sirex's reproductive system.

The symbiotic fungus found in North American sirex is certainly *A. areolatum* but is a different strain from that on which the Kamona nematode is grown in Australia; it not only grows much more slowly than the Australian fungus, but *B. siricidicola* reproduces much more slowly on it (D Williams personal communication). This suggests that the Kamona strain would have some difficulty establishing in sirex-infested trees in North America. Currently the Kamona strain is being passed through many generations on USA fungus (D Williams personal communication) and this should ameliorate the situation.

Even more so than in Australia, using trap trees to introduce nematodes into plantations will be difficult because of a wide variety of bark beetles and other insects that will attack the trees before sirex does; it may therefore be necessary to inoculate naturally-infested trees *in situ* or distribute inoculated, sirex-infested logs, from areas where sirex is common to where it is not.

Both Canada and parts of USA are subject to extremely low temperatures during winter months and it is not yet known what effect this will have on levels of parasitism or even on nematode survival (Williams *et al.* 2007). Obviously there will be no nematode life cycles for a significant part of the year so that total nematode multiplication and migration within the tree will be at least curtailed; to what extent this may affect final levels of parasitism remains to be seen.

12.7 Discussion

Controlling *Sirex noctilio* with nematodes has become much more complicated with the recent arrival of this insect into Canada and the USA, significant problems with nematode biological control in the KwaZulu-Natal region of South Africa and the possibility of sirex eventually establishing in a further 18 countries (Carnegie *et al.* 2006). Without proper nematode control sirex populations are likely to flourish and that can result in billions of dollars of damage to commercial pine plantations around the world.

Although *B. siricidicola* is a highly successful biological control agent it requires continual management and monitoring to be successful. This is particularly so as sirex is still spreading geographically in almost every country into which it has been introduced as well as into new plantations as they reach an age of 10–12 years. In Australia and South America there is also the problem of replacing any defective strain released earlier and in Canada in replacing the strain that entered the country with sirex. Because of this it is important for each country to have a national sirex coordination committee comprised of forest managers and appropriate scientists and a national sirex strategy that is continually updated as necessary.

Australia's national strategy (Haugen *et al.* 1990) and worksheets (National Sirex Co-ordination Committee 2002) are a good beginning but different climatic conditions, tree species and forestry practices will doubtless require various modifications to be made, which may even vary for different regions within a country. Thus, time of trap tree establishment and nematode inoculation should obviously be much earlier in Brazil and KwaZulu-Natal where sirex emergence is at least 2 months earlier than in Australia and the South African Cape peninsula. The spacing of inoculations and the number of nematodes introduced in each inoculation was carefully worked out to achieve nearly 100% parasitism, but also to result in large parasitised sirex, for sirex-infested *P. radiata* in Tasmania (Bedding & Akhurst 1974). However, for different tree species, where trees dry more rapidly or where long cold winters reduce the possible number of free-living generations, more nematodes in more closely-spaced inoculations may be necessary to achieve the same result and this will need to be determined by experimentation. In areas such as KwaZulu-Natal where infested *P. patula* dry out very rapidly, it may even be necessary to inoculate sirex-infested material from areas where this is plentiful, as early as possible, and then collect the inoculated logs at central points for periodical water spraying prior to distributing them where required.

It would be tempting to consider the introduction of new or multiple strains of *B. siricidicola*, perhaps better adapted to certain forest conditions in a particular area. This should be avoided if at all possible. Firstly, the Sopron, and later the Kamona, strains were carefully selected over many years from hundreds of isolates from all over the world at a huge cost (these strains are thought to be equivalent but were isolated from the field at different times and sites). Secondly, most natural strains are likely to be adapted so that they do as little harm to their host's population as possible, which is not what is required. Finally, once an inferior strain has been released it can be very difficult if not impossible to replace it. Nevertheless, if the Sopron and Kamona strains do not enter the eggs or do not significantly reduce egg numbers of a particular strain of sirex (that has been introduced to a region/country or has arisen from mutation of normal sirex) then there is no alternative but to look for another nematode strain that does. The development of such a strain of sirex from "normal" sirex is of course a possibility and that would become a major problem. However, it is more likely that an ineffective nematode strain would develop since nematodes have many generations each year compared with usually only one for sirex.

There is a still a considerable body of applied research required to make nematode control fully effective in the various regions where sirex is or is likely to become a problem. These include the effect of temperature, tree species, fungal interaction, effect of other insects, inoculation strategies for different situations, trap tree protection and molecular probes for sirex and nematode strains.

Presumably *S. noctilio* will remain a major pest of pine trees throughout the world for the foreseeable future but nematodes should be an effective means of controlling it for hundreds of years to come, provided the situation is continually monitored and managed appropriately.

References

Akhurst RJ (1975) A study of the free-living phase of *Deladenus*, nematodes parasitic in woodwasps. MSc Thesis, University of Tasmania

Bedding RA (1967) Parasitic and free-living cycles in entomogenous nematodes of the genus *Deladenus*. Nature London 214:174–175

Bedding RA (1968) *Deladenus wilsoni* n.sp. and *D. siricidicola* n.sp. (Neotylenchidae), entomophagous nematodes parasitic in siricid woodwasps. Nematologica 14:515–525

Bedding RA (1972) Biology of *Deladenus siricidicola* (Neotylenchidae), an entomophagous nematode parasitic in siricid woodwasps. Nematologica 18:482–493

Bedding RA (1975) Five new species of *Deladenus* (Neotylenchidae), entomophagousmycetophagous nematodes parasitic on siricid woodwasps. Nematologica 20:204–225

Bedding RA (1979) Manipulating the entomophagous-mycetophagous nematode, *Deladenus siricidicola* for the biological control of the woodwasp *Sirex noctilio* in Australia. In: Waters WE (ed) Current topics in forest entomology. USDA Forest Service Gen Tech Pap WO-8. pp 144–147

Bedding RA (1984) Nematode parasites of Hymenoptera. In: Nickle WR (ed) Plant and insect parasitic nematodes. Marcel Dekker, New York. pp 755–795

Bedding RA (1992) Strategy to overcome the crisis in control of sirex by nematodes. Austral For Grower Summer 1991/92:15–16

Bedding RA (1993) Biological control of Sirex noctilio using the nematode *Deladenus siricidicola*. In: Bedding RA, Akhurst RJ, Kaya HK (eds) Nematodes and the biological control of insect pests. CSIRO Publ, East Melbourne, Australia. pp 11–20

Bedding RA, Akhurst RJ (1974) Use of the nematode *Deladenus siricidicola* in the biological control of *Sirex noctilio* in Australia. J Aust Entomol Soc 13:129–135

Bedding RA, Akhurst RJ (1978) Geographical distribution and host preferences of *Deladenus* species (Nematoda: Neotylenchidae) parasitic in siricid woodwasps and associated hymenopterous parasitoids. Nematologica 24:286–294

Bedding RA, Iede ET (2005) Application of *Beddingia* for Sirex wood wasp control. In: Grewal PS, Ehlers R-U, Shapiro-Ilan DI (eds) Nematodes as biological control agents. CABI Publ, Cambridge, MA. pp 385–399

Blinova SL, Korenchenko EA (1986) *Phaenopsitylenchus lacicis* g.n. and sp.n. (Nematoda: Phaenopsitylenchidae fam. N.) parasite of *Phenops guttulata* and remarks on taxonomy of nematodes of the superfamily Sphaerularioidea. Acad Nauk SSSR Trudy Gelmint Lab 34:14–23

Calder J, Bedding RA (2002) Standard operating procedures for production, experimental analysis and quality assurance testing of *Beddingia siricidicola*. National Sirex Co-ordination Comm, Canberra, Australia

CanFI2001 Database (2007) Canadian Forest Service, Natural Resources Canada, http://cfs.nrcan.gc.ca/subsite/canfi/index-canfi [accessed March 2008]

Carnegie AJ, Waterson DG, Eldridge RH (2005) The history and management of sirex wood wasp, *Sirex noctilio* (Hymenoptera: Siricidae), in pine plantations in New South Wales, Australia. NZ J For Sci 35:3–24

Carnegie A J, Matsuki M, Haugen DA, Hurley BP, Ahumada R, Klasmer P, Sun J, Iede ET (2006) Predicting the potential distribution of *Sirex noctilio* (Hymenoptera: Siricidae), a significant exotic pest of *Pinus* plantations. Ann For Sci 63:119–128

Coutts MP (1969a) The mechanism of pathogenicity of *Sirex noctilio* on *Pinus radiata*. 1, Effects of the symbiotic fungus *Amylostereum* sp. (Thelophoraceae). Aust J Biol Sci 22:915–924

Coutts MP (1969b) The mechanism of pathogenicity of *Sirex noctilio* on *Pinus radiata*. 2, Effects of *S. noctilio* mucus. Aust J Biol Sci 22:1153–1161

Dyer D (2007) An overview of the national *Sirex* control strategy in South Africa. Internatl Sirex Symposium, Pretoria, South Africa 9–16 May 2007. p 36

Gaut IPC (1969) Identity of the fungal symbiont of *Sirex noctilio*. Aust J Biol Sci 22:905–914

de Groot P (2007) An overview of the *Sirex noctilio* situation in Canada. Internatl Sirex Symposium, Pretoria, South Africa 9–16 May 2007. p 30

Haugen DA, Underdown MG (1990) *Sirex noctilio* control program in response to the 1987 Green Triangle outbreak. Aust Forest 53:33–40

Haugen DA, Bedding RA, Underdown MG, Neumann FG (1990) National strategy for control of *Sirex noctilio* in Australia. Austral For Grower 13 No 2

Hoebeke ER, Haugen DA, Haack R (2005) *Sirex noctilio*: Discovery of a Palearctic siricid woodwasp in New York. Newsl Mich Entomol Soc 50:24–25

Hurley BP, Slippers B, Hatting HJ, Croft PK, Wingfield MJ (2007) The *Sirex* control program in the eastern parts of South Africa: lessons from research efforts between 2004–2006. Internatl Sirex Symposium, Pretoria, South Africa 9–16 May 2007. p 38

Iede ET, Penteado S do RC, Schaitza EG (1998) *Sirex noctilio* in Brazil: detection, evaluation, and control. In: Proc Conference Training in the control of *Sirex noctilio* by the use of natural enemies, Colombo, Brazil. pp 45–52

Iede ET, Klasmer P, Penteado S do R.C (2000) *Sirex noctilio* in South America: distribution, monitoring and control. In: XXI Internatl Congr Entomol. Foz do Iguacu, Aug. 2000. Embrapa Soja Anais, Londrina-PR. p 474

Iede ET, Penteado S do RC, Filho WR (2007) The woodwasp *Sirex noctilio* in Brazil – Monitoring and control. Internatl Sirex Symposium, Pretoria, South Africa 9–16 May 2007. p 20

Madden JL (1968) Physiological aspects of host tree favourability for the woodwasp, *Sirex noctilio* F. Proc Ecol Soc Aust 3:147–149

Madden JL (1974) Oviposition behaviour of the woodwasp, *Sirex noctilio* F. Aust J Zool 22:341–351

Madden JL (1977) Physiological reactions of *Pinus radiata* to attack by the woodwasp, *Sirex noctilio* F. (Hymenoptera: Siricidae). Bull Ent Res 67:405–426

Madden JL, Irvine CJ (1971) The use of lure trees for the detection of *Sirex noctilio* in the field. Aust Forest 35:164–166

National Sirex Co-ordination Committee (2002) National Sirex Control Strategy Operations Worksheets

Poinar Jr GO, Jackson TA, Bell NL, Wahid MB (2002) *Elaeolenchus arthenonema* n. g., n. sp. (Nematoda: Sphaerularioidea: Anandranematidae n. fam.) parasitic in the palm-pollinating weevil **Elaeidobius kamerunicus** Faust, with a phylogenetic synopsis of the Sphaerularioidea Lubbock, 1861. System Parasitol 52:219–225

Popiel I, Vasquez EM (1991) Cryopreservation of *Steinernema carpocapsae* and *Heterorhabditis bacteriophora*. J Nematol 23:432–437

Popiel I, Holtemann KD, Glazer I, Womersley C (1988) Commercial storage and shipment of entomogenous nematodes. International Patent No. WO 88/011344

Remillet M, Laumond C (1991) Sphaerularioid nematodes of importance in agriculture. In: Nickle WR (ed) Manual of agricultural nematology. Marcel Dekker, New York. pp 967–1024

Riding IL (1970) Microvilli on the outside of a nematode. Nature Lond 226:179–180

Schneeberger NF (2007) Response to the recent find of *Sirex noctilio* in the United States. International Sirex Symposium, Pretoria, South Africa 9–16 May 2007. p 20

Spadberry JP, Kirk AA (1978) Aspects of the ecology of siricid woodwasps (Hymenoptera: Siricidae) in Europe, North Africa and Turkey with special reference to the biological control of *Sirex noctilio* F. in Australia. Bull Ent Res 68:341–359

Sutherst RW, Maywald GF, Yonow T, Stevens PM (1999) CLIMEX: predicting the effects of climate on plants and animals. CSIRO Publ, Melbourne, Australia

Tribe GD, Cillié JJ (2004) The spread of *Sirex noctilio* Fabricius (Hymenoptera: Siricidae) in South African pine plantations and the introduction and establishment of its biological control agents. African Entomol 12:9–17

Verleur M (2007) Validation for mass inoculations with *Beddingia siricidicola*, despite apparent low inoculation generated parasitism rates. International Sirex Symposium, Pretoria, South Africa 9–16 May 2007. p 40

Williams D, Mastro V, Downer K (2007) Establishment of *Beddingia siricidicola* for biological control of *Sirex noctilio* in the United States: Questions, issues, and challenges. International Sirex Symposium, Pretoria, South Africa 9–16 May 2007. p 32

Wood MS, Stephens NC, Allison B, Howell CI (2001) Plantations of Australia – a report from the National Plantation Inventory and the National Farm Forest Inventory (abridged version). National Forest Inventory, Bureau of Rural Sciences, Canberra

Zondag R (1962) A nematode disease of *Sirex noctilio* (F.). Interim Res. Rep., New Zealand Forest Service 1–6

Zondag R (1966) Observations on a nematode disease of *Sirex noctilio* (F.). NZ For Res Inst, For Entomol Report No 19

Zondag R (1975) A non-sterilising strain of *Deladenus siricidicola*. NZ For Res Inst Report 1974. Rotorua, NZ. pp 51–52

Chapter 13
Fire Ant Control with Entomopathogens in the USA

David H. Oi and Steven M. Valles

Abstract Fire ants are stinging invasive ants from South America that infest over 129.5 million hectares in the southern United States, where eradication is no longer considered possible. The biological control of fire ants, especially by pathogens, is viewed by some as the only sustainable tactic for suppression. Microscopic-based surveys conducted in South America during the 1970s and 1980s led to the discovery of fungi and microsporidia infecting fire ants. Three of these microorganisms have been studied extensively: *Beauveria bassiana* 447, *Thelohania solenopsae*, and *Vairimorpha invictae*. *B. bassiana* 447 causes fire ant mortality but infections do not spread to queens and intercolony transmission was not evident. *T. solenopsae* has been found in the US and has been shown to spread naturally and debilitate colonies. Colony decline has also been associated with *V. invictae*, which is currently being evaluated for host specificity and possible release in the US. Through the use of molecular techniques, viruses infecting fire ant in the US have been discovered and characterized. *Solenopsis invicta* virus-1 can be transmitted easily to uninfected colonies and colony death often results. This virus apparently causes persistent, asymptomatic infections that actively replicate when the host is stressed. Research on fire ant-specific microsporidia and viruses, as well as other fire ant entomopathogens, is summarized to illustrate the efforts that have been undertaken to understand the biology of these pathogens and to facilitate their utilization in biological control of fire ants.

13.1 Introduction to Fire Ants

Fire ants are stinging invasive ants from South America that plague over 129.5 million hectares in the southern United States (US). "Fire ants" is a name that most commonly refers to *Solenopsis invicta*, which has an official common name in the

---

D.H. Oi
USDA, Agricultural Research Service, Center for Medical, Agricultural and Veterinary Entomology, 1600 SW 23rd Dr., Gainesville, Florida 32608 USA
e-mail: david.oi@ars.usda.gov

US of "red imported fire ant". In addition to *S. invicta*, the names "fire ant" and "imported fire ant" also refer to *Solenopsis richteri*, the black imported fire ant. These closely related species belong to the *Solenopsis saevissima* species complex and even hybridize where they co-occur. Both species were inadvertently introduced separately into the US (ca 1918 and 1933 for *S. richteri* and *S. invicta*, respectively (Tschinkel 2006)). *S. invicta* is the most prevalent species in the US, mainly occurring in the southeastern states with its northern limits in North Carolina, Tennessee, Arkansas, and Oklahoma. This species is now of worldwide concern with infestations reported from Australia, mainland China, Hong Kong, Taiwan, and Mexico (Tschinkel 2006, Sánchez-Peña et al. 2005).

The fire ant inflicts a painful, burning sting and frequently a person will receive numerous stings simultaneously when ants swarm out of their nest to attack an intruder. This greatly intensifies the pain and can cause panic. In addition, it is conservatively estimated that 1% of individuals stung in the US are susceptible to becoming allergic to the venom and at risk for anaphylaxis (Triplett 1976). Deaths from fire ant stings have been reported, and lawsuits have resulted in awards of over $US1 million.

The annual economic impact of fire ants in the US is estimated to be over $US6.5 billion across both urban and agricultural sectors (Pereira et al. 2002). In addition, the dominance of fire ants in natural ecosystems has reduced biodiversity and harmed wildlife (Wojcik et al. 2001). Given the tremendous impact that fire ants have had in the US, incursions into previously non-infested areas have instigated very expensive eradication programs. The cost of a planned, but aborted, 10-year eradication program in California was valued at $US65.4 million (Jetter et al. 2002). The current eradication program in Australia cost more than $US144 million over 7 years (2001–2007, McNicol 2006).

In the southern US, eradication is no longer considered possible, and instead, integrated pest management (IPM) for fire ants is encouraged. While toxicant-based fire ant baits are the major component of fire ant IPM, the inclusion of parasites and pathogens as biological control agents is increasing. The biological control of fire ants, and especially pathogens, is viewed by some as the only sustainable tactic for suppression of the ubiquitous fire ants. In this chapter we discuss the discovery and use of entomopathogens for fire ant control, from the early surveys utilizing microscopy to the more recent use of molecular techniques to advance microbial control of this notorious invasive pest.

13.2 Past and Present Fire Ant Microbial Control Projects

13.2.1 Natural Enemy/Pathogen Surveys

Classical biological control, that is the release, establishment and spread of effective natural enemies of a pest, is one approach that offers the possibility for permanent regional suppression of fire ants. The arrival of exotic species into new continents often occurs without introduction of the natural enemies associated with exotics in their areas of endemism. For the red imported fire ant, over 35 natural enemies have been identified in South America (Williams *et al.* 2003) compared to about seven in the US (Collins & Markin 1971, Cook *et al.* 1997, Valles *et al.* 2004). The lack of effective natural enemies can allow exotic species to attain much higher population densities in newly invaded regions than in their native homelands. Accordingly, fire ant populations in the US are generally 5–10 times higher than in South America (Porter *et al.* 1997).

Interest in natural enemies of both *S. invicta* and *S. richteri* extended back into the era when chemical eradication or control using chemicals was being emphasized (Williams *et al.* 2001). In the 1960s Silveira-Guido *et al.* (1973) conducted extensive studies on the parasitic ant *Solenopsis (Labauchena) daguerrei*, which was found on several *Solenopsis* species in South America. Other parasites, such as *Pseudacteon* phorid flies and *Orasema* eucharitid wasps, have been reported from fire ants. Among these parasitoids, several species of *Pseudacteon* flies have been released as classical biological control agents in the US. Surveys for pathogens of fire ants in South America and the southeastern US have been conducted since the 1970s with several microorganisms eventually being evaluated as biocontrol agents (Williams *et al.* 2003). In 1971 and 1973 surveys of the *Solenopsis saevissima* fire ant complex were conducted in the state of Mato Grosso in Brazil by researchers from the University of Florida. The fungus *Metarhizium anisopliae* and the microsporidium *Thelohania solenopsae* were isolated from *S. invicta* collected in these surveys (Allen & Buren 1974, Knell *et al.* 1977). In contrast to *M. anisopliae*, a cosmopolitan entomopathogenic fungus, the *T. solenopsae* infection was considered to be the first report of a microsporidian infection in ants (Allen & Buren 1974). Subsequent surveys in Uruguay and Argentina in 1974 found *T. solenopsae* in *S. richteri* (Allen & Silveira-Guido 1974). More extensive surveys in Paraguay, Argentina, Uruguay, and Brazil during 1975 and 1976 by researchers from the US Department of Agriculture, the University of Florida, and Ohio State University detected *T. solenopsae* in several other fire ant species in the *S. saevissima* complex (Allen & Knell 1980). Other surveys by USDA researchers, conducted in Brazil in 1976, 1979, and 1981, detected virus-like particles (Avery *et al.* 1977), a neogregarine similar to *Mattesia geminata*, a spore-forming bacterium, and a dimorphic fungus (Jouvenaz *et al.* 1980, 1981). In addition, another microsporidium infecting *S. invicta*, *Vairimorpha invictae*, was discovered and described (Jouvenaz & Ellis 1986), as well as a nematode, *Tetradonema solenopsis* (Nickle & Jouvenaz 1987). Both of these organisms killed infected *S. invicta*, but only *V. invictae* has been evaluated as a biological control agent (see Section 13.2.4). In the mid-1980s, isolates of entomopathogenic fungi from ants collected in Mato Grosso, Brazil were screened for pathogenicity to fire ants. One isolate, *Beauveria bassiana* 447, was selected for laboratory culturing and testing in the US and eventually patented (Stimac & Alves 1994, see Section 13.2.2). In 1987 a project was established at the USDA South American Biological Control Laboratory in Argentina that focused on the ecology of pathogens and other natural enemies of fire ants. This project continues today and has been instrumental in furthering research toward utilization of fire ant pathogens in the US, as exemplified by an updated pathogen survey and the discovery of a new nematode (Briano *et al.* 2006, Poinar *et al.* 2007).

The search for fire ant (i.e. *S. invicta* and *S. richteri*) pathogens also occurred in the US with limited surveys in 1971 and 1972 in northern Florida and in Mississippi, and a more extensive pathogen survey of fire ants in the southeastern US was completed in 1977. These surveys found only ubiquitous, nonspecific organisms, with the exception of a mildly pathogenic fungus, infecting *S. invicta* (Broome 1974, Jouvenaz *et al.* 1977, 1981). In anticipation of *S. invicta* entering western Texas, a survey of microorganisms in ants in this area was conducted 1978 and 1979, with only one ant from over 2,500 nests being infected with an entomophthoralean fungus, indicating a dearth of ant pathogens in this area (Beckham *et al.* 1982). Interest in finding fungal pathogens continued in 1989 with the screening of colony founding *S. invicta* queens (ca. 1000) collected after a mating flight in Texas. *Metarhizium anisopliae* var. *anisopliae* and a *Conidiobolus* species were isolated from some queens with the *M. anisopliae* isolate exhibiting pathogenicity in a laboratory bioassay (Sánchez-Peña & Thorvilson 1992). Ants collected in fire ant studies in 2000–2003 in Alabama, Florida, and Tennessee resulted in the discoveries of a new *Mattesia* sp. (Pereira et al. 2002, Valles & Pereira 2003a), and the fungi *Myrmicinosporidium durum* (Pereira 2004) and possibly *Akanthomyces* sp. (RM Pereira personal communication) infecting *S. invicta*. *Mattesia* sp. infection, designated as yellow-head disease, was associated with mortality in *S. invicta*, but thus far disease transmission has not been accomplished (Pereira *et al.* 2002).

Molecular biology has provided a new approach for searching for fire ant pathogens. Variants of the intracellular bacteria *Wolbachia* have been confirmed in *S. invicta* populations from the US and South America using *Wolbachia*-specific gene sequences (Shoemaker *et al.* 2003, Bouwma *et al.* 2006). However, the impact of *Wolbachia* on the fitness of *S. invicta* colonies has yet to be documented, with a single variant being examined and its effect insignificant (Bouwma & Shoemaker 2007). Other bacteria have been isolated from *S. invicta* midguts and characterized by 16s rRNA gene analysis and sequencing to potentially provide a vehicle for introducing genes into fire ants for their control (Li *et al.* 2005). The discovery and characterization of SINV-1 and other new viruses in fire ants (Valles *et al.* 2004) also illustrate how molecular techniques can enhance the search for pathogens (see Section 13.3).

Besides the surveys of fire ants to search for potential pathogens, various formulated entomopathogens and entomogenous nematodes originally isolated from other insects have been tested on fire ants. In general, these formulations require direct contact with ants to obtain infection and, in theory, the infection would spread to other colony members and possibly other colonies. However, the removal of cadavers or unhealthy colony members from nests, grooming behavior, antimicrobial secretions, and relocation of nesting sites limits the spread of infections (Oi & Pereira 1993). Applications of nematodes to fire ants resulted in excessive grooming and applications to nests often caused colonies to relocate with minimal colony reductions (Drees *et al.* 1992, Jouvenaz *et al.* 1990).

Bypassing these behaviors should facilitate the spread of an effective pathogen, thus pathogens isolated from naturally infected fire ants conceivably would have more transmission potential. In the following sections we discuss pathogens that have been isolated from *S. invicta* and have had significant assessments of their potential as microbial control agents of this invasive ant.

13.2.2 Beauveria bassiana

Beauveria bassiana is an entomopathogenic fungus that infects many insect species. As mentioned previously, several isolates from *S. invicta* collected in Brazil were screened for efficacy against fire ants in the late 1980s, with one isolate, *B. bassiana* 447 (ATCC 20872; Bb447), being selected for further assessment. This isolate could be cultured efficiently on rice to produce spores (Stimac *et al.* 1993, Stimac & Alves 1994). When *S. invicta* adults were sprayed directly with Bb447 suspensions of 10^8 spores/ml, virtually 100% mortality occurred and evidence of infection was observed in 88% of the cadavers (Pereira *et al.* 1993). However, field application of various formulations of Bb447 spores by injection with pressurized CO_2 into *S. invicta* nests, which are mounds of soil above subterranean networks of tunnels, or scattering rice/fungus on top of nests, resulted in 48–100% of the treated nests remaining active or relocating. Infection was confirmed in 52–60% of adult ants sampled from treated nests and piles of dead infected brood were observed (Oi *et al.* 1994). The lack of control under field conditions could be the result of several factors. Transmission within colonies can be limited by the hygienic behavior of ants, such as the removal of spores by grooming and the discarding of cadavers outside of nests before sporulation (Siebeneicher *et al.* 1992, Oi & Pereira 1993). Pereira *et al.* (1993) demonstrated that mortality of *S. invicta* by Bb447 in non-sterile soil was poor, indicating that soil borne antagonists were hindering infection. Fire ant venom also inhibits the germination of *B. bassiana* conidia (Storey *et al.* 1991). Poor transmission of another isolate of *B. bassiana* spores among fire ants has also been reported (Siebeneicher *et al.* 1992). Pereira *et al.* (1993) conjectured that for transmission to occur, localized pockets of large quantities of spores would need to be deposited within a fire ant nest, and then the pockets must be visited by ants.

In addition to direct spore applications to nests, bait formulations have been developed with Bb447 and other isolates. However, spore contact and germination through the ant cuticle must occur for infection. Formulations that promote such contact, such as attractive, dry powders that are difficult to carry by ants may be most suitable for baits (Pereira & Stimac 1997). Conidia of *B. bassiana* originally isolated from the Mexican leaf-cutting ant, *Atta mexicana*, and subsequently reisolated from *S. invicta* were encapsulated in sodium alginate and then dried into pellets. Alginate, commonly used in processed foods, is a polysaccharide gum extract from algae. These pellets were foraged upon when coated with peanut oil, a food readily accepted by fire ants. The pellets were evidently retained in the nest to allow for sporulation and infection. Significant reductions in fire ant colony activity were reported from field plots where the oil-coated mycelia pellets were broadcast

(Thorvilson *et al.* 2002). However, these results could not be replicated by other researchers (Collins *et al.* 1999, DH Oi unpublished data).

Given the inhibition of *B. bassiana* conidial germination with exposure to venom and soil antagonists and the hygienic behavior of ants, efficiently infecting fire ant queen(s) and the majority of nestmates to kill individual colonies is problematic. In addition, natural intercolony transmission and control have not been demonstrated. Thus, the effective use of *B. bassiana* as a microbial pesticide or a biological control agent against fire ants currently seems unlikely.

13.2.3 **Thelohania solenopsae**

Thelohania solenopsae is an obligate intracellular microsporidian pathogen that was first observed in alcohol-preserved specimens of *S. invicta* collected in 1973 during a survey in the city of Cuiabá, Mato Grosso, Brazil (Allen & Buren 1974). It is a relatively common fire ant pathogen in South America, being found in 25% of sites surveyed (Briano *et al.* 1995c, 2006). It was noted during the initial survey that infected colonies were smaller and had less vigor when disturbed (Allen & Buren 1974). Subsequent observations in South America suggested that *T. solenopsae* infection was a chronic, debilitative disease causing fire ant populations to decrease rapidly after 1–2 years during periods of stress such as drought (Allen & Knell 1980). Field experiments in Argentina with *S. richteri* documented 83% fewer nests in *T. solenopsae*-infected plots and infected colonies were significantly smaller (Briano *et al.* 1995a,b). The discovery of *T. solenopsae* in the US in 1996 (Williams *et al.* 2003) and the ability to initiate infections by transferring brood from infected colonies (Willams *et al.* 1999, Oi *et al.* 2001) allowed for further documentation of the pathogen's detrimental effects on *S. invicta* colonies in the US. Laboratory inoculations of colonies containing single or multiple (3–12) queens resulted in 88 and 100% reductions in brood within 29 and 52 weeks after inoculation, respectively. Queens from infected colonies weighed less, had declining oviposition rates, and died earlier than queens from healthy colonies (Willams *et al.* 1999, Oi & Williams 2002). Evidence of transovarial transmission was also reported from both *S. richteri* and *S. invicta* (Briano *et al.* 1996, Valles *et al.* 2002). Reductions of *S. invicta* population indices (= estimates of fire ant populations based on the number of ants and the presence of brood in individual nests) in infected field plots in the US were also reported, ranging from "weak" to 63%. Reductions were often due to the presence of smaller colony sizes instead of the total elimination of colonies and reductions often fluctuated (Cook 2002, Oi & Williams 2002, Fuxa *et al.* 2005a). The 6-month or longer decline in colonies most likely allowed for re-infestations to occur, thus documenting field reductions was less consistent (Oi & Williams 2002). Potentiation was exhibited when *T. solenopsae*-infected *S. invicta* died faster after feeding on fire ant bait containing hydramethylnon (Valles & Pereira 2003b), and also after infection by *B. bassiana* (Brinkman & Gardner 2000).

The level of control associated with *T. solenopsae* infection is insufficient especially in urban areas where tolerance to fire ant stings is very low. Perhaps the most compelling effect of fire ant biological control introductions is the potential delay in re-infestation in areas cleared of fire ants by insecticides. Widespread establishment of *T. solenopsae* and other fire ant biological control agents in unmanaged lands could diminish sources of re-infestations. Infected colonies that split from larger colonies or simply move from untreated to treated areas may eventually die faster. Fire ants can also spread and colonize new areas with newly-mated queens after nuptial flights. *T. solenopsae* was present in fire ant reproductives that initiate nuptial flights as well as in newly-mated queens, which had poorer survivorship and colony founding ability (Oi & Williams 2003). Slower fire ant re-infestation and consistent control was documented in an area where *T. solenopsae* and a fire ant parasitoid, *Pseudacteon tricuspis*, were released and became established (Fig. 13.1, Oi *et al.* 2008). Because the impact of *P. tricuspis* alone is low (Morrison & Porter 2005), *T. solenopsae* may have played the greater role in the slower re-infestation.

S. invicta colonies occur as two social forms: monogyne with one queen per colony and polygyne with multiple queens per colony. Infections of *T. solenopsae* were more prevalent (56–83%) in fire ant populations that consisted of polygyne colonies, and rare (0–2%) in monogyne populations (Oi *et al.* 2004, Milks *et al.* 2008). Polygyne fire ants, including brood and queens, move freely between colonies. Infected individuals can relocate to uninfected nests or healthy ants can move into infected nests, and not all queens within a colony are infected. Thus, infections in polygyne populations can be more persistent (Oi *et al.* 2004). In

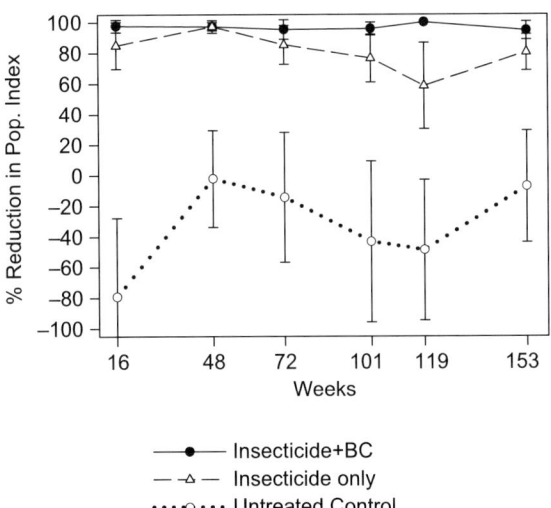

Fig. 13.1 Reductions in *S. invicta* population indices per plot (average % reduction ±95% CI) among areas where (**a**) the fire ant biological control agents *T. solenopsae* and the parasitic fly *Pseudacteon tricuspis* (= BC) were established and insecticide was applied; (**b**) only insecticide was applied; and (**c**) no biological control agents or insecticides were used. Negative reductions represent increases in *S. invicta* populations relative to pretreatment populations. Weeks after insecticide application or biocontrol releases are indicated on the x-axis (adapted from Oi *et al.* 2008)

contrast, monogyne colonies are territorial and there is little intercolony brood or queen exchange. Monogyne colonies infected with *T. solenopsae* apparently succumb without efficient intercolony transmission to maintain or spread the infection. It is also hypothesized that when mixed social forms occur, *T. solenopsae* confers a competitive disadvantage to the social form with higher infection prevalence. In areas of Louisiana where both social forms were living, the social form with more colonies infected with *T. solenopsae* declined more, relative to the other social form (Fuxa *et al.* 2005a, b). In Argentina, *T. solenopsae* infections were found to be nearly equally present in both social forms of *S. invicta* (Valles & Briano 2004). However, the social form assay used may not be applicable to South American fire ant populations and monogyne prevalence needs to be re-evaluated (DD Shoemaker personal communication).

The host range of *T. solenopsae* is apparently restricted to the *Solenopsis saevissima* species group. In South America, besides *S. invicta*, species from which *T. solenopsae* have been reported include *S. richteri* (Allen & Silveira-Guido 1974), *S. saevissima, S. quinquecuspis, S. macdonaghi, S. blumi* [= *S. quinquecuspis*] (Allen & Knell 1980), and *S. interrupta* (Briano *et al.* 2006). Infections in nine other non-*Solenopsis* genera of ants were not detected from samples collected in infected areas in both the US and South America (Williams *et al.* 1998, Briano *et al.* 2002). When infected *S. invicta* brood was introduced to colonies of several ant species from the US, including species from the *Solenopsis geminata* species group (*S. geminata* and *S. xyloni*), infections did not occur. In contrast, introductions of *S. invicta* brood inocula produced infections in the *Saevissima* species group (*S. richteri, S. invicta*, and their hybrid) (Table 13.1).

Currently the only known method of consistent intercolony transmission of *T. solenopsae* is by transfer of live, infected brood with rates of up to 80% transmission being reported. In contrast, 4–25% transmission has been reported when brood was tended by infected adult ants (Allen & Knell 1980, Oi *et al.* 2001), with the possibility of contamination mentioned in Oi *et al.* (2001). While four spore types have been described for *T. solenopsae* (Knell *et al.* 1977, Oi *et al.* 2001, Shapiro *et al.* 2003, Sokolova & Fuxa 2001, Sokolova *et al.* 2003, 2004), infection has yet to be initiated by inoculation with isolated spores despite several attempts using spores mixed with various foods (Allen & Knell 1980, Oi *et al.* 2001, Shapiro *et al.* 2003). Chen *et al.* (2004) hypothesized that spores found in the meconia of pupating larvae were the source for horizontal transmission.

Live, infected brood has been used to successfully initiate *T. solenopsae* field infections in several states in the US (Florida, Louisiana, Mississippi, Oklahoma, and South Carolina). These successful inoculations were conducted in polygyne populations of *S. invicta*, with the social form determined by either PCR or by assessment of nest densities in combination with adult worker sizes (Greenberg *et al.* 1985, Macom & Porter 1996). As mentioned previously, sustained infections are most prevalent in polygyne populations, and it is becoming more difficult to find polygyne populations in the US that are completely free of *T. solenopsae*. Establishing sustained infections in monogyne populations has generally been unsuccessful (Fuxa *et al.* 2005a, DH Oi unpublished data). Similarly, since *S. richteri* in the

Table 13.1 Host range of *T. solenopsae* from surveys in South and North America and laboratory inoculations

Field surveys in Argentina and Brazil[a]			
Ant	# Infected nests/ # nests	Ant	# Traps with infected ants/ # traps
Acromyrmex	0/45	*Brachymyrmex*	0/2
Camponotus	0/1	*Camponotus*	0/46
Pheidole	0/4	*Crematogaster*	0/28
S. macdonaghi	11/19	*Dorymyrmex*	0/2
S. invicta	28/255	*Linepithema*	0/10
S. richteri	38/261	*Paratrechina*	0/2
S. quinquecuspis	+[b]	*Pheidole*	0/66
S. saevissima	+[b]	*Wasmannia*	0/1
S. interrupta	+[b]	*Solenopsis* sp.	1/5
		S. invicta	1/67
		S. richteri	22/75
Field survey: Florida, US[c]		**Lab inoculations with US ants**[e]	
Ant	# Infected nests/ # nests	Ant	# Infected colonies/ # colonies
Camponotus floridanus	0/1	*C. floridanus*	0/1
Brachymyrmex depilis	0/1	*Linepithema humile*	0/3
Dorymyrmex bureni	0/9	*Monomorium floricola*	0/3
Pheidole moerens	0/1	*Solenopsis geminata*	0/5
P. metallescens	0/1	*Solenopsis xyloni*	0/8
Trachymyrmex septentrionalis	0/1	*S. richteri* x *S. invicta* hybrid	3/6
Solenopsis geminata	0/15	*S. richteri*	1/7
Pheidole[d]	0/17	*S. invicta*	12/19

[a] Data from Briano *et al.* (2002).
[b] *T. solenopsae* infections reported without quantification (Allen & Knell 1980, Briano *et al.* 2006); + = present.
[c] Data from Williams *et al.* (1998).
[d] Ants collected in Texas, US (Mitchell *et al.* 2006).
[e] DH Oi unpublished data.

US is monogyne, several inoculations of *S. richteri* populations in Mississippi and Tennessee with live infected *S. invicta* brood have failed (DH Oi unpublished data). It was speculated that failure to establish infections in *S. richteri* could be attributed to poor cross-fostering of the *S. invicta* brood despite initially being carried into nests; *S. richteri* being monogyne in the US; and/or host isolate incompatibility.

The method of using live brood as inocula is labor intensive and inefficient. Brood is obtained from field collected *T. solenopsae*-infected colonies and the process of excavating nests, rearing colonies, and separating brood from adult ants is inconsistent, yielding from <1 to 40 + g brood per colony. Infection rates of inocula per colony are based on the presence of *T. solenopsae* in 10 individual stained slide mounts of fourth instars and/or prepupae, or 10 individual wet mounts of non-melanized pupae. Separated brood (3–5 g) with infection rates of 60–80% is poured into individual nests where ants from the recipient colony would tend the brood. Shipping or transporting live brood is difficult, requiring careful handling and refrigeration, and inoculating individual nests is time consuming. More efficient methods of inoculation, such as broadcasting formulated inocula, are not available currently. Development of an infective spore formulation(s) that can be stored and is compatible with fire ant bait application equipment would facilitate the dissemination of *T. solenopsae* and perhaps increase infection rates in monogyne populations. Nevertheless, the widespread distribution of *T. solenopsae* in polygyne *S. invicta* in the US and its documented field impact demonstrates its utility as a microbial biological control of fire ants.

13.2.4 Vairimorpha invictae

Vairimorpha invictae is another microsporidium that was described from a *S. invicta* colony collected in the state of Mato Grosso, Brazil, in the early 1980s (Jouvenaz & Ellis 1986, Jouvenaz & Wojcik 1981). *V. invictae* is an obligate, intracellular parasite that produces two spore types: unicellular octospores contained in groups of eight within sporophorous vesicles and binucleate free spores (Jouvenaz & Ellis 1986). *V. invictae* was detected in 2.3% of 2528 *S. invicta* and *S. richteri* colonies at 13% of 154 sites surveyed from 1991–1999 in Argentina. In surveys conducted mainly in Argentina and in portions of Bolivia, Chile, Paraguay, and Brazil from 2001–2005, similar percentages of *V. invictae* infected sites were reported (12% of 262 sites), although the percentage of infected colonies was higher (10% of 2064 colonies). Within individual sites, *V. invictae* prevalence can be high, with 23–83% of colonies being infected (Briano & Williams 2002, Briano *et al.* 2006, Porter *et al.* 2007). In comparison, *T. solenopsae* has been reported to be present in 25% of sites (n = 185 and 262) with 8 and 13% (n = 1836 and 2064, respectively) of colonies infected (Briano *et al.* 1995c, 2006).

Laboratory evidence for the pathogenicity of *V. invictae* included faster mortality among naturally infected, starved adult workers and higher infection rates among dead workers than live workers (Briano & Williams 2002). Infections initiated by the introduction of live infected brood or dead infected adults into small laboratory colonies of *S. invicta* resulted in significant reductions (>80%) in colony growth (Oi *et al.* 2005). Declines in field populations of *S. invicta* (69%) were associated with natural *V. invictae* infections and also simultaneous infections with *T. solenopsae* in Argentina (Briano 2005). The persistence of *V. invictae* field infections appears to be more sporadic with wide and abrupt fluctuations in prevalence, whereas *T. solenopsae* maintains a fluctuating yet sustained infection level (Briano *et al.* 2006). Faster colony declines were observed when simultaneous infections of the two microsporidia occurred in the laboratory (Williams *et al.* 2003).

V. invictae can be found in all life stages of *S. invicta* including eggs. However, the low number of infected eggs and queens makes the importance of vertical transmission in the *V. invictae* life cycle uncertain (Briano & Williams 2002). While laboratory colonies of *S. invicta* can be infected through the introduction of live infected brood or infected adults that died naturally, infection by isolated spores has not been achieved (Jouvenaz & Ellis 1986, Briano & Williams 2002, Oi *et al.* 2005). However, larvae can be infected when reared to the pupal stage by infected adult workers (Oi *et al.* 2007).

The host range of *V. invictae* appears to be restricted to ants in the *Solenopsis saevissima* species group. Infections have been reported from *S. invicta*, *S. richteri*, and *S. macdonaghi* in field surveys of nests and bait trapping in Argentina and Brazil. In addition, infections were not observed in 10 non-*Solenopsis* genera (Briano *et al.* 2002). Similarly, infections were not detected in 235 non-ant arthropods (10 orders, 43 families, 80 species), and 947 non-*Solenopsis* ants (12 genera, 19 species) collected at baits from five *V. invictae* sites in Argentina (Porter *et al.* 2007). Inoculations with *V. invictae*-infected brood of laboratory colonies of *Solenopsis geminata* and *Solenopsis xyloni*, two fire ant species in the *Solenopsis geminata* species group found in North America, did not result in infections (Oi *et al.* 2007). Thus, the host specificity of *V. invictae* is favorable for release as a biological control agent in the US and efforts are underway to secure approval for its release.

13.3 Molecular Techniques Facilitate Virus Discovery in Fire Ants

Although viruses are considered important biological control agents for use against insect pests (Lacey *et al.* 2001), they have not been examined for their potential use against fire ants. Indeed, before 2004, the only report present in the literature concerned with virus infections in fire ants was the observation of "virus-like particles" in an unidentified *Solenopsis* species from Brazil (Avery *et al.* 1977). As indicated in Section 9.2.1, efforts to discover microbial infections in fire ants in South America were conducted by either brute force examination of large numbers of colonies, or attempts to identify and examine diseased ant colonies, exclusively by microscopic methods (Jouvenaz *et al.* 1977, 1981, Jouvenaz 1983, Wojcik *et al.* 1987). Despite searches over several decades, no virus had been shown to infect *S. invicta*.

In late 2001, a molecular-based approach was employed in an attempt to discover virus infections in fire ant colonies. An expression library was created, sequenced and analyzed to identify potential viral infections through homologous gene identification (Valles *et al.* 2008). The library was created from all stages (eggs, larvae, pupae, workers, and the queen) of a monogyne *S. invicta* colony and 2,304 clones were sequenced. After assembly and removal of mitochondrial and poor quality sequences, 1,054 unique sequences were identified and deposited into the GenBank database (Accession Numbers EH412746 through EH413799). Six ESTs exhibited significant homology with single-stranded RNA viruses (3B4, 3F6, 11F1, 12G12, 14D5, and 24C10). Subsequent analysis of these putative viral ESTs revealed that 3B4 was most likely a ribosomal gene of *S. invicta*, 11F1 was a positive-strand RNA virus contaminant introduced into the colony from the cricket food source (Valles & Chen 2006), 12G12 appeared to be a plant-infecting tenuivirus also introduced into the colony as a field contaminant, and 3F6, 14D5, and 24C10 were unique and exhibited significant homology (expectation scores $< 10^{-5}$) with the single-stranded Acute Bee Paralysis virus (ABPV) (Valles et al. 2004, Valles & Strong 2005).

13.3.1 Genome Acquisition, Construction, and Characterization of Solenopsis invicta *Virus-1*

Using these ESTs (3F6, 14D5, and 24C10) as a platform, we conducted 5' and 3' Rapid Amplification of cDNA Ends (RACE) to acquire the entire genome of this likely virus. The polyadenylated RNA genome was comprised of 8,026 nucleotides (GenBank Accession Number: AY634314), which encoded two large open reading frames (ORF). These ORFs were flanked and separated by short untranslated regions. The 5' proximal ORF (defined by nucleotides $28-218 =$ ORF 1) encoded a predicted amino acid sequence possessing significant identity with the helicase, protease, and RNA-dependent RNA polymerase (RdRp) regions from positive-strand RNA viruses. The predicted amino acid sequence of the 3' proximal ORF (defined by nucleotides 4390 to $7803 =$ ORF 2) exhibited similarity to virus structural proteins of positive-strand RNA viruses, especially the Acute Bee Paralysis virus. Electron microscopic examination of negatively stained samples from virus-infected fire ants (as determined by RT-PCR) revealed isometric particles with a diameter of 30–35 nm (Fig. 13.2), also consistent with positive-strand RNA viruses.

This new virus, currently named *Solenopsis invicta* virus-1 (SINV-1), represents the first virus to be discovered in *S. invicta*. SINV-1 was easily transmitted to uninfected *S. invicta* by feeding and the replicative genome strand (or negative strand) was consistently present in infected ants indicating that the virus was replicating and that the ant was serving as host (Valles et al. 2004, Hashimoto et al. 2007, SM Valles unpublished data). To date, two forms (based on sequence differences) have been de-

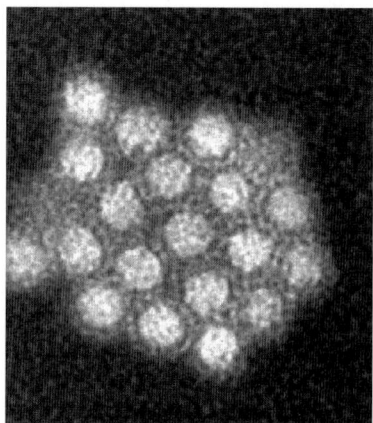

Fig. 13.2 Electron micrograph of *Solenopsis invicta* virus-1 purified from infected fire ants

40 nm

scribed, namely SINV-1 (Valles *et al.* 2004) and SINV-1A (Valles & Strong 2005). These forms are distinct, and can be differentiated by RT-PCR. However, nucleotide changes result in largely synonymous amino acid changes indicating that SINV-1A is likely a genotype of SINV-1.

SINV-1 is found in all fire ant caste members and developmental stages. Worker ants exhibited the highest genome copy number (2.1×10^9 copies/worker ant) and pupae exhibited the lowest (4.2×10^2 copies/pupa) (Hashimoto *et al.* 2007). Mean genome copy number (based on quantitation of the RdRp) was lowest in early larvae and pupae. Overall, SINV-1 genome copy number increased throughout larval development, declined sharply at pupation, then increased in adults. No symptoms were observed among infected nests in the field. However, under certain situations (stressors), infected colonies exhibited extensive brood death and often colonies collapsed as a result. Thus, SINV-1 fits the paradigm for many insect-infecting positive-strand RNA viruses (Chen & Siede 2007). Specifically, infection is chronic and asymptomatic, which can convert to an active-lethal state under certain conditions. After the initial discovery (Valles *et al.* 2004), research has focused on characterizing and understanding SINV-1 in hopes of utilizing it as a control agent.

13.3.2 SINV-1 Phenology, Host Specificity, Distribution, and Tissue Tropism

Valles *et al.* (2007) reported the phenology, geographic distribution, and host specificity of SINV-1. The prevalence of SINV-1 and -1A among fire ant nests in Florida exhibited a distinct seasonal pattern (Fig. 13.3). Infection rates of SINV-1 and -1A were lowest from December to April, increasing rapidly in May and remaining high through August, before declining again in autumn (October). A significant relationship was observed between mean monthly temperature and SINV-1 ($p < 0.0005$, $r = 0.82$) and SINV-1A ($p < 0.0001$, $r = 0.86$) infection rates in *S. invicta* colonies. Relatively higher temperatures were associated with correspondingly higher intercolony SINV-1 infection rates, a relationship that has been noted previously for other RNA viruses and their insect hosts (Bailey 1967, Plus *et al.* 1975).

SINV-1 was reported to be distributed widely among *S. invicta* populations. It was detected in *S. invicta* from all US states examined (except New Mexico) and Argentina (Valles *et al.* 2007). SINV-1 and -1A were also found to infect other *Solenopsis* species. SINV-1 was detected in *S. richteri* and the *S. invicta/richteri* hybrid from northern Alabama and *S. geminata* from Florida, but not Hawaii. SINV-1A was detected in *S. geminata* and *S. carolinensis* collected in Florida and the *S. invicta/richteri* hybrid from Alabama. However, among nearly 2,000 arthropods collected from pitfall traps from north Florida, none except for *S. invicta* tested positive for SINV-1 or SINV-1A. Thus, SINV-1 appears to be specific to the *Solenopsis* genus, with *S. invicta* and *S. richteri* the main hosts. Positive-strand RNA viruses can exhibit exceptionally wide (e.g. Cricket Paralysis virus) or narrow (e.g. Drosophila

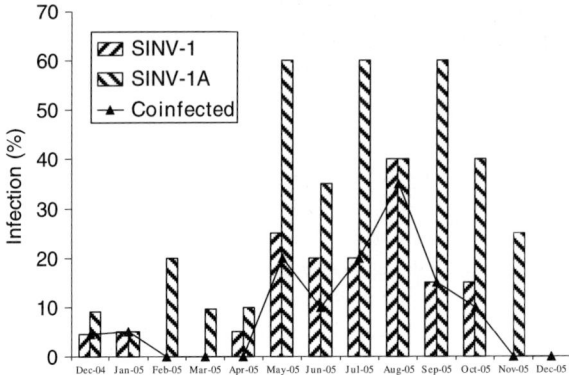

Fig. 13.3 Percentage SINV-1 and SINV-1A infection among field-collected *S. invicta* nests sampled from two locations in Gainesville, Florida. Nests infected with both genotypes are indicated by the line graph. Graph adapted from Valles *et al.* (2007)

C virus complex) host ranges (Christian & Scotti 1998). Although SINV-1 appears to be similar to the Drosophila C virus complex in that it infects only species in a single genus (*Solenopsis*), more direct challenges of other arthropods with SINV-1 in laboratory experiments have not yet been conducted and may indicate otherwise.

Phylogenetic analyses of regions of the SINV-1/-1A genome corresponding to structural proteins indicated significant divergence between viruses infecting North American and South American *S. invicta*. Based on the fact that positive-strand RNA viruses have high mutation rates (Domingo & Holland 1997) in the order of 10^{-4}–10^{-3} per nucleotide site per replication (Holland *et al.* 1982) and recombination does occur via template switching during transcription, the phylogenetic data indicate a significant duration of separation between the virus samples taken from North and South American *S. invicta*. The phylograms also indicate that the North American viral strains have diverged more recently from the common ancestor compared with viral strains from Argentina (i.e. North American strains display fewer nucleotide changes than Argentinean strains). Although hypothetical, these data suggest that SINV-1 was introduced into North America along with founding *S. invicta* or *S. richteri*. This statement is supported by the lack of infection among other *Solenopsis* species (*S. geminata* and *S. xyloni*) in areas apparently devoid (Hawaii) or with incipient infestations of *S. invicta* (Mexico, New Mexico and California). It is further hypothesized that SINV-1 infection of *S. geminata* and *S. carolinensis* (in Florida) may have originated from introduced *S. invicta* and leapt to these native *Solenopsis* species. If this event occurred, it is possible that the SINV-1 infection of native *Solenopsis* species could have provided *S. invicta* with a competitive advantage by reducing the fitness of the native ants relative to *S. invicta*. Further study may provide a more conclusive determination of the origin of SINV-1 and may even provide additional insight into the complex *Solenopsis* phylogenetic relationships. Rapidly-evolving RNA viruses may provide details about host population structure

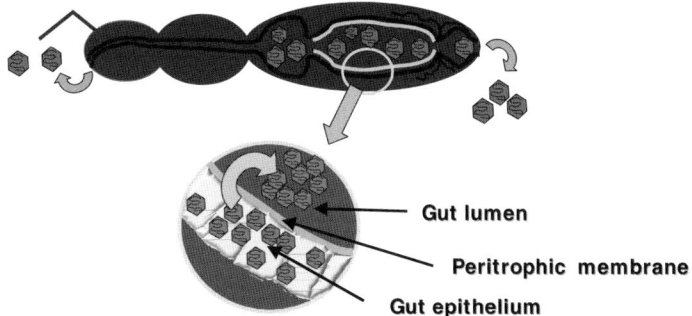

Fig. 13.4 Proposed location of SINV-1 replication and mechanism of transmission. SINV-1 likely replicates in the gut epithelial cells of the midgut. As viral particles are synthesized, they are shed into the gut lumen where they are dispersed to ant nestmates by trophallaxis or substrate contamination by defecation

and demographic history that might not be possible from host genetic data alone (Biek *et al*. 2006).

To determine the susceptible host cell(s), quantitative real-time PCR was employed. This method was also utilized to determine the infection rate among individual ants and colonies of *S. invicta* (Hashimoto & Valles 2007). Among tissues examined from SINV-1-infected ants (larvae and worker ants), the midgut consistently had the highest number of SINV-1 genome copies (>90% of the total). Negative staining and electron microscopy of the supernatant of gut homogenates revealed the presence of spherical virus particles with a diameter of 30–35 nm, consistent with SINV-1 (and positive-strand RNA viruses). Therefore, SINV-1 appears to replicate in gut epithelial cells of *S. invicta*. Viral particles are also found in high abundance in the midgut contents. It is proposed that viral replication occurs in the epithelial cells of the midgut and virus particles are shed into the gut lumen. From there, the particles may be passed to nestmates by trophallaxis or substrate contamination by defecation (Fig. 13.4).

The number of SINV-1 genome copies in infected larvae and workers was often quite high ($> 10^9$ copies/ant). A strong correlation was observed between colony infection rate (number of infected ants/nest) and the number of viral particles (Hashimoto & Valles 2007).

13.3.3 Use of SINV-1 for Controlling S. invicta

Although SINV-1 is currently present in the *S. invicta* US population, we still anticipate its utility as a microbial control agent against this ant pest. SINV-1 appears to fit the paradigm of RNA virus infections in honeybees and other arthropods in that they are present as persistent, asymptomatic infections. However, often when the host experiences certain stressors, the virus begins actively replicating and causes debilitating host symptoms or death. So, there may be a way to emulate these stressors and induce an active-lethal phase of the viral infection. Furthermore, different strains of SINV-1 are likely to exhibit a range of virulence levels. For example, colony collapse disorder in honeybees has recently been associated with Israeli Acute Paralysis virus of bees (Cox-Foster et al. 2007). Investigations on the stress response and conversion to the active-lethal phase of SINV-1 and identification of more virulent genotypes of SINV-1 that could be mass produced for *S. invicta* control are in progress.

13.4 Conclusions and Outlook for Using Pathogens to Control Fire Ants

Over 20 years ago, Jouvenaz (1986) discussed the constraints of using pathogens for the biological control of fire ants. Lacking were knowledge on the biology of fire ant pathogens, such as the mode of intercolony transmission, and efficient methods to screen colonies for pathogens. Examination by microscopy of aqueous extracts has revealed fungi and other spore-forming microorganisms, and now molecular techniques have allowed the discovery and characterization of viruses. Molecular biology will continue to improve our understanding of the biology and ecology of fire ant pathogens and should play a role in solving limitations to pathogen transmission. Constraints such as mass production and dissemination will have to be resolved once the control potential and host specificity of pathogens have been determined. Continuing research on fire ant-specific microsporidia and viruses illustrates the effort to understand their biology and improve their utilization as biological controls for fire ants. Fire ants and their tremendous reproductive capability, mobility, and adaptability, makes their eradication daunting at best. Self-sustaining biological control is one of the few, if not the only, control measure that offers the possibility of long-term regional suppression. Of the known 40+ natural enemies of fire ants, pathogens are a small, but increasing portion as evidenced by the recent discoveries of new viral pathogens infecting *S. invicta* in the US. Additional fire ant pathogens remain to be found, studied, and hopefully introduced from South America. Ongoing research to utilize pathogens for the biological and microbial control of fire ants should eventually yield more measurable benefits and provide a useful model for combating other invasive ants.

References

Allen GE, Buren WF (1974) Microsporidian and fungal diseases of *Solenopsis invicta* Buren in Brazil. J. New York Entomol Soc 82:125–130

Allen GE, Knell JD (1980) Pathogens associated with the *Solenopsis saevissima* complex in South America. Proc Tall Timbers Conf Ecol Animal Control Habitat Mgmt 7:87–94

Allen GE, Silveira-Guido A (1974) Occurrence of microsporida in *Solenopsis richteri* and *Solenopsis* sp. in Uruguay and Argentina. Fla Entomol 57:327–329

Avery SW, Jouvenaz DP, Banks WA, Anthony DW (1977) Virus-like particles in a fire ant, *Solenopsis* sp., (Hymenoptera: Formicidae) from Brazil. Fla Entomol 60:17–20

Bailey L (1967) The incidence of virus diseases in the honeybee. Ann Appl Biol 60:43–48
Beckham RD, Bilimoria SL, Bartell DP (1982) A survey for microorganisms associated with ants in western Texas. Southwest Entomol 7:225–229
Biek R, Drummond AJ, Poss M (2006) A virus reveals population structure and recent demographic history of its carnivore host. Science 311:538–541
Bouwma AM, Shoemaker DD (2007) No evidence for *Wolbachia* phenotypic effects in the fire ant *Solenopsis invicta*. In: Oi DH (ed) Proc 2007 Annual Imported Fire Ant Conf April 23–25, Gainesville, FL. pp 53–54
Bouwma AM, Ahrens ME, DeHeer CJ, Shoemaker DD (2006) Distribution and prevalence of *Wolbachia* in introduced populations of the fire ant *Solenopsis invicta*. Insect Mol Biol 15: 89–93
Briano JA (2005) Long-term studies of the red imported fire ant, *Solenopsis invicta*, infected with the microsporidia *Vairimorpha invictae* and *Thelohania solenopsae* in Argentina. Environ Entomol 34:124–132
Briano JA, Williams DF (2002) Natural occurrence and laboratory studies of the fire ant pathogen *Vairimorpha invictae* (Microsporida: Burenellidae) in Argentina. Environ Entomol 31:887–894
Briano JA, Patterson RS, Cordo HA (1995a) Long-term studies of the black imported fire ant (Hymenoptera: Formicidae) infected with a microsporidium. Environ Entomol 24: 1328–1332
Briano J, Patterson R, Cordo H (1995b) Relationship between colony size of *Solenopsis richteri* (Hymenoptera: Formicidae) and infection with *Thelohania solenopsae* (Microsporidia: Thelohaniidae) in Argentina. J Econ Entomol 88:1233–1237
Briano JA, Jouvenaz DP, Wojcik DP, Cordo HA, Patterson RS (1995c) Protozoan and fungal diseases in *Solenopsis richteri* and *S. quinquecuspis* (Hymenoptera: Formicidae), in Buenos Aires province, Argentina. Fla Entomol 78:531–537
Briano JA, Patterson RS, Becnel JJ, Cordo HA (1996) The black imported fire ant, *Solenopsis richteri*, infected with *Thelohania solenopsae*: intracolonial prevalence of infection and evidence for transovarial transmission. J Invertebr Path 67:178–179
Briano JA, Williams DF, Oi DH, Davis LR Jr (2002) Field host range of the fire ant pathogens *Thelohania solenopsae* (Microsporida: Thelohaniidae) and *Vairimorpha invictae* (Microsporida: Burenellidae) in South America. Biol Control 24:98–102
Briano JA, Calcaterra LA, Vander Meer R, Valles SM, Livore JP (2006) New survey for the fire ant microsporidia *Vairimorpha invictae* and *Thelohania solenopsae* in southern South America, with observations on their field persistence and prevalence of dual infections. Environ Entomol 35:1358–1365
Brinkman MA, Gardner WA (2000) Enhanced activity of *Beauveria bassiana* to red imported fire ant workers (Hymenoptera: Formicidae) infected with *Thelohania solenopsae*. J Agric Urban Entomol 17:191–195
Broome JR (1974) Microbial control of the imported fire ant *Solenopsis richteri*. Ph.D. dissertation, Mississippi State University, Mississippi State, MS
Chen YP, Siede R (2007) Honey bee viruses. Adv Virus Res 70:33–80
Chen JSC, Snowden K, Mitchell F, Sokolova J, Fuxa J, Vinson SB (2004) Sources of spores for the possible horizontal transmission of *Thelohania solenopsae* (Microspora: Thelohaniidae) in the red imported fire ants, *Solenopsis invicta*. J Invertebr Pathol 85:139–145
Christian, PD, Scotti PD (1998) Picorna like viruses of insects. In: Miller LK, Ball LA (eds) The insect viruses. Plenum, New York. pp 301–336
Collins HL, Markin GP (1971) Inquilines and other arthropods collected from nests of the imported fire ant, *Solenopsis saevissima richteri*. Ann Entomol Soc Am 64:1376–1380
Collins H, Callcott AM, McAnally L, Ladner A, Lockley T, Wade S (1999) Field and laboratory efficacy of *Beauveria bassiana* alone or in combination with imidacloprid against RIFA. In: 1999 Accomplishment Report Gulfport Plant Protection Station, CPHST, PPQ, USDA, Gulfport Mississippi. pp 96–99 http://cphst.aphis.usda.gov/sections/SIPS/annual_reports.htm

Cook TJ (2002) Studies of naturally occurring *Thelohania solenopsae* (Microsporida: Thelohaniidae) infection in red imported fire ants, *Solenopsis invicta* (Hymenoptera: Formicidae). Environ Entomol 31:1091–1096

Cook JL, Johnston JS, Vinson SB, Gold RE (1997) Distribution of *Caenocholax fenyesi* (Strepsiptera: Myrmecolacidae) and the habitats most likely to contain its stylopized host, *Solenopsis invicta* (Hymenoptera: Formicidae). Environ Entomol 26:1258–1262

Cox-Foster DL, Conlan S, Holmes EC, Palacios G, Evans JD, Moran NA, Quan P, Briese T, Hornig M, Geiser DM, Martinson V, VanEngelsdorp D, Kalkstein AL, Drysdale A, Hui J, Ahai J, Cui L, Hutchison SK, Simons JF, Egholm M, Pettis J, Lipkin WI (2007) A metagenomic survey of microbes in honey bee colony collapse disorder. Science 318:283–287

Domingo E, Holland JJ (1997) RNA virus mutations and fitness for survival. Ann Rev Microbiol 51:151–178

Drees BM, Miller RW, Vinson SB, Georgis R (1992) Susceptibility and behavioral response of red imported fire ants (Hymenoptera: Formicidae) to selected entomogenous nematodes (Rhabditida: Steinernematidae and Heterorhabditidae). J Econ Entomol 85:365–370

Fuxa JR, Sokolova YY, Milks ML, Richter AR, Williams DF, Oi DH (2005a) Prevalence, spread, and effects of the microsporidium *Thelohania solenopsae* released into populations with different social forms of the red imported fire ant (Hymentoptera: Formicidae). Environ Entomol 34:1139–1149

Fuxa JR, Milks ML, Sokolova YY, Richter AR (2005b) Interaction of an entomopathogen with an insect social form: an epizootic of *Thelohania solenopsae* (Microsporidia) in a population of the red imported fire ant, *Solenopsis invicta*. J Invertebr Pathol 88:79–82

Greenberg L, Fletcher DJC, Vinson SB (1985) Differences in worker size and mound distribution in monogynous and polygynous colonies of the fire ant *Solenopsis invicta* Buren. J Kans Entomol Soc 58:9–18

Hashimoto Y, Valles SM (2007) *Solenopsis invicta* virus-1 tissue tropism and intra-colony infection rate in the red imported fire ant: A quantitative PCR-based study. J Invertebr Pathol 96:156–161

Hashimoto Y, Valles SM, Strong CA (2007) Detection and quantitation of *Solenopsis invicta* virus in fire ants by real-time PCR. J Virolog Methods 140:132–139

Holland J, Spindler K, Horodyski F, Grabau E, Nichol S, VandePol S (1982) Rapid evolution of RNA genomes. Science 215:1577–1585

Jetter KM, Hamilton J, Klotz JH (2002). Red imported fire ants threaten agriculture, wildlife, and homes. Calif Agric 56(1):26–34

Jouvenaz DP (1983) Natural enemies of fire ants. Fla Entomol 66:111–121

Jouvenaz DP (1986) Diseases of fire ants: problems and opportunities. In Lofgren CS, Vander Meer RK (eds) Fire ants and leaf-cutting ants: biology and management. Westview Press, Boulder, CO. pp 327–338

Jouvenaz DP, Ellis EA (1986) *Vairimorpha invictae* n. sp. (Microspora: Microsporida), a parasite of the red imported fire ant, *Solenopsis invicta* Buren (Hymenoptera: Formicidae). J Protozool 33:457–461

Jouvenaz DP, Wojcik DP (1981) Biological control of the imported fire ants. Semi Annual Report of Research Conducted on Imported Fire Ants. SEA, Agric. Res. Gainesville, FL and Gulfport, MS. Report 81(1):54–55

Jouvenaz DP, Allen GE, Banks WA, Wojcik DP (1977) A survey for pathogens of fire ants, *Solenopsis* spp. in the southeastern United States. Fla Entomol 60:275–279

Jouvenaz DP, Banks WA, Atwood JD (1980) Incidence of pathogens in fire ants, *Solenopsis* spp., in Brazil. Fla Entomol 63:345–346

Jouvenaz DP, Lofgren CS, Banks WA (1981) Biological control of imported fire ants; a review of current knowledge. Bull Entomol Soc Am 27:203–208

Jouvenaz DP, Lofgren CS, Miller RW (1990) Steinernematid nematode drenches for control of fire ants, *Solenopsis invicta*, in Florida. Fla Entomol 73:190–193

Knell JD, Allen GE, Hazard EI (1977) Light and electron microscope study of *Thelohania solenopsae* n. sp. (Microsporida: Protozoa) in the red imported fire ant, *Solenopsis invicta*. J Invertebr Pathol 29:192–200

Lacey LA, Frutos R, Kaya HK, Vail P (2001) Insect pathogens as biological control agents: Do they have a future? Biol Control 21:230–248

Li H, Medina F, Vinson SB, Coates CJ (2005) Isolation, characterization, and molecular identification of bacteria from the red imported fire ant (*Solenopsis invicta*) midgut. J Invertebr Pathol 89:203–209

Macom TE, Porter SD (1996) Comparison of polygyne and monogyne red imported fire ant (Hymenoptera: Formicidae) population densities. Ann Entomol Soc Am 89:535–543

McNicol C (2006) Surveillance methodologies used within Australia. Various methods including visual surveillance and extraordinary detections; above ground and in ground lures. In: Graham LC (ed) Proc Red Imported Fire Ant Conf March 28–30, 2006, Mobile, Alabama. pp 69–73

Milks ML, Fuxa JR, Richter AR (2008) Prevalence and impact of the microsporidium *Thelohania solenopsae* (Microsporidia) on wild populations of red imported fire ants, *Solenopsis invicta*, in Louisiana. J. Invertebr Pathol 97:91–102

Mitchell FL, Snowden K, Fuxa JR, Vinson SB (2006) Distribution of *Thelohania solenopsae* (Microsporida: Thelohaniidae) infecting red imported fire ant (Hymenoptera: Formicidae) in Texas. Southwest Entomol 31:297–306

Morrison LW, Porter SD (2005) Testing for population-level impacts of introduced *Pseudacteon tricuspis* flies, phorid parasitoids of *Solenopsis invicta* fire ants. Biol Control 33:9–19

Nickle WR, Jouvenaz DP (1987) *Tetradonema solenopsis* n. sp. (Nematoda: Tetradonematidae) parasitic on the red imported fire ant, *Solenopsis invicta* Buren, from Brazil. J Nematol 19:311–313

Oi DH, Pereira RM (1993) Ant behavior and microbial pathogens. Fla Entomol 76:63–74

Oi DH, Williams DF (2002) Impact of *Thelohania solenopsae* (Microsporidia: Thelohaniidae) on polygyne colonies of red imported fire ants (Hymenoptera: Formicidae). J Econ Entomol 95:558–562

Oi DH, Williams DF (2003) *Thelohania solenopsae* (Microsporidia: Thelohaniidae) infection in reproductives of red imported fire ants (Hymenoptera: Formicidae) and its implication for intercolony transmission. Environ Entomol 32:1171–1176

Oi DH, Pereira RM, Stimac JL, Wood LA (1994) Field applications of *Beauveria bassiana* for the control of the red imported fire ant (Hymenoptera: Formicidae). J Econ Entomol 87:623–630

Oi DH, Becnel JJ, Williams DF (2001) Evidence of intra-colony transmission of *Thelohania solenopsae* (Microsporidia: Thelohaniidae) in red imported fire ants (Hymenoptera: Formicidae) and the first report of a new spore type from pupae. J Invertebr Pathol 78:128–134

Oi DH, Valles SM, Pereira RM (2004) Prevalence of *Thelohania solenopsae* (Microsporidia: Thelohaniidae) infection in monogyne and polygyne red imported fire ants (Hymenoptera: Formicidae). Environ Entomol 33:340–345

Oi DH, Briano JA, Valles SM, Williams DF (2005) Transmission of *Vairimorpha invictae* (Microsporidia: Burenellidae) infections between red imported fire ant (Hymenoptera: Formicidae) colonies. J Invertebr Pathol 88:108–115

Oi DH, Valles SM, Briano JA (2007) Laboratory host specificity testing of the microsporidian pathogen *Vairimorpha invictae*. In: Oi DH (ed) Proc 2007 Annual Imported Fire Ant Conf April 23–25, 2007, Gainesville, Florida. pp 90–91

Oi DH, Williams DF, Pereira RM, Horton PM, Davis T, Hyder A, Bolton H, Zeichner B, Porter SD, Boykin S, Hoch A, Boswell M, Williams G (2008) Combining biological and chemical controls for the management of red imported fire ants (Hymenoptera: Formicidae). Amer Entomol 54:46–55

Pereira RM (2004) Occurrence of *Myrmicinosporidium durum* in *Solenopsis invicta* and other new host ants in eastern United States. J Invertebr Pathol 86:38–44

Pereira RM, Stimac JL (1997) Biocontrol options for urban pest ants. J Agric Entomol 14:231–248

Pereira RM, Stimac JL, Alves SB (1993) Soil antagonism affecting the dose response of workers of the red imported fire ant, *Solenopsis invicta*, to *Beauveria bassiana* conidia. J Invertebr Pathol 61:156–161

Pereira RM, Williams DF, Becnel JJ, Oi DH (2002) Yellow head disease caused by a newly discovered *Mattesia* sp. in populations of the red imported fire ant, *Solenopsis invicta*. J Invertebr Pathol 81:45–48

Plus N, Croizier G, Jousset FX, David J (1975) Picornaviruses of laboratory and wild *Drosophila melanogaster*: Geographical distribution and serotypic composition. Ann Microbiol 126A: 107–121

Poinar G, Porter SD, Hyman BC, Tang S (2007) *Allomermis solenopsi* n. sp. (Nematoda: Mermithidae) parasitizing the fire ant *Solenopsis invicta* Buren (Hymenoptera: Formicidae) in Argentina. Syst Parasitol 68:115–128

Porter SD, Williams DF, Patterson RS, Fowler HG (1997) Intercontinental differences in the abundance of *Solenopsis* fire ants (Hymenoptera: Formicidae): escape from natural enemies? Environ Entomol 26:373–384

Porter SD, Valles SM, Davis TS, Briano JA, Calcaterra LA, Oi DH, Jenkins RA (2007) Host specificity of the microsporidian pathogen *Vairimorpha invictae* at five field sites with infected *Solenopsis invicta* fire ant colonies in northern Argentina. Fla Entomol 90:447–452

Sánchez-Peña SR, Thorvilson HG (1992) Two fungi infecting red imported fire ant founding queens from Texas. Southwestern Entomol 17:181–182

Sánchez-Peña SR, Patrock RJW, Gilbert LA (2005) The red imported fire ant is now in Mexico: documentation of its wide distribution along the Texas-Mexico border. Entomol News 116: 363–366

Shapiro AM, Becnel JJ, Oi DH, Williams DF (2003) Ultrastructural characterization and further transmission studies of *Thelohania solenopsae* from *Solenopsis invicta* pupae. J Invertebr Pathol 83:177–180

Shoemaker DD, Ahrens M, Sheill L, Mescher M, Keller L, Ross KG (2003) Distribution and prevalence of *Wolbachia* infections in native populations of the fire ant *Solenopsis invicta* (Hymenoptera: Formicidae). Environ Entomol 32:1329–1336

Siebeneicher SR, Vinson SB, Kenerley CB (1992) Infection of the red imported fire ant by *Beauveria bassiana* through various routes of exposure. J Invertebr Pathol 59:280–285

Silveira-Guido A, Carbonell J, Crisci C (1973) Animals associated with the *Solenopsis* (fire ants) complex, with special reference to *Labauchena daguerrei*. Proc Tall Timbers Conf Ecol Animal Control Habitat Management 4:41–52

Sokolova Y, Fuxa JR (2001) Development of *Thelohania solenopsae* in red imported fire ants, *Solenopsis invicta* from polygynous colonies results in the formation of three spore types. In: Suppl J Eukaryot Microbiol Proc Seventh Internat Worksh Opportunistic Protists 48:85S

Sokolova YY, McNally LR, Fuxa JR (2003) PCR-based analysis of spores isolated from smears by laser pressure catapult microdissection confirms genetic identity of spore morphotypes of a microsporidian *Thelohania solenopsae*. J Eukaryot Microbiol 50:584–585

Sokolova YY, McNally LR, Fuxa JR, Vinson SB (2004) Spore morphotypes of *Thelohania solenopsae* (microsporidia) described microscopically and confirmed by PCR of individual spores microdissected from smears by position ablative laser microbeam microscopy. Microbiology 150:1261–1270

Stimac JL, Alves SB (1994) Ecology and biological control of fire ants. In: Rosen D, Bennett FD, Capinera JL (eds), Pest management in the subtropics: biological control – a Florida perspective. Intercept Ltd., Andover, England, UK. pp 353–380

Stimac JL, Pereira RM, Alves SB, Wood LA (1993) *Beauveria bassiana* (Balsamo) Vuillemin (Deuteromycetes) applied to laboratory colonies of *Solenopsis invicta* Buren (Hymenoptera: Formicidae) in soil. J Econ Entomol 86:348–352

Storey GK, Vander Meer RK, Boucias DG, McCoy CW (1991) Effect of fire ant (*Solenopsis invicta*) venom alkaloids on the in vitro germination and development of selected entomogenous fungi. J Insect Pathol 58:88–95

Thorvilson HD, Wheeler B, San Francisco M (2002) Development of *Beauveria bassiana* formulations and genetically marked strains as a potential biopesticide for imported fire ant control. Southwestern Entomol Suppl no 25:19–29

Triplett RF (1976) The imported fire ant: health hazard or nuisance? South Med J 69:258–259
Tschinkel WR (2006) The fire ants. Belknap Press of Harvard Univ. Press, Cambridge, MA
Valles SM, Chen Y (2006) Serendipitous discovery of an RNA virus from the cricket, *Acheta domesticus*. Fla Entomol 89:282–283
Valles SM, Pereira RM (2003a) Use of ribosomal DNA sequence data to characterize a neogregarine pathogen of *Solenopsis invicta* (Hymenoptera: Formicidae). J Invertebr Pathol 84:114–118
Valles SM, Pereira RM (2003b) Hydramethylnon potentiation in *Solenopsis invicta* by infection with the microsporidium *Thelohania solenopsae*. Biol Control 27:95–99
Valles SM, Briano JA (2004) Presence of *Thelohania solenopsae* and *Vairimorpha invctae* in South American populations of *Solenopsis invicta*. Fla Entomol 87:625–627
Valles SM, Strong CA (2005) *Solenopsis invicta* virus–1A (SINV-1A): distinct species or genotype of SINV-1? J Invertebr Pathol 88:232–237
Valles SM., Oi DH, Pereira OP, Williams DF (2002) Detection of *Thelohania solenopsae* (Microsporidia: Thelohaniidae) in *Solenopsis invicta* (Hymenoptera: Formicidae) by multiplex PCR. J Invertebr Pathol 81:196–201
Valles SM, Strong CA, Dang PM, Hunter WB, Pereira RM, Oi DH, Shapiro AM, Williams DF (2004) A picorna-like virus from the red imported fire ant, *Solenopsis invicta*: initial discovery, genome sequence, and characterization. Virology 328:151–157
Valles SM, Strong CA, Oi DH, Porter SD, Pereira RM, Vander Meer RK, Hashimoto Y, Hooper-Bui LM, Sanchez-Arroyo H, Davis T, Karpakakunjaram V, Vail KM, Graham LC, Briano JA, Calcaterra LA, Gilbert LE, Ward R, Ward K, Oliver JB, Taniguchi G, Thompson DC (2007) Phenology, distribution, and host specificity of *Solenopsis invicta* virus. J Invertebr Pathol 96:18–27
Valles SM, Strong CA, Dang PM, Hunter WB, Pereira RM, Oi DH, Shapiro AM, Williams DF (2008) Expressed sequence tags from the red imported fire ant, *Solenopsis invicta*: annotation and utilization for discovery of new viruses. J Invertebr Pathol 99:74–81
Williams DF, Knue GJ, Becnel JJ (1998) Discovery of *Thelohania solenopsae* from the red imported fire ant, *Solenopsis invicta*, in the United States. J Invertebr Pathol 71:175–176
Williams DF, Oi DH, Knue GJ (1999) Infection of red imported fire ant (Hymenoptera: Formicidae) colonies with the entomopathogen *Thelohania solenopsae* (Microsporidia: Thelohaniidae). J Econ Entomol 92:830–836
Williams DF, Collins HL, Oi DH (2001) The red imported fire ant (Hymenoptera: Formicidae): a historical perspective of treatment programs and the development of chemical baits for control. Amer Entomol 47:146–159
Williams DF, Oi DH, Porter SD, Pereira RM, Briano JA (2003) Biological control of imported fire ants (Hymenoptera: Formicidae). Amer Entomol 49:150–163
Wojcik DP, Allen CR, Brenner RJ, Forys EA, Jouvenaz DP, Lutz RS (2001) Red imported fire ants: impact on biodiversity. Amer Entomol 47:16–23
Wojcik DP, Jouvenaz DP, Banks WA, Pereira AC (1987) Biological control agents of fire ants in Brazil. In: Eder J, Rembold H (eds) Chemistry and biology of social insects. Peperny, Munich. pp 627–628

Chapter 14
Biological Control of the Cassava Green Mite in Africa with Brazilian Isolates of the Fungal Pathogen *Neozygites tanajoae*

Italo Delalibera Júnior

Abstract Cassava green mite (CGM), *Mononychellus tanajoa* (Acari: Tetranychidae), became one of the most important pests of cassava in Africa after its introduction from the Neotropics in the early 1970s. Exploration for potential natural enemies to be introduced in Africa began in Brazil in 1988 and soon revealed that the pathogenic fungus *Neozygites tanajoae* is an important natural enemy of CGM in Brazil and causes epizootics in CGM populations in a broad range of climates. During the last two decades, a series of studies was conducted to understand the biology and ecology of this pathogen in order to facilitate its use in Africa. A collection of *N. tanajoae* and *N. floridana* isolates from Brazil, Colombia and Benin was established, and *N. tanajoae* isolates from Brazil were selected based on efficient mortality and mummification levels. The selected isolates are specific to CGM, kill all active stages of the host with median lethal times of <5 days, and infections are initiated by even a single capilliconidium. Using living, infected mites, experimental releases of two isolates of *N. tanajoae* from Brazil and one isolate from Benin were carried out in the Adjohoun district, Republic of Benin, in January 1999. Post-release monitoring conducted up to 48 weeks after the releases revealed that higher infection rates were achieved only by introductions of Brazilian isolates and not where the native isolate or non-infected mites were released. Using molecular probes it was confirmed that the fungus recovered after the releases was from Brazil.

14.1 Introduction

Cassava, *Manihot esculenta* (Euphorbiaceae), is a perennial woody shrub native to the Americas that was probably domesticated along the southern border of the Amazon basin (Olsen & Schaal 1999). Cassava can tolerate drought and grow in low-nutrient soils and for these reasons this crop plays an important role for food

---
I. Delalibera Jr
Department of Entomology, Plant Pathology and Agricultural Zoology, ESALQ/University of São Paulo, Av. Pádua Dias 11, C.P. 9, 13418-900 Piracicaba, São Paulo Brasil
e-mail: italo@esalq.usp.br

security in many developing countries. This plant was introduced to Africa from the Neotropics in the sixteenth century (Jones 1959). Cassava has become one of the most important staple crops in sub-Saharan Africa, particularly in the poorest regions where it ensures a food reserve for low-income farmers (Yaninek & Schulthess 1993). Today, cassava is cultivated in all tropical regions of the world (Bellotti et al. 1987). Over 50% of the current global cassava production is in Africa (IFAD-FAO 2000). The roots are the main part of the plant that is consumed but leaves can also be used for human consumption (Megevand et al. 1987). It is the world's fourth most important staple after rice, wheat and maize and is an important component in the diet of over one billion people (IFAD-FAO 2000). It is also used commercially for the production of animal feed and starch-based products.

The cassava green mite (CGM), *Mononychellus tanajoa* (Acari: Tetranychidae), was first reported in NE Brazil in 1938 (Bellotti et al. 1999). This species was discovered attacking cassava in Uganda in 1971 (Yaninek & Herren 1988), and it has since become one of the most important pests of this crop in Africa, causing up to 80% reduction in cassava yield (Yaninek et al. 1989).

Cassava is cultivated in Africa by resource-poor farmers within traditional farming systems on a small scale (Yaninek et al. 1989). This crop is mostly grown for subsistence and growers cannot afford to buy pesticides. Therefore, intervention to control CGM must be simple, appropriate, adaptive and inexpensive (Elliot et al. 2000). For these reasons, a continent-wide biological control program was conceived in 1979 to introduce natural enemies of CGM from South America to Africa (Yaninek & Herren 1988).

Molecular comparisons among CGM biotypes from Africa, Brazil and Colombia suggested that a single introduction of this pest species occurred in Africa from CGM populations closer to Colombia than to Brazil (Navajas et al. 1994). In the early 1980s, CIAT (International Center for Tropical Agriculture) began intensive explorations of natural enemies in Latin America for introduction into Africa. However, none of the phytoseiid predatory mites released from Colombia and other South American countries (not including Brazil) from 1984–1988 became established (Bellotti et al. 1999). Exploration for potential natural enemies began in Brazil in 1988 and soon revealed that the pathogenic fungus *Neozygites tanajoae* (= *Neozygites* sp. in early publications) (Zygomycetes: Entomophthorales) (Delalibera et al. 1992) and phytoseiid predatory mites (Moraes et al. 1990, 1991) were the most important natural enemies of CGM in northeastern Brazil. Initially, research was focused on phytoseiids. The predatory mite *Typhlodromalus aripo* introduced from Brazil demonstrated excellent establishment and spread in West Africa (IITA 1998). By the end of 1998, *T. aripo* had established in 16 African countries, reducing CGM populations by an average of two-thirds and increasing cassava yields by a third. In places where *T. aripo* did not establish and/or had low impact on host populations, releases of Brazilian isolates of *N. tanajoae* were considered.

There have been few attempts at classical biological control using entomophthoralean fungi (Elliot et al. 2000, Hajek et al. 2007). Difficulties with in vivo and in vitro production, preservation, and bioassay standardization are some reasons associated with the scarce utilization of this group of pathogens. Although the steps for introducing a biological control agent are generally similar for most pathogens, *Neozygites* presented unique challenges. Bioassay standardization using capilliconidia produced from mummified CGM was one challenge faced during selection of strains. *N. tanajoae* is a particularly fastidious fungus and the difficulty associated with isolating and producing this fungus in vitro was a factor that limited other important studies, e.g. molecular characterization for pathogen detection and confirmation of establishment. During the last two decades a series of studies was conducted to understand the biology and ecology of this pathogen in order to facilitate its use in Africa. Some of these studies are summarized below.

14.2 Surveying Natural Enemies in Brazil

During the first surveys of natural enemies of CGM in Brazil in August 1988, considerable numbers of CGM infected with *N. tanajoae* were found in the State of Bahia (Delalibera et al. 1992). Further investigations were conducted to determine the distribution of *N. tanajoae* within CGM populations in all nine states of northeastern Brazil during February 1989 and June 1990. Those regions were chosen because of their similarities with areas where CGM was of major concern in Africa, i.e. both of these areas are relatively hot and dry for part of the year. *N. tanajoae*-infected CGM were found in 41 out of 82 cassava fields infested with CGM in six of the states. In many fields, the CGM populations were low when the surveys were conducted, thus reducing the chances of finding the pathogen. Therefore, the distribution range of the fungus was probably underestimated.

With the intention of collecting *N. tanajoae* isolates, explorations were carried out from August 1993 to June 1994, again in all nine states of northeastern Brazil. Explorations were conducted during various seasons and at different elevations and climates to increase chances of finding higher variability among strains of the pathogen. During the dry season, August–September 1993, *N. tanajoae*-infected CGM were only detected in the central region of the State of Bahia. The pathogen was frequently found during the winter in June 1994, when CGM populations were very low. During this intensive survey, *N. tanajoae* was collected in all Northeastern states, except Rio Grande do Norte. Later, isolates were collected from the more southern states of Minas Gerais, Paraná and São Paulo. All explorations were conducted by or with Brazilian scientists in accordance with specific permits for exploration required by Brazilian legislation.

N. tanajoae has also been reported attacking CGM in Venezuela (Agudelo-Silva 1986), Palmira, Colombia, (Alvarez Afanador 1990) and in Benin, Africa (Yaninek et al. 1996). Considering the widespread occurrence of *N. tanajoae* in South America, the logical next step for the CGM project was to determine strain variability in relation to virulence as well as other fungal attributes.

14.3 Selecting Efficient Isolates

The first effort toward use of this pathogen as a biological control agent was to select potential isolates from South America to be released in Africa. Major differences exist between *N. tanajoae* and the most common fungal pathogens used for biological control, anamorphs of Hypocreales, e.g. *Beauveria bassiana*. The most substantial difference regarding selection of efficient isolates is that many entomopathogenic Hypocreales grow and sporulate well on artificial culture media and *N. tanajoae* does not. Hypocrealean isolates can be screened against hosts using conidia produced in vitro. Isolation and laboratory maintenance of *N. tanajoae* can be challenging (Delalibera *et al.* 2003, 2004), and for this reason most investigations working toward the classical biological control of CGM in Africa have used mummified cadavers collected either in the field or produced in the laboratory. This in vivo bioassay method, using mummified mites, is technically difficult and less reliable than bioassays using hypocrealean conidia (Delalibera & Hajek 2004), but it is the best alternative for studying *Neozygites*.

A collection of 27 *N. tanajoae* isolates and two *N. floridana* isolates from Brazil, Colombia and Benin was established. This collection has been preserved in vivo, in mummified cadavers at $-10\,°C$ and, under these conditions, the fungus retains viability for up to ten years, but with decreased sporulation over time (Wekesa & Delalibera in press).

Comparisons of virulence of all isolates against CGM demonstrated that most isolates from Brazil were highly virulent (Delalibera & Hajek 2004). Sixteen *N. tanajoae* isolates caused more than 89% mortality and more than 62% of the CGM became mummified. A mummified CGM is characteristically a swollen, brown fungus-killed mite that has great potential to produce conidia. The median mummification time ranged from 4.4 to 6.7 days. Five Brazilian isolates caused >75% mummification with a median mummification time of <5 days. Isolates that cause high mummification in a short period of time would be more likely to cause epizootics and to become established in the new environment.

Conidial discharge was quantified from cadavers of CGM that died from infections with 14 isolates of *N. tanajoae* collected from diverse climates of Brazil to help select potential candidate strains for introduction to Africa (Delalibera *et al.* 2006). These studies were aimed at identifying isolates with lower requirements of humidity for sporulation and isolates that discharged more conidia during short periods of moisture. High levels of sporulation are critical for optimal transmission of the pathogen. At $96 \pm 0.5\%$ RH, production of conidia was very low and even isolates from the semi-arid Brazilian regions produced few conidia. We therefore could not select Brazilian isolates with lower moisture requirements for sporulation. However, significant differences in the numbers of conidia produced by diverse Brazilian isolates were observed at 100% RH. The isolate sporulating least, BIN21, discharged only 45.7% of the number of conidia produced by isolate BIN1, one of the isolates producing the most conidia. Decisions on which isolates to choose for release are complex, because many isolates performed very well in the laboratory against CGM. Bioassays comparing important attributes of the pathogen can be used first to eliminate the less appropriate isolates (e.g. those that do not infect or sporulate well). An example of an isolate not adapted to CGM is BIN21. This isolate was collected from CGM but caused very low mummification (Delalibera & Hajek 2004) and sporulated very poorly (Delalibera *et al.* 2006). The *N. floridana* isolate BIN34 from Colombia was initially considered a candidate for biological control of CGM (Bellotti *et al.* 1999). Later, this isolate proved to be less pathogenic to CGM than *N. tanajoae* isolates from Brazil (Delalibera & Hajek 2004). Results from these studies demonstrated that mummification and differences in production of conidia among isolates should be considered when selecting *Neozygites* isolates for biological control introductions.

14.4 Epizootiology, the Impact of *N. tanajoae* in the Host Population

N. tanajoae causes epizootics in CGM populations in a broad range of climates in Brazil (Gondim Jr. 1992, Delalibera Jr. *et al.* 1999), with infection levels often exceeding 85% (Delalibera Jr. *et al.* 2000, Elliot *et al.* 2000). Epizootics are characterized by a short period between detection of infection and rapid declines in CGM densities. Although climatic conditions in northeastern Brazil regions where epizootics occur are similar to those of Africa, only very low levels of infection by *N. tanajoae* had been observed in CGM populations in Africa (Yaninek *et al.* 1996), indicating that endemic African strains have limited impact on populations of this introduced host.

Elliot *et al.* (2000) used multiple regression analyses to demonstrate that patterns of prevalence of *N. tanajoae* in Brazil corresponded positively with rainfall levels and negatively with CGM populations, possibly implying population regulation. By modeling *N. tanajoae* epizootics, Oduor *et al.* (1997) reached the same conclusion that the fungal pathogen can reduce the population growth of CGM. In both studies, it was assumed that the pathogen alone is not capable of controlling the pest population, but a useful role in combination with other factors is indicated. Elliot *et al.* (2000) therefore proposed that *N. tanajoae* might be of particular use in drier cassava-growing areas where rainfall at the outset of the wet season is not sufficiently intense to cause heavy CGM mortality but may be sufficient to stimulate epizootics of the fungal pathogen, protecting the flush of new cassava growth. Therefore, the introduction of efficient isolates from Brazil to Africa has great potential for reducing losses caused by CGM.

Oduor *et al.* (1995a, b) conducted extensive studies on the effects of abiotic factors on the development of *N. tanajoae* (= *Neozygites* cf. *floridana*). They found that the main attributes of *N. tanajoae* for success in biological control are its short life cycle and its ability to kill all active stages of CGM, with infections initiated by even a single capilliconidium.

14.5 Risk Analysis and Non-Target Effects

When considering concerns about non-target effects and ecological host specificity of introduced exotic entomopathogens, attention should be paid to releasing more specific pathogens or releasing pathogens that have negligible impact on non-target organisms (Hajek et al. 2000). To address this issue, infectivity of *N. tanajoae* was initially tested against a species of the red spider mite, *Tetranychus bastosi*, and also against two species of phytoseiid predatory mites that occur in the target host's habitat in Brazil (Moraes & Delalibera 1992). Houtondji et al. (2002b) determined the safety of a Brazilian isolate of *N. tanajoae* for a variety of non-target insect and mite species, including natural enemies of cassava pests and standard beneficial non-target arthropods. These tests were conducted in accordance with recommendations of the Food and Agriculture Organization of the United Nations, before this pathogen could be introduced in Benin. Test species were the predators of CGM: *Euseius concordis* (Acari: Phytoseiidae), *Euseius citrifolius* (Acari: Phytoseiidae) and *Stethorus* sp. (Coleoptera: Coccinellidae); the phytophagous cassava mealybug, *Phenacoccus herreni* (Hemiptera: Coccidae) and its parasitoid *Apoanagyrus diversicornis* (Hymenoptera: Encyrtidae); a whitefly pest of cassava, *Aleurothrixus aepim* (Hemiptera: Aleyrodidae); and the silkworm *Bombyx mori* (Lepidoptera: Bombycidae). These studies indicated no effects of Brazilian isolates of *N. tanajoae* against the tested organisms.

Twenty-seven isolates of *N. tanajoae* and two isolates of *N. floridana* from Brazil, Colombia, and Benin were assayed against CGM and two mite species associated with cassava in Africa, the red mite, *Oligonychus gossypii*, and the twospotted spider mite, *Tetranychus urticae* (Delalibera & Hajek 2004). The host specificity of *Neozygites* isolates from CGM in Brazil and Benin was used, along with other characteristics, to separate these isolates from *N. floridana* isolates pathogenic to spider mite species, and the CGM isolates from Brazil and Benin were then named *N. tanajoae* (Delalibera et al. 2004). *N. tanajoae* can readily be distinguished from *N. floridana* based on 18S rDNA sequences, host ranges, nutritional requirements for growth in vitro, tolerance to cold (4 °C) and ability to withstand specific cryopreservation techniques. So far, *N. tanajoae* is not known to infect or cause any harm to another host, while assays using *N. floridana* isolates indicate that this species has a broader host range.

14.6 Importation and Quarantine

Quarantine measures to prevent the movement of pests, became a important issue in light of what occurred in Africa after introduction of the cassava mealybug, *Phenacoccus manihoti*, and the cassava green mite, both originally from South America, and other pests. There are several pests and secondary pest in the neotropics at present that could become major pests in Africa, where native natural enemies or tolerant cassava cultivars more resistant to these pests are not available (Herren & Neuenschwander 1991).

Two Brazilian isolates collected during epizootics in 1995, in Cruz das Almas and Piritiba (Alto Alegre village) state of Bahia, Brazil, were sent from EMBRAPA Meio Ambiente (Empresa Brasileira de Pesquisa Agropecuária) to IITA (International Institute of Tropical Agriculture) in Benin via quarantine at the University of Amsterdam, The Netherlands. This introduction was covered by export permits from Brazil and import permits issued by the national authorities of Benin. Brazilian legislation states that specific permits from the Ministry of Science and Technology were required for shipment of collections by Brazilians scientists to other countries. In order to exclude unwanted microorganisms, it is critical to obtain a purified isolate of the pathogen for exportation (Hajek *et al.* 2000). However, at that time, *N. tanajoae* had never been isolated in vitro, and the fungus was shipped as CGM mummified cadavers on dry cotton wool on top of a layer of cotton wool soaked in glycerol at the bottom, inside plastic photographic film canisters. It was assumed that the field collected mummified cadavers would include a greater genetic diversity compared to fungal inoculum produced in the laboratory (in vivo or in vitro), and this could allow for enhanced virulence and ability to adapt to conditions in the new land.

14.7 Pathogen Identification: Development of Molecular Probes

N. tanajoae occurs in Africa, but the endemic pathotypes are inefficient at controlling CGM populations. While in Brazil epizootics of *N. tanajoae* are frequently observed in CGM populations, in Benin, Africa, only low levels of infection have been documented (Yaninek *et al.* 1996). Precise discrimination among the closely related *N. tanajoae* isolates from Brazil and Africa was not possible using classical taxonomic criteria. *N. tanajoae* isolates from Brazil were released in Benin in 1998 and 1999, but molecular probes to distinguish exotic from native strains were still not available at that time, making it impossible to monitor the success of introductions of exotic isolates in regions where endemic strains of the pathogen already existed. Samples of indigenous strains collected before the releases as well as isolates from Brazil were used in molecular characterization of the pathogen. For the development of molecular probes to differentiate among *N. tanajoae* strains, RAPD (Random Amplification of Polymorphic DNA) markers were converted into SCARs (Sequence Characterized Amplified Regions) and specific primers were designed for the detection of indigenous and exotic isolates (I Delalibera *et al.* unpublished data). The protocol for differentiation between isolates of *N. tanajoae* consists of PCR reactions using specific primers to evaluate DNA extracted from single mummified mites collected directly from the field. Each mummified mite is placed inside one well of a 96-well PCR plate, where DNA extraction is performed using 50–100 µl of InstaGeneTM Matrix (Bio-Rad Laboratories, Hercules, CA) according to the manufacturer's directions. PCR amplification is conducted using two pairs of primers: *Neozygites* spp. positive control primers and a set of primers that only amplifies a fragment from Brazilian *Neozygites* isolates. These primers have been validated using samples collected from many states of Brazil and from African countries where the Brazilian isolates had never been released (BV Agboton personal communication). Using these molecular probes, establishment of Brazilian isolates in Benin was confirmed (see below). These molecular probes will be used to evaluate the outcome of *N. tanajoae* releases conducted in Benin and future releases planned for other countries in Africa.

14.8 Field Releases and Post-Release Follow-Up Monitoring

For long term establishment of fungal pathogens an important decision is the number of release sites, the amount of inoculum and which stage of the pathogen to release. These issues are usually determined in part by the ease of pathogen production (Hajek *et al.* 2000). Resting spores could be the first choice for the spore stage of *N. tanajoae* to release, because resting spores are environmentally more resistant than the delicate conidial stages. However, the factors that induce formation of this type of spore are unknown so it was not possible to produce resting spores for release. Based on the difficulties of working with *N. tanajoae*, the two options for stages of release of *N. tanajoae* were living, infected mites or mummified cadavers.

An in vivo release procedure was developed for *N. tanajoae* and tested in the field, releasing different isolates of this fungus in the Adjohoun district, Republic of Benin, in February 1998. Infection levels in 1998 were low so the experimental release protocol was improved and these isolates were released in the Adjohoun district again in January 1999. In 1999, evaluation of infection levels 3 days prior to the releases indicated the absence of *N. tanajoae* (reported as *N. floridana*) infection in nearly all selected fields with mean infection levels lower than 0.08% in some fields (Hountondji *et al.* 2002a). Releases were conducted using cassava leaf discs containing 30 live mites previously exposed to *N. tanajoae* conidia. Leaf discs were individually attached with pins on the tops of fully developed leaves of plants with over 100 active CGM per leaf. Releases were carried out in five locations at least 3 km apart, considered as five replicate blocks. Within each location, four cassava fields located approximately 0.5 km apart were used for treatments, totaling 20 cassava fields. These cassava fields were between 4–5 months old and 0.04–0.5 ha. Treatments consisted of release of mites infected by three isolates of *N. tanajoae* (two Brazilian and one Beninese) and release of uninfected mites as a control treatment.

Post-release monitoring conducted at 8, 14, 22, 36, and 48 weeks after the release revealed the presence of infected mites in most inoculated fields (Hountondji *et al.* 2002a). Higher infection levels were observed in fields inoculated with the Brazilian isolates compared with fields inoculated with the local isolate during all post-release monitoring periods, except 8 weeks post-release. Highest infection levels were observed 48 weeks after the releases, mainly in fields inoculated with the Brazilian isolates from Cruz das Almas (36.5%) and from Piritiba (34.0%). *N. tanajoae* was recovered in 63% of the fields sampled, 48 weeks after the releases (one field inoculated with a Brazilian isolate was harvested and not replanted by 48 weeks). In contrast, *N. tanajoae* was recovered in 10% of all 20 fields before the releases. In control fields, infection by *N. tanajoae* was absent during all but the first and the second monitoring dates when only one infected mite was found in a single field on each occasion. These results demonstrated that higher infection rates can be achieved by release of Brazilian isolates. Whether these releases resulted in permanent establishment remains to be demonstrated. However, there are persuasive lines of evidence to indicate that the introduced pathogen recycled in the new area:(1) in some fields where the Brazilian isolates were released, high infection levels were observed after the crop had been harvested and replanted; (2) resting spores, important for long term survival of the pathogen, were detected in infected mites after the releases, although in very low prevalence; and (3) using the molecular probes it was confirmed that the fungus recovered after the releases was from Brazil. The relative contribution of each introduced isolate to the prevalence of *N. tanajoae* and the impact on the host population in the release areas remains to be determined. The results from these releases in Benin have encouraged planning for further dissemination of the Brazilian isolates in other countries in sub-Saharan Africa.

References

Agudelo-Silva P (1986) A species of *Triplosporium* (Zygomycetes: Entomophthorales) infecting *Mononychelus progressivus* (Acari: Tetranychidae) in Venezuela. Fla Entomol 69:444–446

Alvarez Afanador JM (1990) Estudios de patogenicidad de un hongo asociado con *Tetranychus urticae* Koch y *Mononychellus tanajoa* (Bondar), ácaros plaga de la yuca *Manihot esculenta* Crantz. Undergraduate Thesis. Universidad Nacional de Bogota, Bogota

Bellotti AC, Hershey CH, Vargas O (1987) Recent advances in resistance to insect and mite pests of cassava. Cassava Breeding: A multidisciplinary review. In: Hershey CH (ed) Workshop in the Philippines, 1985. CIAT, Cali, Colombia. pp 117–146

Bellotti AC, Smith L, Lapointe SL (1999) Recent advances in cassava pest management. Annu Rev Entomol 44:343–370

Delalibera Jr I, Hajek AE (2004) Pathogenicity and specificity of *Neozygites tanajoae* and *Neozygites floridana* (Zygomycetes: Entomophthorales) isolates pathogenic to the cassava green mite. Biol Control 30:608–616

Delalibera Jr I, Gomez DRS, Moraes GJ de, Alencar JA de, Araújo WF (1992) Infection of *Mononychellus tanajoa* (Acari: Tetranychidae) by the fungus *Neozygites* sp (Zygomycetes: Entomophthorales) in northeastern Brazil. Fla Entomol 75:145–147

Delalibera Jr I, Moraes GJ de, Gomez DRS (1999) *Neozygites floridana* (Zygomycetes, Entomophthorales) epizootics and population dynamics of phytoseiid predatory mites of *Mononychellus tanajoa* (Acari, Phytoseiidae and Tetranychidae) in Bahia. Rev Brasil Entomol 43:287–291

Delalibera Jr I, Moraes GJ de, Lapointe SL, Silva CAD da, Tamai MA (2000) Temporal variability and progression of *Neozygites* sp. (Zygomycetes: Entomophthorales) in populations of *Mononychellus tanajoa* (Bondar) (Acari: Tetranychidae). An Soc Entomol Brasil 29:523–535

Delalibera Jr I, Hajek AE, Humber RA (2004) *Neozygites tanajoae* sp. nov., a pathogen of the cassava green mite. Mycologia 96:1002–1009

Delalibera Jr I, Hajek AE, Humber RA (2003) Use of cell culture media for cultivation of the mite pathogenic fungi *Neozygites tanajo*ae and *N. floridana*. J Invertebr Pathol 84:119–127

Delalibera Jr I, Humber RA, Hajek AE (2004) Preservation of in vitro cultures of the mite pathogenic fungus *Neozygites tanajoae*. Can J Microbiol 50:579–586

Delalibera Jr I, Demetrio CGB, Manly BFJ, Hajek AE (2006) Effect of relative humidity and origin of isolates of *Neozygites tanajoae* (Zygomycetes: Entomophthorales) on production of conidia from cassava green mite, *Mononychellus tanajoa* (Acari: Tetranychidae) cadavers. Biol Control 39:489–496

Elliot SL, Moraes G J de, Delalibera Jr I., Silva CAD da, Tamai MA, Mumford JD (2000). Potential of the mite-pathogenic fungus *Neozygites floridana* (Entomophthorales: Neozygitaceae) for control of the cassava green mite *Mononychellus tanajoa* (Acari: Tetranychidae). Bull Ent Res 90:191–200

Gondim Jr MGC (1992) Efeito da vegetação nativa no controle biológico de *Mononychellus tanajoa* (Bondar) (Acari: Tetranychidae), na zona da mata de Pernambuco e biologia de *Neoseiulus anonymus* (Chant & Baker) (Acari: Phytoseiidae). Masters thesis. Universidade Federal Rural de Pernambuco, Recife, PE, Brazil

Hajek AE, Delalibera Jr I, McManus ML (2000) Introduction of exotic pathogens and documentation of their establishment and impact. In: Lacey LA, Kaya HK (eds) Field manual of techniques in invertebrate pathology: Application and evaluation of pathogens for control of insects and other invertebrate pests. Kluwer Academic Publishers, Dordrecht, NL. pp 339–369

Hajek AE, McManus ML, Delalibera Jr I (2007) A review of introductions of pathogens and nematodes for classical biological control of insects and mites. Biol Control 41:1–13

Herren HR, Neuenschwander P (1991) Biological control of cassava pests in Africa. Annu Rev Entomol 36:257–283

Hountondji FCC, Lomer CJ, Hanna, R, Cherry AJ, Dara, SK (2002a) Field evaluation of Brazilian isolates of *Neozygites floridana* (Entomophthorales: Neozygitaceae) for the microbial control of cassava green mite in Benin,West Africa. Biocontr Sci Technol 12:361–370

Hountondji FCC, Yaninek JS, Moraes GJ de, Oduor GI (2002b) Host specificity of the cassava green mite pathogen *Neozygites floridana*. BioControl 47:61–66

IFAD-FAO (2000) International fund for agricultural development (IFAD); Food and Agriculture Organization of the United Nations (FAO). The world cassava economy, facts, trends and outlook. IFAD-FAO, Rome. Food-Agr-Prod-38. 46 pp

IITA (1998) Annual report. International Institute for Tropical Agriculture, Cotonou, Benin. pp 73–81

Jones WO (1959) Manioc in Africa. Stanford Univ. Press, California

Megevand B, Yaninek JS, Friese DD (1987) Classical biological control of the cassava green mite. Insect Sci Appl 8:871–874

Moraes GJ, Delalibera Jr I (1992) Specificity of a strain of *Neozygites* sp. (Zygomycetes: Entomophthorales) to *Mononychellus tanajoa* (Acari: Tetranychidae). Exp Appl Acarol 14:89–94

Moraes GJ de, Alencar JA, Wenzel Neto F, Mergulhao SMR (1990) Explorations for natural enemies of the cassava green mite in Brazil. In: Howeler RH (ed) Symposium of the International Society of Tropical Root Crops. International Society for Tropical Root Crops, Department of Agriculture,.Bangkok. pp 351–353

Moraes GJ de, Mesa NC, Braun A (1991) Some phytoseiid mites of Latin America (Acari: Phytoseiidae). Internat J Acarol 17:117–139

Navajas M, Gutierrez J, Bonato O, Bolland HR, Mapangou Divassa S (1994) Intraspecific diversity of the cassava green mite *Mononychellus progresivus* (Acari: Tetranychidae) using comparisons of mitochondrial and nuclear ribosomal DNA sequences and cross- breeding. Exp Appl Acarol 18:351–360

Oduor GI, Moraes GJ de, Yaninek JS, Van der Geest LPS (1995a) Effect of temperature, humidity and photoperiod on mortality of *Mononychellus tanajoa* (Acari: Tetranychidae) infected by *Neozygites* cf. *floridana* (Zygomycetes: Entomophthorales). Exp Appl Acarol 19:571–579

Oduor GI, Yaninek JS, Van der Geest LPS, Moraes GJ de (1995b) Survival of *Neozygites* cf. *floridana* (Zygomycetes: Entomophthorales) in mummified cassava green mites and the viability of its primary conidia. Exp Appl Acarol 19:479–488

Oduor GI, Sabelis MW, Lingeman R, Moraes de GJ, Yaninek JS (1997) Modelling fungal (*Neozygites* cf. *floridana*) epizootics in local populations of cassava green mites (*Mononychellus tanajoa*). Exp Appl Acarol 21:485–506

Olsen KM, Schaal BA (1999) Evidence on the origin of cassava: Phylogeography of *Manihot esculenta*. Proc Natl Acad Sci USA 96:5586–5591

Wekesa VW, Delalibera Jr I (2008) Long-term preservation of *Neozygites tanajoae* (Entomophthorales: Neozygitaceae) in cadavers of *Mononychellus tanajoae* (Acari: Tetranychidae). Biocontr Sci Technol (in press)

Yaninek JS, Herren HR (1988) Introduction and spread of the cassava green mite *Mononychellus tanajoa* (Bondar) (Acari: Tetranychidae) an exotic pest in Africa and the search for appropriate control methods: a review. Bull Ent Res 78:1–13

Yaninek JS, Schulthess F (1993) Developing an environmentally sound plant protection for cassava in Africa. Agric Ecosys Environ 46:305–324

Yaninek JS, Moraes GJ de, Markham RH (1989) Handbook on the cassava green mite (*Mononychellus tanajoa*) in Africa. Alphabyte, Rome

Yaninek JS, Saizonou S, Onzo A, Zannou I, Gnanvossou D (1996) Seasonal and habitat variability in the fungal pathogens, *Neozygites* cf. *floridana* and *Hirsutella thompsonii*, associated with cassava mites in Benin, West Africa. Biocontr Sci Technol 6:23–33

Chapter 15
Microbial Control for Invasive Arthropod Pests of Honey Bees

Rosalind R. James

Abstract Honey bees are critical to world agriculture because of their role in crop pollination. Unfortunately, the sustainability of this bee is threatened by an increasing number of invasive pests, particularly the tracheal mite, varroa mite, and small hive beetle. Integrated pest management has not been well utilized by beekeepers, partly due to a lack of biological control agents. Microbial control strategies have been investigated for varroa mites using fungal pathogens, but have produced variable results. Difficulties have arisen because bees maintain hives at temperatures that are detrimental to the fungi, and the immature stages of the mites can avoid the fungi. It is also difficult to mass produce highly virulent and persistent fungal spores, and products are not available for use. One option to investigate further is the search for pathogens of the pests in their native range, as has been done in the introduction of biological control agents to field crop pests. Also, pests that have part of their life cycle outside the hive, such as small hive beetles, may be more amenable to biological control.

15.1 How it All Started

I first met Vaclav (Bill) Ruzika in 1998 in Weslaco, Texas, which is located just north of the US–Mexican border in the Lower Rio Grande Valley. I was working there as a researcher for the USDA Agricultural Research Service. Ruzika is a character to remember, born and raised in Czechoslovakia, but when I knew him, he lived in Canada where he was, and still is, a retired engineer-turned-entrepreneur, running a ski resort and a string of honey bee hives. I think that at the time he was selling both honey and packaged bees. He came to Weslaco to meet with Bill Wilson and me about a fungus-based product called Mycar® (Abbott Laboratories, North Chicago, IL) and claimed that it had the ability to control varroa mites (*Varroa destructor*

R.R. James
USDA, Agricultural Research Service, Pollinating Insects—Biology, Management, Systematics Research Unit, 5310 Old Main Hill, Logan, Utah 84322-5310 USA
e-mail: Rosalind.James@ars.usda.gov

(Acari: Varroidae)) in honey bees. Mycar® was composed of dried hyphae of the fungus *Hirsutella thompsonii* and was originally marketed for control of the citrus rust mite (*Phyllocoptruta oleivora* (Acari: Eriophyidae)) (McCoy & Couch 1982), although by the time of our meeting, it was no longer on the market. Ruzika said that a honey bee researcher in Czechoslovakia, Oldrich Haragsim, had found that Mycar® reduced the reproduction of varroa mites in honey bee hives. So Ruzika had done his homework and obtained a live culture of the Mycar®-strain from Clayton McCoy in Florida (where Ruzika spent his winter holidays). McCoy had done the original research that lead to the development of Mycar® (McCoy & Couch 1982). Ruzika's idea brought Bill Wilson and me together because he was an experienced honey bee researcher, and I specialized in microbial control of insect pests, and we had previously discussed the idea of trying some kind of microbial control for honey bee pests. This visit was the impetus we needed to set the idea in motion. However, we were only one stop on Ruzika's way home to Alberta from Florida; he stopped at every honey bee research facility he could, including ones as out of the way as ours, and as a result, a few other people caught on to his idea and began looking into the possibility of microbial control for varroa mites. In addition, as it turned out, research on the same topic had already begun at the Rothamsted Experiment Station in the United Kingdom.

What was so appealing to me about this idea, besides my general interest in both microbial control and bees, was the potential to help fulfill the great need for biological control in honey production. At the time, no alternatives to chemical control were available for varroa mites, yet honey often appeals to people because it is seen as a pure and natural product. Furthermore, one of the biggest problems we have in developing microbials for field crops is poor persistence of the microbes after they are applied to crops and exposed to sunlight and drying conditions. The honey bee hive is dark and humid. Are these not the perfect conditions for a fungal microbial control agent? Initially, the problems I was concerned about were finding a microbe pathogenic to the mites, yet not to the honey bees, and making sure that honey bee housekeeping practices were not so thorough that they cleaned out all the spores applied to the hive. As it turned out, there were more complications than these three factors, but first, let me give you a little background in honey beekeeping and the invasive pests that have come into North America in the last few decades.

15.2 Honey Beekeeping

Honey bees are generally known for three things: they produce honey, they sting, and they pollinate flowers. Managed bees are indeed vital to world agriculture, providing pollination service for more than 90 crops, including legumes, rape, almonds, apples, berries, and numerous seed crops. However, honey bees have also played an important role in European and American culture as well, where their industriousness and social behaviors have been repeatedly used as metaphors for human political and social agendas (Horn 2005). The honey bees belong to the genus *Apis*, and the most commonly managed bees are *A. mellifera* and *A. cerana*. In the US, farmers rent more than two million *A. mellifera* hives a year (Morse & Calderone 2000), but the sustainability of this bee is being threatened by an increasingly large complex of problems, making them less and less economical to manage.

For the region of North and Central America, somewhere between 2600 and 4900 species of bees have been described, depending on the taxonomic classification you choose (Krombein *et al.* 1979, Michener *et al.* 1994); however, none of the bees native to this region are honey bees. *Apis mellifera* was first brought into North America in the 17th century with the early European settlers, in part to help fulfill the ideal of the New World being the land of milk and honey (Horn 2005), and the bees then spread across the continent with the advancement of European settlements. According to Thomas Jefferson, Native Americans called honey bees the white man's fly, and "considered their approach as indicating the approach of the settlements of the whites" (Jefferson 1785). As agriculture has industrialized and increased in scale, the importance of the honey bee for pollination, rather than solely for honey production, has increased. For example, Burgett (2004) evaluated the economics of beekeeping in Oregon for the year 2003, and found that renting hives for pollination was essential for most beekeepers because honey sales were not profitable enough to off-set expenses. Of course, the profitability of honey varies from year to year depending on the current market price, but also, the costs of production have increased over the years due to increasing needs for pest management.

15.3 Exotic, Invasive Invertebrate Pests of Honey Bee Hives

Prior to 1984, the greatest pests in honey beekeeping were diseases, and American foulbrood was the disease that caused the greatest concern. In 1984, a mite parasitic on honey bees, the tracheal mite (*Acarapis woodi* (Acari: Tarsonemidae)), entered the US from Mexico (Shimanuki *et al.* 2006). This mite is called the tracheal mite because it infests and reproduces in the tracheae, or breathing tubes, of honey bees. The mites feed on the hemolymph of the bee by puncturing the tracheal wall. Female mites enter the tracheae of young adult bees, those less than 1 week old, and lay eggs. It takes 3–4 days for the eggs to hatch, and about 12 days for the young mites to mature and mate. Newly mature, mated female mites disperse by leaving the trachea and attaching to the hairs of the host, where they wait for a new host to come along, and the lifecycle is then repeated. Heavy mite infestations are most likely to cause high mortality of the adult bees during the early spring, before the bees begin their spring population growth.

In 1987, another mite parasitic to honey bees was detected in the US. This new invader, the varroa mite (*Varroa destructor*), was first found in Wisconsin on a package of bees from Florida. The varroa mite was originally a parasite of the Asian honey bee (*Apis cerana*) but was already well adapted to *A. mellifera* when it came into contact with this new host. It was probably introduced at least twice into the US, first from Korea, most likely by way of Europe (where varroa was already widespread) into Florida, and then later from Japan or Thailand (de Guzman *et al.* 1997, 1999, Anderson & Trueman 2000). The mite is now widespread in *A. mellifera* colonies throughout North America. The adult females are much larger than tracheal mites, large enough to see with the naked eye, and they ride around on adult worker bees as a means of phoresy, and they also feed on the hemolymph of adults. I have found these mites to occur most commonly on the intersegmental membranes of adult honey bees, between the second and third sternites of the abdomen, but sometimes between the first and second sternites, and often embedding themselves partially under the sternal plate.

Honey bees raise their young in a wax comb and all stages of immature bees are referred to as "brood." The queen bee lays eggs individually in the cells of the comb, and the attending adult bees then feed the developing larvae until they become pre-pupae and begin to spin cocoons, at which time the attending adults produce a cap over each cell, sealing the brood inside. Worker brood (non-reproductive females) become pre-pupae 6 days after the eggs have hatched and drone brood (males) become pre-pupae 8 days after hatching. When a female adult varroa mite detects a cell containing 5 day old brood, it leaves the adult host and enters the brood cell. After the cell is capped, this mother mite feeds on the developing brood and lays an egg. This first egg will develop into a female mite. The second egg laid by the mother mite will become a male and subsequently the mother mite only lays female eggs, but no more than one per day. After the mite eggs hatch, the developing immatures feed on the host in the capped cell. The cells remain capped for approximately 10.5 days for workers and 12 days for drones. When the host completes development, the young adult bee emerges from the cell, and the mature female mites emerge with it. These females will have mated with their brother before emergence. The mother mite also emerges, being able to infest more than one cell during her lifetime. The male and any immature female mites probably die after the host emerges from the cell, although males can occasionally be found wandering around on the comb. This parasite has proven to be a much more devastating pest than the tracheal mite, to the extent that many beekeepers have somewhat forgotten about tracheal mites and concentrate their efforts on controlling varroa mites. It is possible that some of the chemical control measures for varroa mites also control the tracheal mite secondarily. I will discuss the varroa mite impact and control efforts in more detail later.

The newest invasive invertebrate pest is the small hive beetle (*Aethina tumida* (Coleoptera: Nitidulidae)). It was first found in the US in 1996 in a colony of *A. mellifera* in Florida (Shimanuki *et al.* 2006). The beetle now has been found throughout the US, but it is primarily a problem in the southeastern US. Adult beetles invade honey bee colonies in large numbers and feed on the pollen, honey and brood. They lay eggs inside the hive, hiding the eggs in areas where the bees either will not detect them or cannot get access to them. The larvae also develop in the pollen, honey and brood, and produce a large amount of a slimy substance with a putrefied smell. A yeast associated with this slime produces compounds that attract more beetles to the hive (Torto *et al.* 2007). The beetle can often be controlled by maintaining strong bee colonies and using hive entrance reducers so that the bees can defend themselves more easily. The beetle can also be a pest of stored combs.

15.4 Integrated Pest Management and Honey Beekeeping

After World War II and the invention of DDT, insect control in the US took a dramatic turn. This new chemical pesticide that had low plant and mammalian toxicity was a wonder drug, and its success stimulated the development of other synthetic insecticides. However, the glamour died away as the shortcomings of this approach began to present themselves. Initially, the chemicals that were favored by farmers were those with long environmental persistence and broad spectrum activity. However, they also killed the invertebrate natural enemies, and this trait led to outbreaks of secondary pests that had previously been controlled via natural biological control. The most persistent insecticides also bio-accumulated, that is, they increased in concentration through the food chain, leading to high concentrations in fish, predatory birds and mammals. Thus, even though the mammalian toxicity may have been low, some animals at the top of the food chain ended up getting exposed to concentrations that were detrimental to their reproduction. In addition, the extensive use of these highly effective chemicals sometimes induced pest populations to develop resistance to the pesticides, and with no natural enemies around, this again led to severe pest outbreaks.

Integrated pest management (IPM) is a pest control strategy that emerged as a result of the crises generated by pesticides (Perkins 1982). From one perspective, the base for IPM is biological control, and IPM replaced control strategies where pesticides were at the core. In IPM programs, pesticides are used, but they are used in a strategic manner to avoid disrupting benefits that can be gleaned from biological control. For example, pesticides were developed that were less broad in their target range and were applied more selectively, such as only during the time when they would be most effective, and only when the benefits outweighed the costs based on the population levels of the pests (these critical population levels are termed "thresholds"). Biological control became an important area of research for entomologists, and researchers began "foreign explorations" for the natural enemies of exotic pests for the purpose of releasing them in the US. In addition, other strategies began to be employed, such as the development of insect pheromones to trap pests, monitor their populations, or disrupt their mating. Thus the name "integrated" pest management, because the objective was to integrate the use of several strategies to control a complex of pests in each crop.

The impetus for IPM for field crops and forestry began in the 1970s (Perkins 1982), but the idea of using IPM strategies for honey bee hives is much newer. I remember the idea being discussed at a national honey bee meeting in 2000. Many beekeepers had not heard of IPM at that time, and if they had, they did not really know what it was. Still today, no real IPM strategies are being used in beekeeping. A few research papers exist that attempt to develop some of the basic data needed for an IPM program. For example, an economic threshold for varroa mite control has been described (Delaplane & Hood 1999), the most effective time to apply miticides has been evaluated (Strange & Sheppard 2001, Gatien & Currie 2003), and strategies for coping with miticide resistance have been evaluated, but a general IPM strategy for hive pests has not been widely implemented.

Part of the problem has been a lack of any natural biological control in the beekeeping ecosystem. This problem may not be unique for honey bee colonies, but rather, typical of both highly artificial agricultural systems and invasive pests. Pest control options for beekeepers are limited mainly by two factors. First, the bees are arthropods and easily killed by many pesticides, and second, it is critical not to contaminate the honey and other hive products. Beekeepers try to maintain colonies vigorous enough that the bees can manage their own pest problems, which perhaps is one form of biological control. For example, some genetic lines of honey bees can detect and remove brood that is infested with varroa (Spivak & Reuter 2001), or corral and disable small hive beetles (Neumann *et al.* 2001). However, the general lack of natural biological control means that no natural enemies exist to either be released or protected in the hive. Furthermore, even if we are able to find more natural enemies, the hive environment poses special problems that are very different from those encountered in developing biological control for a forest or an agricultural field. In particular, the hive is a somewhat contained system with a carefully controlled environment that is not always accessible (or amenable) to human manipulations. Thus, some of the IPM approaches taken in the past will not work for bees.

Microbial control may be more adaptable to bee systems than the introduction of parasitoids or predators because honey bees are less likely to remove a microbe from the hive than they would an insect or mite that was introduced as a biological control agent. Parasites and predators of honey bees must be able to deceive the host into letting them into the hive environment, often using chemically-mediated behaviors, or by being too small to be detected and removed. For this reason, microbial control of invasive hive pests probably holds the most promise for future biological control, and is a reason for the considerable interest that has been generated in developing a microbial control for varroa mite, a pest that easily ranks number one in the honey bee world.

15.5 Microbial Control of Varroa Mite

15.5.1 The Identification of Potential Biological Control Agents

Several chemical and cultural controls have been developed for varroa mites, but none of them are entirely effective, and the mites have developed resistance to those that were most effective. Thus, varroa still remains a significant pest of honey bees, and is central to any pest management program. New strategies for its control are constantly being sought. This brings me back to the visit from Bill Ruzika in Texas. The fungus he brought to us was *H. thompsonii*, and we tested it in the laboratory and found it to be a very poor pathogen of varroa mites. However, we also tested several other strains of *H. thompsonii* that originally came from the citrus rust mite, and some of these strains had very high virulence. Our best strain (ARSEF[1] 5858)

---

[1] ARSEF is the USDA Agricultural Research Service's Entomopathogenic Fungus Collection in Ithaca, NY.

had an LC_{90} of 6.3×10^2 conidia/mm^2, 4 days after a spray application (Kanga & James 2002). Compare this to our best strain of *M. anisopliae* (isolated from a product called Bioblast®), that had an LC_{90} of 1.6×10^4 conidia/mm^2 (Kanga & James 2002).

Much of the initial search for an appropriate fungal species and strain focused on finding a fungal pathogen of any arthropod that might show high virulence towards varroa. Many pathogens are known to infect mites (Poinar & Poinar 1998, Chandler *et al*. 2000, Van der Geest *et al*. 2000). The most frequently encountered mite pathogens are fungi, and among the fungi, the most common are the Entomophthorales (Poinar & Poinar 1998). Other common fungal pathogens are in the genus *Hirsutella*, and *M. anisopliae* has been tested for microbial control of ticks (Zhioua *et al*. 1997, 1999, Guedes Frazzon 2000). Rickettsiae, protozoa, microsporidia, and viruses have also been shown to infect mites, but in our initial approach to biological control we needed to focus our efforts because even laboratory-based bioassays were very labor intensive and time consuming. Varroa mites had to be collected directly from honey bee hives just prior to each bioassay. This meant maintaining several varroa-infested honey bee colonies. Keeping the hives infested, yet still viable, required constant monitoring of infestation levels and moving frames of brood from strong colonies to weaker ones. Also, once the mites were collected for a bioassay, they had to be fed live bee pupae several times a week. A few anamorphs of hypocealean fungi emerged as the preferred biocontrol candidates for the following reasons:

(1) The spores of these fungi infect their hosts when they come into contact with the cuticle. Varroa are parasites with sucking mouthparts, and so it would be difficult to apply any pathogen or toxin that needs to go through the gut, such as most bacteria and viruses, because the mites are not likely consume a sufficient number of infective propagules to initiate infection.
(2) These fungi are relatively easy to mass culture. Production methods for infective propagules of Entomophthorales on artificial media do not exist.
(3) Many of the insect-pathogenic Hypoceales are very heat sensitive, but some *Hirsutella* and *Metarhizium* species are more heat tolerant and would be able to survive and infect in the honey bee hive environment. Honey bees regulate the temperature of the hive at approximately 34 °C, which is too high for many entomopathogenic fungi.
(4) Drion Boucias and Clayton McCoy (University of Florida) had a large collection of *H. thompsonii* strains isolated from the citrus rust mite, and many species of *M. anisopliae* have been isolated and stored in general collections, providing a large stock of potential strains to screen.

Although it is generally true that entomopathogenic Hypoceales are easy to mass produce, not all are, and *Hirsutella* is one of the more difficult genera, although the strain of *H. thompsonii* that Ruzika provided us had once been produced on a commercial scale, so it is possible. Stephan Jaronski, a former Abbott Laboratory production scientist (now with the USDA, Agricultural Research Service), said that he had difficulty finding an adequate formulation for *H. thompsonii* spores, and as a result, Mycar® was primarily a hyphal product, and spores were produced by the fungus after the hyphae were applied in the field. This was to prove to be an insurmountable problem later. Like Abbott Laboratories, I was never able to produce a stable spore product that could be applied in the field, even though some of the *H. thompsonii* strains were highly virulent. One of the problems may have been that *H. thompsonii* spores do not readily adhere to varroa cuticle, other than on their feet as they walk around contaminated surfaces (Peng *et al.* 2002). In addition, we could not obtain a reasonable shelf life for the spores once they were harvested from the culture medium, and if spores did not survive well under controlled storage, survival was not likely to improve after being released into the hive environment.

Our studies, and those of others at the time, focused on fungal pathogens known from other species of mites, or from insects. None of the pathogens initially tested had been found to naturally occur on varroa. Our reasoning was that many of these fungi have broad host ranges and could potentially be more pathogenic to varroa than to their natural host, in much the same way that varroa has a much greater impact on the survival of *A. mellifera* than it has on its native host, *A. cerana*. Sometimes when a pathogen switches to a new host, it can be more virulent than it is on its native host. This approach proved fruitful, initially. In general, varroa mites were found to be more susceptible than other mite species to fungal pathogens, and Shaw *et al.* (2002) speculate that this may be due to the mites evolving in a relatively pathogen-free environment in the bee hive.

Meikle *et al.* (2006, 2007) looked for pathogens of varroa mites in bee hives and obtained six isolates of *Beauveria bassiana* (Meikle *et al.* 2006), and then tested two of these naturally occurring pathogens in the field to see if they could get them established in the hive (Meikle *et al.* 2007). I had not previously considered using *B. bassiana* as a biocontrol agent for varroa because this species has been shown several times to be very susceptible to high temperatures, with a significant inhibition of germination and growth above 32 °C (Walstad *et al.* 1970, Hywel-Jones & Gillespie 1990, James *et al.* 1998). However, contrary to expectations, it was found to occur in the hive environment (Meikle 2006) and to be pathogenic to varroa (Shaw *et al.* 2002, Meikle 2006).

15.5.2 Field Trials

Once potential biocontrol agents had been identified, the next logical step was to begin field trials. Several things are necessary to successfully accomplish a field trial, including:

(1) Propagating a sufficient amount of the control agent, or finding a partner in commercial fungal fermentation who can supply spore material;
(2) Developing an application strategy and formulation, or several strategies;
(3) Establishing the amount of control agent material that will be needed, based on the laboratory bioassays;
(4) Developing, at least a method for evaluating varroa populations inside a bee colony, but in addition, methods to evaluate infection rates, honey bee colony strength, and spore survival rates, would be ideal.

Our first experiments on honey bee colonies were done using small observation hives (two-frame hives behind glass) (Kanga & James 2002). These experiments tested the effects of *H. thompsonii* spores in an aqueous suspension sprayed on the bees in the observation hives, and then we monitored varroa mortality (based on the number of dead mites that fell to the bottom of the hive each day). Queen egg-laying and bee death rates were also monitored. In these experiments, approximately 50% of the dead mites were found to be infected with the biocontrol agent after application, but the increases in mortality rates were small, and not statistically greater than the controls (Kanga & James 2002). I repeated these small hive tests in the field, but could not induce infection or any measurable effects on the mite population levels, probably due to poor activity of the spores in the beehives. In these later experiments, we used a dry spore and mycelia powder, and this formulation may not have been as effective as spraying fresh spores. Unlike our spray applications to observation hives, we were not able to recover any viable spores in the hive after application.

While we were experimenting with *H. thompsonii*, two companies (Sylvan and Earth Bioscience) expressed interests in producing *M. anisopliae* spores for biocontrol in the US. Sylvan wanted to produce a strain of the fungus that had previously been used in a product for termite control (called Bioblast) that had been produced by EcoScience. Earth Biosciences was interested in producing a fungus that had been developed by Bayer (called Bio 1020 by Bayer) for biocontrol of several insects (including ticks). Earth Bioscience referred to this as Strain F52. Several field experiments have been conducted with these strains, but with mixed results, as described below.

15.5.2.1 Experiments with the Bioblast Strain of *M. anisopliae*

In the spring of 2002, we conducted a small scale field trial using "nuclear colonies" of honey bees. Nuclear colonies are essentially mini-colonies created by shaking adult bees from full-sized colonies and placing them in a small hive that consists of 5-frames of comb foundation (a standard brood box in the US contains 10 frames of comb) and a new queen. For our experiments, we used 0.9 kg of adult bees and Italian queens. The nuclear colonies were treated with a spore preparation containing 1.07×10^9 spores/g, which was produced at a commercial facility using rice as the substrate. The spores were applied at a rate of 46.8 g/hive (5×10^{10} spores/hive) as either a dry spore powder, or by adhering the spores to plastic strips with vegetable oil. These strips were then inserted between the comb frames. These application methods reduced varroa mite levels to below that of control hives, and to a level that was similar to hives treated with the miticide fluvalinate (Kanga *et al.* 2003). This field trial was conducted in southern Texas, which has a subtropical climate. Kanga *et al.* (2006) repeated these experiments using 3.0×10^{11} spores per hive and a similar number of bees. These spores were produced in-house on Sabouraud dextrose agar (SDA). With this higher dosage, Kanga *et al.* (2006) again obtained significant control of the mites, especially if the colonies were broodless. When colonies are broodless, all the mites in the hive are on adult bees and thus exposed to the fungus.

Furthermore, with no brood present, the mites are not able to reproduce. A broodless condition, or nearly so, may occur during late fall and winter in northern climates, and sometimes during mid-summer in sub-tropical climates. In these experiments, however, they did not use a control that was broodless, so mite levels were compared to those in a reproducing colony, which may be somewhat misleading.

In October 2002, I applied a dry spore powder to colonies of bees obtained from a commercial bee keeper in Logan, Utah. These colonies consisted of two 10-frame hive boxes (called "supers") with approximately 3.5–4.5 kg of bees. Logan has a dry, temperate climate with winter high temperatures usually below freezing between December and March. Spores were applied using talc as an inert ingredient. *Metarhizium anisopliae* was again produced on SDA. The night before treatments were to be applied, the spores were scraped off the culture plates and air dried in a laminar-flow hood over night. The next morning, 30 g of talc powder was mixed with 57 g of spores, yielding a mix that was 1.75×10^{10} spore/g. This mix was then measured into three different treatment rates: H (5×10^{10} spores, 2.86 g), M (2.5×10^{10} spores, 1.43 g), and L (1.0×10^{10} spores, 0.6 g). Controls were not treated, and extra talc was not added to the M and L rates to balance the bulk of material added to the hives, a technique we used later (see below). The spore treatments were applied to the hives by sprinkling half the spores between the frames in each of the supers. Spore viability at the time of treatment was 98.6%, as determined by plating the spores on SDA and then counting the ratio of germinated to ungerminated spores under a microscope (for 500 spores) after the spores had been incubated 18 hours at 25 °C. The hives were then left for the winter. The resulting number of mites per bee did not differ significantly between treatments the next spring, however, the proportion of colonies that were still alive at the end of the winter increased as the application rate increased (Table 15.1).

In September 2003, we repeated the overwintering experiments that we had conducted in Logan, but relocated the hives from the mouth of a canyon (where they were located in 2002) to the floor of Cache Valley, with the hopes that we could improve overwintering survival since the valley was not as cold and windy as the previous location. Also, this time we cultured the fungus on sterile rice, and used

Table 15.1 Effect of *Metarhizium anisopliae* treatments on the overwintering survival of honey bee colonies in 2002

	Spore application rate (per hive)			
	Control	Low 1.0×10^{10}	Medium 2.5×10^{10}	High 5.0×10^{10}
Hives treated in the fall	4	5	7	6
Hives alive in the spring	1	2	5	5
% Survival	25	40	71	83
Mites per 100 bees pre-treatment (mean ± SE)	9.2 ± 2.3	23 ± 5.1	7.6 ± 1.5	8.2 ± 2.3
Mites per 100 bees in the spring (mean ± SE)	–	3.4 ± 1.6	7.0 ± 2.9	5.0 ± 1.0

rice flour as the inert ingredient. To harvest the spores from the rice, the cultured rice was dried overnight in a laminar-flow hood, and the spores were removed by shaking the rice in a sieve. The yield was 5.50×10^9 spores/g. Due to the poor yield of spores, the treatments were altered and included: H (2.48×10^{10} spores, 4.5 g), M (1.24×10^{10} spores, 2.5 g spores +3.25 g rice flour), and the control (6.5 g rice flour), but these treatments were applied twice, once on September 29 and then again on October 8. On September 29, spore viability was 97% and percent moisture of the spores harvested was 67.8%. For the second treatment, a new batch of spores was produced in the same manner. Spore viability of the second batch was 96.5% with 30% moisture. To test how well spores prepared in this manner stored, unused portions of the second batch were stored at 10 °C. After 17 days viability was still 95%. Thus, the high moisture content did not adversely affect spore survival. In fact, we found that for this strain, dry spores did not survive as well in storage at 4 and 35 °C as non-dried spores (Fig. 15.1). Eight hives were treated for the control and M treatments, and seven hives for the H treatment.

Unfortunately, the treatments failed to increase winter survival in this second trial. Mite infestation levels at the beginning of the experiment (September 29, 2003) were lower this second year, with the mean number of mites per 100 bees being 9.69 (SE = 2.34), 7.93 (SE = 1.04) and 10.82 (SE = 2.30) for the control, M, and H treatments, respectively. The numbers of hives still alive on March 23, 2004, were 4 in the control, 1 in treatment M, and 3 in treatment H, yielding the following winter mortality rates: 50% in the control, 87.5% in the M treatment, and 57.1% in the H treatment. Due to the weak condition of the remaining colonies, we did not sample bees in the remaining hives to estimate mite infestation levels.

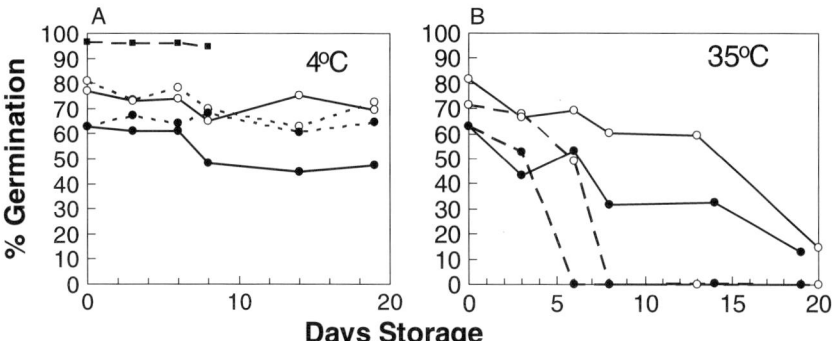

Fig. 15.1 Survival of *Metarhizium anisopliae* spores when stored at different temperatures. To determine viability, spores were tested for an ability to germination on nutrient agar after 24–48 hours incubation at 25 °C. Open circles are spores that were air-dried for 24 hours before being harvested from rice, and solid circles are spores that were air-dried for 48 hours. Dotted lines are spores that were stored as a powder (no formulation), and solid lines are spores that were attached to plastic strips using vegetable oil (cooking spray). Squares are spores that were not dried. Spores were stored at ambient RH (∼ 15%) (800–1200 spores were counted for a given point)

15.5.2.2 Experiments with the Earth Bioscience Strain of *M. anisopliae*

In 2004, we began testing strain F52 in beehives. This experiment was conducted near Gainesville, Florida, which has a subtropical climate. The experiments were conducted in June when the bees experience a period of low bloom availability, and thus brood production is reduced. The fungal spores were produced commercially on rice and the spore concentration was 1.1×10^{10} spores/g. Spore viability at the time of treatments was 86% after 48 hours incubation on SDA. We treated commercial honey bee hives (of a similar size to those treated in Logan) with four different treatments: H (2.87×10^{10} spores/hive, 1.4 g spores +5.1 g rice flour), M (1.44×10^{10} spores/hive, 0.7 g spores +5.8 g rice flour), a rice flour control (6.5 g rice four), and an untreated control. We used ten hives for each treatment. Approximately 200 adult bees were sampled from one frame in each hive just before the treatments were applied, and then 3, 6, and 13 days after the treatments were applied. A repeated measures ANOVA was used to determine whether treatment had a significant effect on the mite infestation levels, after the data had been transformed using the arcsine-square root of the mean number of mites per bee. Again, treatments had no significant effect on mite infestation levels. This time, none of the hives died from mites, probably because the experiments were conducted over a shorter period than in the overwintering experiments, and the bees did not have to endure winter conditions.

In May 2005, we went back to our Logan, Utah, location to determine if we could improve varroa control using different application strategies (James *et al.* 2006). We ran these experiments for 62 days to see if there were any long term effects of the fungus. Using one dose, we tested four different application/packaging strategies and an untreated control group. The five treatments were (a) a removable hive frame that contained live, sporulating fungus cultured on a non-woven material, a preparation sometimes referred to as "fungal bands" (Xu *et al.* 2003, Dubois *et al.* 2004) (Fig. 15.2 A), (b) paper packets of dry spores laid across the tops of the frames (Fig. 15.2 B, the bees were expected to tear apart the packets in an attempt to remove them from the hive, and in this way, release the spores over a period of days), (c) dry spores fixed to plastic strips with vegetable oil, and inserted between two hive frames (Fig. 15.2 C,D), (d) dry spore powder dusted onto the bees between the frames; and (e) untreated controls. Treatments were applied to hives obtained from a commercial bee keeper (of approximately the same size as the first Logan experiments), using five hives per treatment. The dry spores were always applied using 1.4 g per hive, and were obtained from Earth Biosciences (produced on rice substrate). The preparation had 3.0×10^{10} spores/g, with a viability of 84% based on 24 hour germination counts, thus we applied 3.5×10^{10} viable spores per hive. The "fungal bands" were produced in-house and contained 2.1×10^7 spores/cm^2, for a total of 3.69×10^{10} spores per band (James *et al.* 2006). Spore viability was 95% at the time of application.

We measured the number of mites per adult bee, and per brood cell, and estimated the total number of bees in the hive (adults and brood separately) so that we could

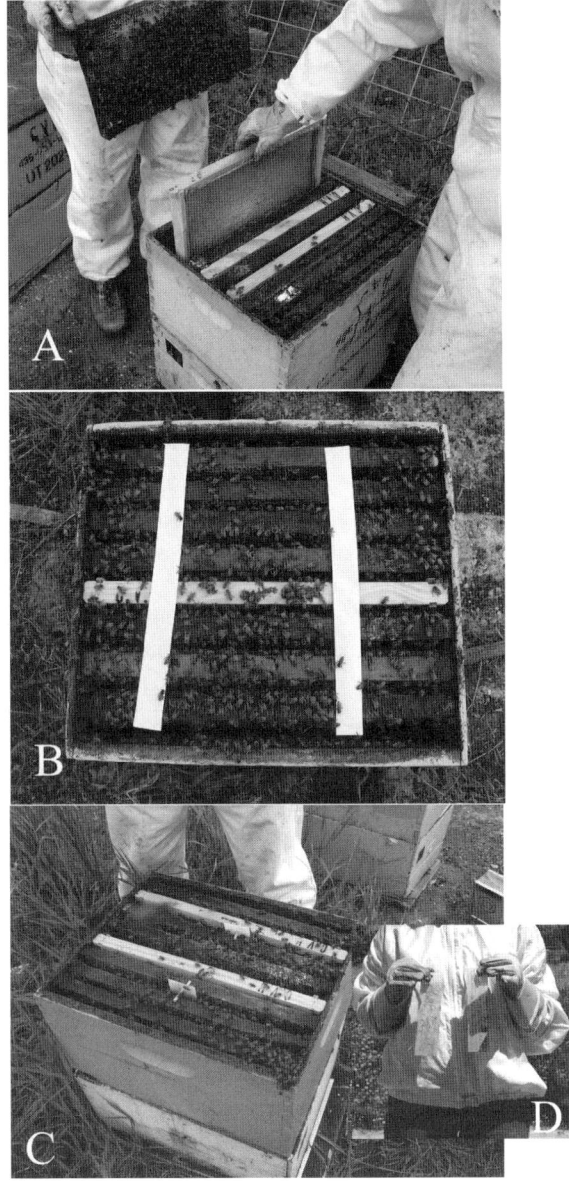

Fig. 15.2 A few *Metarhizium anisopliae* application methods, including (**A**) "fungal bands," (**B**) paper packets containing spores, (**C**) plastic strips coated with oil and spores, then placed in the hives, and (**D**) strips before being placed in the hives

then estimate the total number of mites in the hives. These estimates were made two days before treatments, and again 15, 32, and 62 days post treatment. A repeated measures ANOVA was used to determine whether treatment had a significant effect on the total number of mites and bees in the hive.

Mite numbers increased over the course of the experiment in the packets treatment, and in the plastic-strips treatment (P = 0.07 for packets and P = 0.023 for strips), but did not change significantly in the other treatments, including the controls. The only treatment that showed a decrease in mite numbers was the "fungal bands," but this effect was small and not statistically significant. The populations of adult bees and brood did not change significantly over the course of the experiment, or between treatments (James et al. 2006).

To determine viability of the spores in the hives, samples were taken from the "fungal bands" and from the paper packets on the day of application, and at regular intervals during the course of the experiment. Spore viability in the packets dropped from 84% to 40% in the first day, and then spores slowly lost all viability over the next 20 days. Spores in the "fungal bands" lost viability more rapidly, dropping from 95 to 20% in the first day, and losing all viability by day 14 (James et al. 2006).

With the failure of our creativity in trying to develop a more effective application strategy, we decided to try one high dose of spores mixed with water and sprayed into the hives (Fig. 15.3). This method might not be practical for a beekeeper, but we could at least test the fungus under the best conditions that we could create for it – a high dose with moisture added. This experiment was conducted near Gainesville, Florida, during the winter (November). Twenty hives were randomly assigned to two treatments: treated and control (ten hives per treatment). Fungal spores were produced on rice at the USDA-ARS lab in Stoneville, MS, and spore viability was 96% at the time of treatments. The spore preparation contained 2.9×10^{10} spores/g. For the treated hives, the spores were mixed in a pressurized sprayer with 0.01% Tween 20 (to help disperse and wet the spores) in water at a rate of 17.65 g/L, and applied into the hives at a rate of 12 g/hive (3.34×10^{11} viable spores/hive). The control hives were treated with an equal volume (680 ml) of 0.01% Tween 20. Unfortunately, results from this experiment were no more encouraging. The number of mites per 100 adult bees increased over the course of the experiment, and was unaffected by the treatment (James et al. 2006).

Fig. 15.3 A wand sprayer being used to spray *Metarhizium anisopliae* spore suspension into a honey bee hive

15.5.2.3 Field Trials with *B. bassiana* Strains Isolated from Honey Bee Hives

Meikle *et al.* (2007) applied *B. bassiana* spores to hives. These spores were produced on SDA with yeast extract, then mixed with a silica powder and an electrostatic powder. They used 3.0×10^9 and 8.0×10^9 spores per hive, respectively, in two different experiments, and were able to increase "mite drop" (the number of mites that fall to the bottom board of the hive) after treatments. Significantly more of these trapped mites were infected with *B. bassiana* than in the controls. Unfortunately, the treatments did not have any measurable effect on varroa population levels in the hives. However, it is worth noting that the dosages Meikle *et al.* (2007) tested were several times lower than in all the experiments described above, yet he still saw infections in the hive.

15.5.2.4 General Conclusions Regarding Microbial Control of Varroa

The most frustrating aspect of all these experiments has been that the results are so variable; good control has occasionally been achieved, but not consistently, even under seemingly similar conditions. The substrate on which the spores are produced (rice versus nutrient agar) may have had some effect on the results; spores produced on the nutrient agar were associated with the best results above. This may have to do with how the spores were handled as much as the substrate itself, particularly with how the spores are dried (Hong *et al.* 2000).

I suspect that three main problems plagued this application of microbial control. First, it was difficult to direct applications to the target pest. Many of the mites are enclosed in sealed brood cells and avoid hive applications. Even when the mites are on the bees, we are not certain if they came into contact with the spores. I have found that spores disperse readily in the hive on the adult bees, but I do not know how many spores actually come into contact with the mites. The second problem is that the conditions in the hives are very warm and humid, to the extent that they reduce spore survival. As described above, spore survival in the hive after application is very poor for unformulated spores. Oil has been shown to increase the survival rate of spores under warm conditions (Hong *et al.* 1999), and that may explain the slightly greater success of the spores applied to hives on plastic strips treated with oil. The third problem is dose. The best results have occurred with the highest dosages. However, it has been difficult to obtain such large numbers of spores. To put this into perspective, the highest rates tested for varroa were 3×10^{11} spores per hive (Kanga *et al.* 2006), as compared to, say, the recommended rates of *B. bassiana* for whitefly control in greenhouses where 3×10^{13} spores are used for an entire hectare, although they must be applied every 5–7 days (Faria & Wraight 2001). The dose applied by Kanga *et al.* (2006) is so high that it cannot feasibly be applied all at once; as for whitefly control, it would have to be applied repeatedly. Furthermore, the colonies in those experiments were what beekeepers call "nuclear hives." That is, they only contained the small number of bees used to initiate a new colony, similar to what are sold as packages of bees. They are not the size of a normal, established colony. Thus, much higher doses are probably required for more typical colonies.

15.6 General Conclusions About Biological Control in Honey Bee Hives

Beekeeping seems to be vulnerable to invasion by a myriad of pests, as described earlier in this chapter. What makes the honey bee so susceptible? Is it the lack of any natural biological control agents? Even within the native range of honey bees (Africa, Asia, and Europe), not many natural enemies for honey bee pests have been reported. This is partly due to the social behavior of this bee. Honey bees protect their nests very aggressively, and are on careful watch for invaders, even of the same species. Any beneficial organism would have to be able to avoid this detection system, much as the pests themselves do. For this reason, microbial control seems the most promising biological control strategy, but very few studies have been conducted to identify microbes pathogenic to hive pests. However, the same problems encountered during our attempts to develop microbial control of varroa will be encountered again: high hive temperatures, an inability to easily access the target pest, and the resulting need for very high dosages.

The small hive beetle also spends time outside the bee colony, pupating in the soil and attacking stored comb. It may be easier to target this pest outside of the active bee colony. For example, pupae of the small hive beetle are occasionally killed by fungi in the soil (Ellis *et al*. 2004).

A troublesome area for microbial control of honey bees is availability of products. The number of honey bee hives in any country is relatively small and the value of these hives is also not especially great. As a result, beekeeping does not provide a very large market to support all of the costs associated with developing and registering new microbial control products. New products are most likely to be borrowed from other industries, such as the *M. anisopliae* strains we tested that had been developed for the control of termites and lawn pests. Alternatively, perhaps a more classical biological control approach could be achieved if the beekeeping industry or government agencies paid for the exploration and dispersal of agents with the intention of getting them established in hives, similar to the way the government supports biological control of weeds in rangelands and pastures.

Acknowledgments I am grateful to all the people who encouraged me to pursue the idea of microbial control of varroa mites, and in particular Bill Ruzika, Gerald Hayes, Thomas Corell, and John Cascino. Enthusiasm for the idea has been undaunted, even when I struggled with variable results, and even as some of these people traversed different jobs. Jarrod Leland prepared the "fungal bands" for us. Darren Cox is a commercial beekeeper who provided bees for experiments on several occasions, and who is an excellent research partner. The Latners at a Dadant supply house in Florida also provided us with honey bee colonies and a place to serve as base when we worked in Florida; they too, were great to work with. I had technical assistance in setting up the experiments from several people, in particular from Craig Huntzinger, but also David Barnes, Jose Diaz, Thomas Dowda, Abigail Fox, Sonia Gallegos, Carlos Gracia, Henry Graham, Lambert Kanga, Ellen Klinger, Deanna Larson, and Brian McElroy. Some of these folks are old hands with bees, but others learned to appreciate this intriguing insect, despite a few stings, during the course of the experiments.

References

Anderson D, Trueman JWH (2000) *Varroa jacobsoni* (Acari: Varroidae) is more than one species. Exp Appl Acarol 24:165–189

Burgett M (2004) 2003 Pacific Northwest honey bee pollination survey. The Speedy Bee 33:8–9

Chandler D, Davidson G, Pell JK, Ball BV, Shaw K, Sunderland KD (2000) Fungal biocontrol of acari. Biocontr Sci Technol 10:357–384

de Guzman LI, Rinderer TE, Stelzer JA (1997) DNA evidence of the origin of *Varroa jacobsoni* Oedumans in the Americas. Biochem Genet 35:327–335

de Guzman LI, Rinderer TE, Stelzer JA (1999) Occurrence of two genotypes of *Varroa jacobsoni* Oud. in North America. Apidologie 30:31–36

Delaplane KS, Hood WM (1999) Economic threshold for *Varroa jacobsoni* Oud. in the southeast USA. Apidologie 30:383–395

Dubois T, Li Z, Hu J-F, Hajek AE (2004) Efficacy of fiber bands impregnated with *Beauveria brongniartii* cultures against the Asian longhorned beetle, *Anoplophora glabripennis* (Coleoptera: Cerambycidae). Biol Control 31:320–328

Ellis JD Jr, Rong IH, Hill MP, Hepburn HR, Elzen PJ (2004) The susceptibility of small hive beetle (*Aethina tumida* Murray) pupae to fungal pathogens. Amer Bee J 144:486–488

Fargues J, Goettel MS, Smits N, Ouedraogo A, Vidal C, Rougier M (1997) Effects of temperature on vegetative growth of *Beauveria bassiana* isolates from different origins. Mycologia 89:383–392

Faria M de, Wraight SP (2001) Biological control of *Bemisia tabaci* with fungi. Crop Protection 20:767–778

Gatien P, Currie RW (2003) Timing of acaracide treatments for control of low-level populations of *Varroa destructor* (Acari: Varroidae) and implications for colony performance of honey bees. Can Entomol 135:749–763

Guedes Frazzon AP, da Silva Vaz IJ, Masuda A, Schrank A, Henning Vainstein M (2000) *In vitro* assessment of *Metarhizium anisopliae* isolates to control the cattle tick *Boophilus microplus*. Vet Parasitol 94:117–125

Hong TD, Jenkins NE, Ellis RH (1999) Fluctuating temperature and the longevity of conidia of *Metarhizium flavoviride* in storage. Biocontr Sci Technol 9:165–176

Hong TD, Jenkins NE, Ellis RH (2000) The effects of duration of development and drying regime on the longevity of conidia of *Metarhizium flavoviride*. Mycol Res 104:662–665

Horn T (2005) Bees in America: How the honey bee shaped a nation. University Press of Kentucky, Lexington, KY

Hywel-Jones NL, Gillespie AT (1990) Effect of temperature on spore germination in *Metarhizium anisopliae* and *Beauveria bassiana*. Mycol Res 94:389–392

James RR, Croft BA, Shaffer TB, Lighthart B (1998) Impact of temperature and humidity on host-pathogen interactions between *Beauveria bassiana* and a coccinellid. Environ Entomol 27:1506–1513.

James RR, Hayes GW, Leland JE (2006) Field trials on the microbial control of varroa with the fungus *Metarhizium anisopliae*. Amer Bee J 146:968–972

Jefferson, T 1785. Notes on the State of Virginia, as cited in Horn T (2005) Bees in America: How the honey bee shaped a nation. University Press of Kentucky, Lexington, KY

Kanga LHB, James RR (2002) *Hirsutella thompsonii* and *Metarhizium anisopliae* as potential microbial control agents of *Varroa destructor*, a honey bee parasite. J Invertebr Pathol 81:175–184

Kanga LHB, Jones WA, James RR (2003) Field trials using the fungal pathogen, *Metarhizium anisopliae* (Deuteromycetes: Hyphomycetes) to control the ectoparasitic mite, *Varroa destructor* (Acari: Varroidae) in honey bee, *Apis mellifera* (Hymenoptera: Apidae) colonies. J Econ Entomol 96:1091–1099

Kanga LHB, Jones WA, Gracia C (2006) Efficacy of strips coated with *Metarhizium anisopliae* for control of *Varroa destructor* (Acari: Varroidae) in honey bee colonies in Texas and Florida. Exp Appl Acarol 40:249–258

Krombein KV, Hurd PD, Smith DR, Burks BD (1979) Catalog of Hymenoptera in America north of Mexico, Vol. 2. Smithsonian Institution Press, Washington, DC

McCoy CW, Couch TL (1982) Microbial control of the citrus rust mite with the mycoacaricide, Mycar. Fla Entomol 65:116–126

Meikle WG, Mercadier G, Girod V, Derouané F, Jones WA (2006) Evaluation of *Beauveria bassiana* (Balsamo) Vuillemin (Deuteromycota: Hyphomycetes) strains isolated from varroa mites in southern France. J Apicul Res 45:219–220

Meikle WG, Mercadier G, Holst N, Nansen C, Girod V (2007) Duration and spread of an entomopathogenic fungus, *Beauveria bassiana* (Deuteromycota: Hyphomycetes), used to treat varroa mites (Acari: Varroidae) in honey bee (Hymenoptera: Apidae) hives. J Econ Entomol 100:1–10

Michener CD, McGinley RJ, Danforth BN (1994) The bee genera of North and Central America. Smithsonian Institution Press, Washington, DC

Morse RA, Calderone NW (2000) The value of honey bees as pollinators of U.S. crops in 2000. Bee Culture 132:1–19

Neumann P, Solbrig AJ, Hepburn HR, Elzen PJ, Baxter J, Ratnieks FLW (2001) Behavioural resistance mechanisms of African honey bees toward the small hive beetle (*Athina tumida*, Coleoptera: Nitidulidae). Apidologie 32:495–496

Peng C, Zhu X, Kaya H (2002) Virulence and site of infection of the fungus, *Hirsutella thompsonii*, to the honey bee ectoparasitic mite, *Varroa destructor*. J Invertebr Pathol 81:185–195

Perkins JH (1982) Insects, experts, and the insecticide crisis. Plenum Press, New York

Poinar G Jr, Poinar R (1998) Parasites and pathogens of mites. Annu Rev Entomol 43:449–469

Shaw K, Davidson G, Clark SJ, Ball BV, Pell JK, Chandler D, Sunderland KD (2002) Laboratory bioassays to assess the pathogenicity of mitosporic fungi to *Varroa destructor* (Acari: Mesostigmata), an ectoparasitic mite of the honeybee, *Apis mellifera*. Biol Control 24:266–276

Shimanuki H, Flottum K, Harman A (2006) The ABC and XYZ of bee culture. A. I. Root Company, Medina, OH

Spivak M, Reuter GS (2001) *Varroa destructor* infestation in untreated honey bee (Hymenoptera: Apidae) colonies selected for hygienic behavior. J Econ Entomol 94:326–331

Strange JP, Sheppard WS (2001) Optimum timing of miticide applications for control of *Varroa destructor* (Acari: Varroidae) in *Apis mellifera* (Hymenoptera: Apidae) in Washington State, USA. J Econ Entomol 94:1324–1331

Torto B, Arbogast RT, Alborn H, Suazo A, van Engelsdorp D, Boucias D, Tumlinson JH, Teal PEA (2007) Composition of volatiles from fermenting pollen dough and attractiveness to the small hive beetle *Aethina tumida*, a parasite of the honeybee *Apis mellifera*. Apidologie 38:1–10

Van der Geest LPS, Elliot SL, Breeuwer JAJ, Beerling EAM (2000) Diseases of mites. Exp Appl Acarol 24:497560

Walstad JD, Anderson RF, Stambaugh WJ (1970) Effects of environmental conditions on two species of muscardine fungi (*Beauveria bassiana* and *Metarhizium anisopliae*). J Invertebr Pathol 16:221–226

Xu, JZ, Fan, MZ, Li, Z-Z (2003) Comparative studies on different ways for releasing *Beauveria brongniartii* against Asian longicorn beetles. Chin J Biol Control 19:27–30

Zhioua E, Browning M, Johnson PW, Ginsberg HS, LeBrun RA (1997) Pathogenicity of the entomopathogenic fungus *Metarhizium anisopliae* (Deuteromycetes) to *Ixodes scapularis* (Acari: Ixodidae). J Parasitol 83:815–818

Zhioua E, Ginsberg HS, Humber RA, LeBrun RA (1999) Preliminary survey for entomopathogenic fungi associated with *Ixodes scapularis* (Acari: Ixodidae) in southern New York and New England, USA. J Med Entomol 36:637

Part V
Safety and Public Issues

Chapter 16
Human Health Effects Resulting from Exposure to *Bacillus thuringiensis* Applied during Insect Control Programmes

David B. Levin

Abstract Products based on *Bacillus thuringiensis* (*Bt*) such as Foray 48B, which contains *Bt kurstaki* strain HD-1, must meet rigorous standards required by the US Environmental Protection Agency, the US Food and Drug Administration, the Canadian Pesticide Management and Regulatory Agency, and Health Canada, before they are approved for commercial use in Canada and the US. These agencies consider *Bt*-based products to be neither toxic nor pathogenic to mammals, including humans. Despite these approvals, there remains widespread public concern about negative health effects associated with aerial applications of *Btk* during insect control programmes. Major health impact assessment studies in the US and Canada suggested there were no negative short-term human health effects associated with aerial applications of Foray 48B. A similar health impact assessment conducted in New Zealand reported short term irritant effects and some worsening of pre-existing conditions such as allergies and asthma. These findings warrant further investigation following aerial applications of commercial *Bt* products in populated urban areas.

16.1 Introduction

Bacillus thuringiensis (*Bt*) is a ubiquitous, soil-dwelling, spore-forming, gram-positive bacterium that produces intracellular "crystal proteins" during sporulation. These crystal proteins are precursors to toxins that specifically attack certain groups of insects (Schnepf *et al*. 1998). Different subspecies or varieties of *Bt* produce different crystal proteins. For example, one subspecies of *Bt* produces a crystal protein that is a precursor to a toxin that kills beetles (Order Coleoptera). Another subspecies of *Bt* produces a protein that kills mosquitoes (Order Diptera). *Bt kurstaki* (*Btk*) is commonly used as a biological insecticide to control larvae of moths and butterflies (Order Lepidoptera) that defoliate trees and shrubs. Spores and crystal

---

D.B. Levin
Department of Biosystems Engineering, E2-376 EITC, University of Manitoba, Winnipeg, Manitoba R3T 5V6 Canada
e-mail: levindb@cc.umanitoba.ca

proteins lie dormant in the soil and on surfaces of plant roots and foliage. When insect larvae feed on the contaminated plant material, the crystal proteins are proteolytically cleaved within the insect mid-gut to become potent toxins. The activated proteins insert into the cytoplasmic membrane of the larval mid-gut epithelial cells causing these cells to rupture, enabling the bacteria to penetrate and multiply within the insect, ultimately resulting in death.

Bt is part of a cluster of bacterial species known as the *Bacillus cereus*-group (see Section 16.1.1) that includes *Bacillus cereus*, a well-known food-poisoning bacterium, and *Bacillus anthracis*, the bacterium that causes anthrax (Logan & Turnbull 1999, Carlson *et al.* 1994). Genetic analyses have confirmed that *Bt* is "very closely related" to *B. cereus* (Carlson & Kolsto 1993, Carlson *et al.* 1994).

Commercial preparations of *Btk* have been used in aerial spray programmes to eradicate and control populations of a variety of lepidopteran insects in the United States, Canada and New Zealand. These aerial applications are often conducted in or near densely populated urban areas. Because of the similarities between *Bt*, *B. cereus* and *B. anthracis*, *Bt*-based products are subjected to extensive evaluation both prior to and subsequent to becoming commercially available (McClintock *et al.* 1995). Government agencies, such as the United States Environmental Protection Agency, the US Food and Drug Administration, the Canadian Pesticide Management and Regulatory Agency and Health Canada, consider *Bt* products to be "safe", i.e. they are neither toxic nor pathogenic to mammals, including humans, and their potential negative impacts on other "non-target" species are within acceptable limits. Despite these approvals, there has been widespread public concern about human exposures to *Btk* after aerial applications (Swadner 1994, Ginsberg 2006). This chapter discusses the evidence for and against harmful health effects posed by aerial applications of *Btk* during insect eradication and control programmes.

16.1.1 Genetic Analyses of Bacillus cereus-*Group Bacteria*

The genus *Bacillus* has been divided into five groups (Ash *et al.* 1991). Group I contains a large number of soil-dwelling species, including *B. thuringiensis*, *B. sphaericus*, *B. subtilis*, *B. anthracis* and *B. cereus*. Some of these bacilli are very closely related and form separate groups within the Group I *Bacillus* species. The *B. cereus*-group consists of *B. cereus*, *B. anthracis*, *B. mycoides* and *Bt*, with *B. cereus*, *B. mycoides* and *Bt* being so closely related that some taxonomists consider the three species to be subspecies of *B. cereus* "*sensu lato*" (Leonard *et al.* 1997).

Analysis of the genomes of various strains of *B. cereus* and *Bt* by pulse-field electrophoresis revealed that some strains of *B. cereus* are more similar to certain strains of *Bt* than they are to other strains of *B. cereus*, suggesting that these two species may actually be variants of the same species (Carlson *et al.* 1994). Indeed, early attempts to distinguish between *B. cereus* and *Bt* using nucleotide sequence analysis of genes that encode 16S ribosomal RNA were unsuccessful (Xu & Côté 2003). However, further sequence analyses of 16S rRNA genes, in combination with sequence analyses of 23S rRNA and gyrB genes from a large number of *B. cereus*-group species, revealed that the three most common species of the *B. cereus*-group, *B. cereus*, *B. mycoides* and *Bt*, were each heterogeneous in all three gene loci, while all analysed strains of *B. anthracis* were found to be homogeneous (Bavykin *et al.* 2006). Therefore, based on the analyses of 16S and 23S rRNA sequence variations, the bacteria within the *B. cereus*-group may be divided into seven subgroups: *B. anthracis*, *B. cereus* A and B, *B. thuringiensis* A and B, and *B. mycoides* A and B. Although "*B. cereus*-like" strains were identified in six of the seven subgroups and "*Bt*-like" strains were identified in three of the subgroups, phylogenetic analyses indicated that the majority of *B. cereus* and *Bt* strains were associated with discrete clades (groups or clusters of organisms with a common genetic ancestor).

These analyses do not support the proposed unification of *B. cereus* and *Bt* into one species, and further indicate that rRNA and gyrB gene sequences may be used for discriminating *B. anthracis* from other bacteria in the *B. cereus* group (Bavykin *et al.* 2006). Moreover, complete genome sequence of *B. cereus* strain ATCC 14579 together with the gapped genome of *B. anthracis* strain A2012 enabled comparative analyses that have identified genes that are conserved between *B. cereus* and *B. anthracis*, and genes that are unique for each species. These unique genes have been used to clarify relationships among the *B. cereus*-group bacteria (Ivanova *et al.* 2003).

Bacteria commonly contain small, circular, extra-chromosomal DNA molecules called "plasmids", which can encode a variety of genes. The plasmids contained within *B. cereus*-group species are essential in defining the phenotypic traits associated with the biology, ecology and pathogenicity of these bacteria. For example, *B. anthracis* contains two plasmids, pXO1 and pXO2, encoding toxin production and encapsulation, respectively. These plasmids define the mammalian-specific pathogenic potential of this species, whereas the presence of plasmids encoding *cry* genes, which produce the "crystal protein" endotoxins specific to insects, defines *Bt* isolates. Detailed sequence analyses of plasmids isolated from *Bt*, *B. cereus*, and *B. anthracis* strains have revealed conserved sequence fragments from the *B. anthracis* virulence plasmid pXO1 in plasmids isolated from *B. cereus* and *Bt* (Hu *et al.* 2006), suggesting either a common origin of these plasmids or the possibility of genetic exchange between all three species. Studies of gene transfer between *B. cereus*-group bacteria have confirmed that plasmids can pass freely among *B. cereus* and *Bt* strains, but that plasmid transfer takes place only within dead insect larvae among vegetatively growing bacteria and at very low frequencies (Yuan *et al.* 2007).

Although commercial products such as Foray 48B are based on defined strains of *Btk* that are cultivated under strict quality-controlled conditions, it is possible that a commercial *Bt* strain could acquire genes from *B. cereus* or *B. anthracis* through conjugation within the environment of a dead insect, or intestinal tract of an animal. Although this is theoretically possible, the frequency of this type of recombination in nature would be extremely low, and it is highly unlikely that a *Btk* product could become contaminated with *B. cereus* or *B. anthracis*, or with plasmids derived from these bacteria, during commercial production.

16.2 Does *Btk* Cause Primary Infections in Humans?

Animal experimentation has shown that interperitoneal injection of *B. thuringiensis* can cause death in guinea pigs (Fisher & Rosner 1959), and that pulmonary infection can result in death of immunocompromised mice (Hernandez *et al.* 1999). These experiments, however, reflect neither the types nor the consequences of human exposure to *Btk*. Exposures to *Btk* have been associated with elevated immune response and gastrointestinal illness, but cases in which *Btk* has been demonstrated to be the causal agent of illness are extremely rare (see references below).

A study of farm workers who harvested *Btk*-treated vegetables found that while there was no evidence of occupation-related respiratory symptoms, positive skin-prick tests to several spore extracts were seen in exposed workers (Bernstein *et al.* 1999). In particular, there was a significant increase in the number of positive skin tests to spore extracts at 1 and 4 months after exposure to *Btk* spray. The number of positive skin test responses was also significantly higher in workers with high levels of exposure compared with those with medium or low levels of exposure to *Btk*. Nasal lavage cultures from exposed workers were positive for the commercial *Btk* organism, as demonstrated by specific molecular genetic probes. Specific IgE antibodies were present in more high-exposure workers than in the low and medium groups. Specific IgG antibodies were detected at a greater frequency in the high than in the low-exposure group. Specific IgG and IgE antibodies to *Bt* were present in all groups of workers. The presence of IgG and IgE antibodies indicates that workers were exposed to levels of *Btk* sufficient to stimulate an immune response. A high level of IgE is an indication of "sensitization", the most extreme result of which could be anaphylactic shock when the individual is exposed to the antigen at a later date.

More recently a longitudinal respiratory health study was conducted among > 300 Danish greenhouse workers (Doekes *et al.* 2004). Serum IgE was measured by enzyme immunoassay (EIA) designed to detect *Btk* or the fungus *Verticillium lecanii*. Many sera had detectable IgE specific to *Btk* (23–29%) or *Verticillium* (9–21%). IgE titres from the 2- and 3-year follow-up (n = 230) were highly correlated with the original exposures, and in some individuals (< 15%), IgE titres increased in response to exposure to different *Btk* or *Verticillium* products (both r > 0.70). Thus, exposure to one type of *Btk* product may confer a risk of IgE-mediated sensitisation to a variety of *Btk* products. This study, however, did not determine or identify the allergenic components in the preparations. Moreover, control studies on IgE levels in non-exposed farm workers were not conducted. Thus, a direct correlation between IgE sensitisation and *Btk* is tenuous. Moreover, it must be emphasised that no *Btk*-related illnesses were reported by either Bernstein *et al.* (1999) or Doekes *et al.* (2004).

Btk has been isolated from the faeces of workers who applied *Btk* in greenhouses (Jensen *et al.* 2002a, b). Faecal samples from 12 Danish greenhouse workers were collected for microbial analysis. Seven workers were using *Bt*-based insecticides, whereas five were employed at greenhouses that did not use *Bt*. Bacteria were isolated on *B. cereus*-specific solid substrate, and colonies were further identified using the polymerase chain reaction (PCR). The PCR method was used for the identification of the enterotoxin genes HblA and BceT. The expression of enterotoxins was detected with two commercial serological kits. Primers specific for 16S–23S spacer region were used to identify the bacteria as members of the *B. cereus*-group. Several primers towards insecticidal genes were used to further characterise the isolates as subspecies of *Bt*. Two faecal samples from the *Bt*-exposed greenhouse workers were positive for *B. cereus*-like bacteria. One isolate that was positive for intracellular crystalline inclusions characteristic of *Bt*, and for *cry* genes by PCR, was also positive for *B. cereus* enterotoxins. RAPD profiles of the faecal isolate were identical to those of strains isolated from a commercial product. The methods used in this study verified that the faecal isolate was identical to the *Bt* isolate found in the biopesticide used. Again, it must be emphasised that the person from whom this bacteria was isolated displayed no symptoms or signs of illness (Jensen *et al.* 2002a, b).

Bt has been associated with intestinal illness (Jackson *et al.* 1995). During investigation of a gastroenteritis outbreak in a chronic care institution, Norwalk virus was found in stool specimens from two individuals and bacterial isolates, presumptively identified as *B. cereus* were isolated from four individuals (including one with Norwalk virus). All clinical isolates were subsequently identified as *Bt*. All *B. cereus* and *Bt* isolates showed cytotoxic effects characteristic of enterotoxin-producing *B. cereus*. An additional 20 isolates each of *B. cereus* and *Bt* from other sources were tested for cytotoxicity, and all, with the exception of one *B. cereus* isolate, showed cytotoxic activity. It must be emphasised, however, that Norwalk virus, and not *Bt*, was identified as the causal agent of the gastrointestinal illness.

B. cereus-like enterotoxin genes have been identified in natural isolates of *Bt* strains (Perani *et al.* 1998, Swiecicka *et al.* 2006), as well as in commercial strains of *Bt* (Damgaard 1995). Strains of *B. cereus* and *Bt* were tested by the Tecra VIA kit for the ability to produce a diarrhoeal enterotoxin (Damgaard 1995). The strains of *Bt* were isolated from commercial *Bt*-based insecticides (Bactimos, Dipel, Florbac FC, Foray 48B, Novodor FC, Turex, VecTobac and XenTari). The production of diarrhoeal enterotoxin varied by a factor of more than 100 among the different strains tested. *B. cereus* (F4433/73) produced the highest amount of enterotoxin and the *Bt* strain isolated from DiPel the lowest. The products were tested for their content of diarrhoeal enterotoxin and all products, except MVP, which does not contain viable *Bt* spores, contained diarrhoeal enterotoxins. These results indicate a potential risk for gastroenteritis caused by *Bt*.

Proven cases of *Btk* causing clinical disease in humans are extremely rare, if indeed any exist. In one case, a corneal ulcer developed in a previously healthy 18 year-old farmer who accidentally splashed a commercial formulation of *Btk* into his eye (Samples & Buettner 1983). Siegel and Shadduck (1990) suggested the possibility that while spores may have persisted in the eye and were recovered, they may not have actually caused the ulcer. More recently multiple thigh and knee abscesses containing a different subspecies of *Bt*, *Bt konkukian*, were found in a previously healthy soldier who was severely wounded by a landmine-explosion (Hernandez *et al.* 1998). It was unclear whether *Bt* was the causative agent or a contaminant. Numerous reviews of mammalian safety of *Btk* report that the risk to humans from *Btk* is extremely small (Siegel 2001, O'Callaghan & Glare 2003).

16.3 Exposure of Humans during Aerial Spraying

Large-scale applications of commercial *Btk* preparations have been applied in aerial spray programmes within or very near to densely populated urban populations. Surveillance for human health impacts was conducted during aerial sprays in the United States (Green *et al.* 1990), Canada (Noble *et al.* 1992, 1994, Pearce *et al.* 2002a, b), and New Zealand (Anonymous 1993, Aeraqua Medical Services 2001, 2005a, b). Details concerning the bioaerosols generated during aerial applications of Foray 48B, which contains *Btk* strain HD-1 (*Btk* HD1), and the extent of human exposures to *Btk* during these applications, are available from studies conducted in conjunction with a European gypsy moth, *Lymantria dispar*, eradication programme conducted in Victoria, British Columbia (BC) Canada (de Amorim *et al.* 2001, Teschke *et al.* 2001).

Aerial applications of Foray 48B were carried out in Victoria, BC, on May 9–10, May 19–21, and June 8–9, 1999. The spray was applied by aircraft between 5:00 am and 7:00 am (Pacific time), when the wind was at 10 kilometres/hour (approximately 6 miles/hour) or less. The spray was applied at 4 litres/hectare (approximately 0.25 U.S. gallons/acre) at a rate of 70 litres/minute (approximately 20 U.S. gallons/minute), and a droplet size of between 110 and 130 µm. The objective was to apply the bacteria so maximum amounts would adhere to tree foliage. The application of Foray 48B resulted in 99% mortality of the gypsy moth population (British Columbia Ministry of Forest and Range 2007).

Two hundred and fifty-six bulk air samples were taken during aerial application at outdoor locations, both within and down-wind of the spray zone, to monitor the presence and concentration of airborne *Btk* HD1 spores during and after the spray application. These samples were taken by pulling air through 0.5 µm pore-size Teflon filters mounted in 37 mm close-face cassettes using constant flow battery-powered pumps. To characterise the size distributions of the *Btk* HD1-containing aerosols, air samples were taken using a size-selective, six-stage cascade impactor (Andersen plates) mounted with trypticase-soy-agar plates (Teschke *et al.* 2001). These air samples were drawn through 0.5 µm filters so that particles were separated and settled on the agar plates on the basis of their airborne size (0.7–1.1 µm). Air was drawn at a rate of 28.3 litres/min with a high volume pump maintained by an orifice placed after the sixth stage of the agar plates. Samples were taken both outside and inside the homes of participating residents during and after the spray application.

Nasal swabs were obtained from asthmatic children whose families volunteered to perform these tests. These families were located both within the spray zone and outside the spray zone (down-wind). Post-spray swabs were taken the morning after the spray, before any doors or windows were opened. Environmental (air and water) and human (nasal swab) samples, collected before and after aerial applications of Foray 48B, were analysed for the presence of *Btk* HD1 (de Amorim *et al.* 2001). Random Amplified Polymorphic DNA (RAPD) analysis, *cry* gene specific polymerase chain reaction (PCR), and dot-blot DNA hybridisation techniques were used to identify over 11,000 isolates of bacteria. Identification of *Btk* HD1 in air samples collected on the filters of the bulk air samplers and on the agar plates of the

Andersen samplers permitted calculation of the concentrations of airborne *Btk* HD1 spores both outside and inside homes.

Identification of *Btk* HD1 in nasal swab samples permitted an estimation of the human exposure to the airborne spores. *Bacillus* isolates with genetic patterns consistent with those of *Btk* HD1 used in the spray were identified in 85.4% of the isolates obtained from the air samples and 76.6% isolates obtained from the human nasal swab samples (de Amorim *et al.* 2001). The average concentration of airborne *Btk* HD1 measured outside of residences during the 30-minute period of the spraying within the spray zone was 739 colony forming units per cubic meter of air (CFU/m^3), where in this case the number of CFUs is essentially equivalent to number of spores (Teschke *et al.* 2001). Additional outdoor air samples were taken up to 9 days post-spray. The outdoor air concentrations of *Btk* HD1 diminished in two phases. In the first few hours, the airborne *Btk* HD1 concentration declined by half the original concentration. Numbers of CFUs then declined more slowly over the following 8 days.

During and immediately after the spray application, the average concentration of airborne *Btk* HD1 measured inside residences within the spray zone was 2.3–4.6 times lower than outside. Data from the Andersen samplers revealed that *Btk* HD1 aerosols with a medium diameter of 4.3–7.3 µm were present in air outside residences within 15 minutes after the spray application began. The aerosols were not visible to the human eye and were small enough to penetrate airways of the respiratory system (Teschke *et al.* 2001).

Approximately 5–6 hours after the spray application began, indoor concentrations of airborne *Btk* HD1 increased and then exceeded the outdoor concentrations, with an average of 245 CFU/m^3 of indoor air. During this time period, outdoor air concentrations had begun to fall to an average of 77 CFU/m^3. Given that the rate of inhalation of an adult at rest is approximately 20 m^3/day or 0.833 m^3/hour, at 5–6 hours post-spray, people in the spray zones can be estimated to have inhaled approximately 203 colony forming units of *Btk* HD1 (spores)/hour. While there are insufficient data to estimate the average quantity of spores inhaled by people residing in homes that were part of this study, it is clear that spore counts of *Btk* HD1 did rise in homes inside the spray zone following spraying, and that a substantial number of nasal swabs from people residing inside the spray zone were positive for *Btk* HD1. In addition, a significant amount of drift was detected up to 1000 metres (0.6 miles) outside the spray zone, so that more people were exposed than was anticipated (Teschke *et al.* 2001).

16.4 Evidence of Human Illness Associated with Aerial Applications of *Btk*

Studies to determine human health impacts resulting from aerial applications of Foray 48B typically consist of passive surveillance with laboratory follow-up of *Bacillus* isolates submitted to diagnostic laboratories, evaluation of hospital admissions and/or physician visits, collection of public self-reports of adverse effects, and monitoring specific symptoms of exposed groups. A surveillance study for human infections caused by *Btk* applications, among residents of Lane County, Oregon, was conducted during two seasons of *Btk* spraying for gypsy moth control (Green *et al.* 1990). *Bacillus* isolates from cultures obtained for routine clinical purposes were tested for presence of *Btk*. Detailed clinical information was obtained for all *Bt*-positive patients. About 80,000 people lived in the spray area in the first year of spraying, and 40,000 in the area sprayed in the second year. Fifty-five *Btk*-positive cultures were identified. The cultures had been taken from 18 different body sites or fluids. Fifty-two (95%) of the *Btk* isolates were assessed to be probable contaminants and not the cause of clinical illness. For three patients, *Btk* could not be definitively determined to be the causal agent of the infection, as each of these *Btk*-positive patients had pre-existing medical problems.

A major health impact assessment to determine possible mental and physical health effects as a consequence of human exposures to aerial applications of Foray 48B was conducted in conjunction with the Victoria (British Columbia, Canada) gypsy moth control programme in 1999 (Pearce *et al.* 2002a). A randomly selected population of adults (1009 individuals) living in the community of Victoria on Southern Vancouver Island was interviewed pre- and post-spray utilising a symptom survey and a health status survey tool called the Short Form 12 Health Status Profile (SF-12).

Analysis of symptom reporting showed no significant differences in symptom reporting following the aerial spray application either when comparing pre- and post-spraying time frames or whether the location of residence was inside or outside the spray zone. Based on the SF-12, there were no significant changes in physical health scores and a small improvement in the average mental health score post-spray for residents both inside and outside the spray zone. The data suggest that aerial exposure to *Btk*-based insecticides did not result in measurable acute health effects in the general population. No differences in reported symptoms between those inside and outside the spray zone or before and after the spraying were found. For those living inside the spray zone, there was no change in reported symptoms after spraying except for an improvement in the category "other." For those living outside the spray zone, there was an improvement in "unexplained tiredness" after the spraying occurred. There were no differences in other reported symptoms. Based on multivariate analysis, it was seen that the best predictor for the presence of specific symptoms post-spray was if a person reported the same symptoms in the pre-spray interview. Living inside the spray zone was not a predictor for any of the self-reported symptoms. Based on the standardised measure of health status, there was a small improvement in the average mental health score after the spray period for residents inside and outside the spray zone. There were no significant changes in the physical health scores.

An additional study conducted during the Victoria spray programme endeavoured to determine if aerial spraying of Foray 48B was associated with an increase in the symptoms or change in the Peak Expiratory Flow Rate of children with asthma (Pearce *et al.* 2002b). Children with a medical history of severe asthma living in the spray zone were matched with children with severe asthma living outside of the spray zone. Peak Expiratory Flow Rates, asthma symptoms and non-asthma symptoms were recorded in diaries, both prior to and following aerial applications of Foray 48B. No statistical differences in asthma symptom scores was found between subjects and controls, either before or after the spray were observed; nor were there significant changes in Peak Expiratory Flow Rates for subjects after the spray period. Thus, this study found no evidence of adverse effects resulting from aerial applications of Foray 48B.

A similar study was undertaken in New Zealand. Between January 2002 and May 2004, a series of aerial sprays of Foray 48B containing *Btk* HD1 was made to areas of Auckland and Hamilton, New Zealand. The purpose of the sprays was to eradicate the painted apple moth (*Teia anartoides*; PAM) from Auckland and the Asian gypsy moth (*Lymantria dispar*; AGM) from Hamilton. Eight aerial sprays were applied at approximately weekly intervals between 8 October 2003 and 29 November 2003. Foray 48B was applied at 5 litres/ha for PAM and AGM sprays 1 and 2, and 7 litres/ha for AGM sprays 3 to 8 (Richardson *et al.* 2005).

Two hundred and ninety two residents within the Ministry of Agriculture and Forestry (MAF) West Auckland spray zone were recruited by a door-to-door survey of homes within the most intensively sprayed area ten weeks prior to the first aerial spraying (Petrie *et al.* 2003). Participants completed a symptom checklist and a questionnaire measuring health perceptions. Three months after the start of spraying, 181 (62%) of the original participants responded to a similar postal questionnaire. Symptom reports, health perceptions and visits to healthcare providers were compared between the baseline and the follow-up questionnaire. Rates of symptom complaints in respondents with previously diagnosed asthma, hay fever or other allergies were compared to those for respondents without these prior health conditions.

Petrie *et al.* (2003) reported that symptom complaints increased significantly following the aerial spraying, in particular sleep problems, dizziness, difficulty concentrating, irritated throat, itchy nose, diarrhoea, stomach discomfort and gas discomfort. However, there was no non-exposed control group used in the study. The two surveys were carried out nearly 3 months apart and several of the symptoms included in the study have seasonal variation (New Zealand Ministry of Health 2003). Analyses showed an increase in symptoms in those participants with a previous history of hay fever, which is not unexpected; it was expected that a small percentage of population would experience some effect from the spray programme (New Zealand Ministry of Health 2004). Most residents saw their health as unaffected by the spray programme, and there was no significant increase in visits to general practitioners or alternative healthcare providers.

The New Zealand study and subsequent investigations (Aeraqua Medical Services 2001, 2005a, b) suggest that aerial spraying with Foray 48B caused short term irritant effects and possibly a worsening of pre-existing conditions such as allergies and asthma, although Petrie *et al.* (2003) found no significant increase in asthma symptoms among asthmatics (New Zealand Ministry of Health 2003, 2004). However, the only epidemiological study addressing asthma and respiratory conditions that included a control group from outside the spray area is from Vancouver Island (Pearce *et al*. 2002b), and, as discussed above, this study reported no adverse effects.

That no significant increase in symptoms was observed in the Canadian health impact assessment studies, while some community responses were observed in the New Zealand studies may be a function of the amounts and timing of spray applications. In the Victoria study, Foray 48B was applied at 4 litres/ha between 5:00 am and 7:00 am (while the majority of the population was still indoors). In Auckland and Hamilton, New Zealand, Foray 48B was applied at 5 litres/ha and 7 litres/ha between 8:00 am and 4:00 pm, while the general population was outdoors. While detailed studies of the levels of human exposure to *Btk* were not recorded during the Auckland and Hamilton aerial applications, it is conceivable that greater levels of direct exposure to the spray could account for the increase in reported symptoms.

16.5 Conclusions

Bt products such as Foray 48B, which contains *Btk* strain HD-1, must meet rigorous standards required by the US Environmental Protection Agency, the US Food and Drug Administration, the Canadian Pesticide Management and Regulatory Agency, and Health Canada, before they are approved for commercial use in Canada and the US. These agencies consider *Btk* products to be neither toxic nor pathogenic to mammals, including humans. Despite these official approvals, reports associating *Btk* with human illness can occasionally be found in the literature. Data generated by major health impact assessment studies in the US and Canada suggests there are no negative short-term human health effects associated with aerial applications of Foray 48B. A similar health impact assessment conducted in New Zealand reported short term irritant effects and worsening of hay fever and allergic response following sprays, which was not unexpected. Further health impact assessments should be conducted in conjunction with aerial applications of commercial *Btk* products over urban areas.

References

Aeraqua Medical Services (2001) Health surveillance following Operation EverGreen: A programme to eradicate the white-spotted tussock moth from the eastern suburbs of Auckland. Report to the Ministry of Agriculture and Forestry. 60 pp + Appendices

Aeraqua Medical Services (2005a) A study of presentations of householder concerns to the painted apple moth (PAM) health service and Auckland summer symptom survey. Report to Agri-Quality Ltd. 131 pp http://www.biosecurity.govt.nz/files/pests/painted-apple-moth/pam-health-report-appendix.pdf

Aeraqua Medical Services (2005b) A comparison of presentations of householder concerns to the painted apple moth (PAM) and Asian gypsy moth (AGM) health services. Report to Agri-Quality Ltd. 68 pp http://www.biosecurity.govt.nz/files/pests/painted-apple-moth/pam-health-report-appendix.pdf

Anonymous (1993) Report of health surveillance activities: Gypsy Moth control program. Health Promotion and Disease Prevention, Washington State Department of Health

Ash, C, Farrow JAE, Wallbanks S, Collins MD (1991) Phylogenetic heterogeneity of the genus *Bacillus* revealed by comparative analysis of small subunit ribosomal RNA sequences. Lett Appl Micro 13:202–206

Bavykin SG, Lysov YP, Zakhariev V, Kelly JJ, Jackman J, Stahl DA, Cherni A (2006) Use of 16S rRNA, 23S rRNA, and gyrB gene sequence analysis to determine phylogenetic relationships of *Bacillus cereus* group microorganisms. J Clin Micro 42:3711–3730

Bernstein IL, Bernstein JA, Miller M, Tierzieva S, Bernstein DI, Lummus Z, Selgrade MK, Doerfler DL, Seligy VL (1999) Immune responses in farmworkers after exposure to *Bacillus thuringiensis* pesticides. Environ Health Persp 107:575–582

British Columbia Ministry of Forest and Range (2007) History of Gypsy Moth Infestations in British Columbia. http://www.for.gov.bc.ca/hfp/gypsymoth/history.htm [accessed 17 March 2008]

Carlson CR, Kolsto AB (1993) A complete physical map of a *Bacillus thuringiensis* chromosome. J Bacteriol 175:1053–1060

Carlson CR, Caugant DA, Kolsto A-B (1994) Genotypic diversity among *Bacillus cereus* and *Bacillus thuringiensis* strains. Appl Environ Microbiol 60:1719–1725

Damgaard PH (1995) Diarrheal enterotoxin production by strains of *Bacillus thuringiensis* isolated from commercial *Bacillus thuringiensis*-based pesticides. FEMS Immunol. Medical Microbiol 12:245–250

de Amorim GV, Whittome B, Shore B, Levin DB (2001) Identification of *Bacillus thuringiensis* subspecies *Kurstaki* strain HD1-like bacteria from environmental and human samples after aerial spraying of Victoria, British Columbia, Canada with Foray 48B. Appl Environ Microbiol 67:1035–1043

Doekes G, Larsen P, Sigsgaard T, Baelum J (2004) IgE sensitization to bacterial and fungal biopesticides in a cohort of Danish greenhouse workers: the BIOGART study. Am J Indust Med 46:404–407

Fisher R, Rosner L (1959) Toxicology of microbial insecticide Thuricide. Agric Food Chem 17:686–688

Ginsberg C (2006) Aerial spraying of *Bacillus thuringiensis kurstaki* (Btk). J Pestic Reform 20:13–16

Green M, Heumann M, Sokolow R, Foster LR, Bryant, R, Skeels M (1990) Public health implications of the microbial pesticide *Bacillus thuringiensis*: an epidemiological study, Oregon, 1985–1986. Am J Pub Health 80:848–852

Hernandez E, Ramisse F, Cruel T, Ducoureau JP, Alonso JM, Cavallo JD (1998) *Bacillus thuringiensis* serovar H34-*konkurkian* superinfection: report of one case and experimental evidence of pathogenicity in immunosuppressed mice. J Clin Microbiol 36:2138–2139

Hernandez E, Ramisse F, Cruel T, le Vagueresse R, Cavallo JD (1999) *Bacillus thuringiensis* H34 isolated from human and insecticidal sero-types 3a3b and H14 can lead to death of immunocompetent mice after pulmonary infection. FEMS Immunol Med Microbiol 24:43–47

Hu X, Hansen BM, Hendriksen NB, Yuan Z (2006) Detection and phylogenic analysis of one anthrax virulence plasmid pXO1 conservative open reading frame ubiquitous presented within *Bacillus cereus* group strains. Biochem Biophys Res Comm 349:1214–1219

Ivanova N, Sorokin A, Anderson I, Galleron N, Candelon B, Kapatral V, Bhattacharyya A, Reznik G, Mikhailova N, Lapidus A, Chu L, Mazur M, Goltsman E, Larsen N, D'Souza M, Walunas T, Grechkin Y, Pusch G, Haselkorn R, Fonstein M, Ehrlich SD, Overbeek R, Kyrpides N (2003) Genome sequence of *Bacillus cereus* and comparative analysis with *Bacillus anthracis*. Nature 423(6935):87–91

Jackson SG, Goodbrand RB, Ahmed R, Kasatiya S (1995) *Bacillus cereus* and *Bacillus thuringiensis* isolated in a gastroenteritis outbreak investigation. Lett Appl Microbiol 21:103–105

Jensen GB, Larsen P, Jacobsen BL, Madsen B, Wilcks A, Smidt L, Andrup L (2002a) Isolation and characterization of *Bacillus cereus*-like bacteria from faecal samples from greenhouse workers who are using *Bacillus thuringiensis*-based insecticides. Internat Arch Occup Environ Health 75:191–196

Jensen GB, Larsen P, Jacobsen BL, Madsen B, Smidt L, Andrup L (2002b) *Bacillus thuringiensis* in fecal samples from greenhouse workers after exposure to *B. thuringiensis*-based pesticides. Appl Environ Microbiol 68:4900–4905

Leonard C, Yahua C, Mahilion J (1997) Diversity and distribution of IS231, IS232, and IS240 among *B. cereus*, *B. thuringiensis*, and *B. mycoides*. Microbiol 143:2537–2547

Logan NA, Turnbull PCB (1999) *Bacillus* and recently derived Genera. In: Murray PR, Baron EJ, Pfaller MA, Turnover FC, Yolken RH (eds) Manual of clinical microbiology. ASM Press, Washington, D.C. pp 357–369

McClintock JT, Schaffer CR, Sjoblad RD (1995) A comparative review of the mammalian toxicity of *Bacillus thuringiensis*-based pesticides. Pestic Sci 45:95–105

New Zealand Ministry of Health (2003) Human health considerations in the use of Btk-based insecticide Foray 48B for Asian gypsy moth in Hamilton. Summary report prepared for the Ministry of Health, Ministry of Agiculture and Forestry, and Waikato DHB Public Health Unit. October 2003. Auckland Regional Public Health Service. Report can be accessed at http:///www.moh.govt.nz [accessed March 17 2008]

New Zealand Ministry of Health (2004) Report on the effects of the painted apple moth spray programme. Media release, April 27 2004 http://www.moh.govt.nz/moh/nsf/pagesmh/3019?Open. [accessed 17 March 2008]

Noble MA, Riben PD, Cook GJ (1992) Microbiological and epidemiological surveillance programme to monitor the health effects of Foray 48B Btk spray. Report to the British Columbia Ministry of Forests, September 30 1992

Noble MA, Kandola P, Amos M, Riben P, Cook G, Shaw C (1994) Cluster analysis of community retrieved isolates of *Bacillus thuringiensis* var *kurstaki* (BTK). 93rd General Meeting for the American Society of Microbiology. Las Vegas, Nevada. May 4 1994

O'Callaghan M, Glare TR (2003) Mammalian safety of *Bacillus thuringiensis*. In: Akhurst RJ, Beard CE, Hughes P (eds) Proc 4th Pacific Rim Conf. CSIRO pp 254–261

Pearce M, Behie G, Chappell N (2002a) The effects of aerial spraying with *Bacillus thuringiensis kurstaki* on area residents. Environ Health Rev 46:19–22

Pearce M, Habbick B, Williams J (2002b) The effects of aerial spraying with *Bacillus thuringiensis kurstaki* on children with asthma. Can J Public Health 93:21–25

Petrie K, Thomas M, Broadbent E (2003) Symptom complaints following aerial spraying with biological insecticide Foray 48B. J NZ Med Assoc 14th March 116(1170):1–7

Perani M, Bishop AH, Vaid A (1998) Prevalence of β-exotoxin, diarrheal toxin and specific δ-endotoxin in natural isolates of *Bacillus thuringiensis*. FEMS Microbiol Lett 160:55–60

Richardson B, Kay MK, Kimberley MO, Charles JG, Gresham BA (2005) Evaluating the benefits of dose-response bioassays during aerial pest eradication operations. NZ Plant Prot 58:17–23

Samples JR, Buettner H (1983) Ocular infection caused by a biological insecticide. J Infect Dis 148: 614

Schnepf E, Crickmore N, Van Rie J, Lereclus D, Baum J, Feitelson J, Zeigler DR, Dean DH (1998) *Bacillus thuringiensis* and its pesticidal crystal proteins. Microbiol Molec Biol Rev 62:775–806

Siegel JP (2001) The mammalian safety of *Bacillus thuringiensis*-based insecticides. J Invertebr Pathol 77:13–21

Siegel JP, Shadduck JA (1990) Clearance of *Bacillus sphaericus* and *Bacillus thuringiensis* ssp. *israelensis* from mammals. J Econ Entomol 83:347–355

Swadner C (1994) Insecticide fact sheet. J Pestic Reform 14:13–20

Swiecicka I, Van der Auwera GA, Mahillon J (2006) Hemolytic and nonhemolytic enterotoxin genes are broadly distributed among *Bacillus thuringiensis* isolated from wild mammals. Micro Ecol 52:544–551

Teschke K, Chow Y, Bartlett K, Ross A, van Netten C (2001) Spatial and temporal distribution of airborne *Bacillus thuringiensis* var. *kurstaki* during an aerial spray program for gypsy moth eradication. Environ Health Persp 109:47–54

Xu D, Côté JC (2003) Phylogenetic relationships between *Bacillus* species and related genera inferred from comparison of 3′ end 16S rDNA and 5′ end 16S–23S ITS nucleotide sequences. Internat J Syst Evol Micro 53:695–704

Yuan YM, Hu XM, Liu HZ, Hansen BM, Yan JP, Yuan ZM (2007) Kinetics of plasmid transfer among *Bacillus cereus* group strains within lepidopteran larvae. Arch Microbiol 87:425–431

Chapter 17
Environmental Impacts of Microbial Control Agents Used for Control of Invasive Pests

Maureen O'Callaghan and Michael Brownbridge

Abstract A range of bacteria, viruses, fungi, protists and nematodes has been used for control and eradication of invasive pests. Insect pathogens vary in key characteristics such as specificity, mode of action and persistence, all of which determine their safety profile with respect to impacts on non-target species. Laboratory testing against beneficial species and post-application monitoring of impacts support the view that, while they are not entirely free of hazards to non-target organisms, in comparison with other control methods these microbial control agents are environmentally benign. When considering actions against new pest incursions, where the potential for damage to crops and/or impacts on indigenous ecosystems is enormous, rapid steps must be taken to mitigate or eradicate these pests. In such situations, biopesticides can be attractive control options, and their use is likely to have minimal impact on beneficial and other non-target species. Such advantages have been clearly demonstrated with *Bacillus thuringiensis*, which has not precipitated any major ecological disturbances, even when used in very intensive and prolonged eradication programmes.

17.1 Introduction

Bacteria, viruses, fungi, protists and nematodes comprise the major groups of arthropod pathogens that have been used in control and eradication attempts on invasive pests. Microbial agents are generally considered as environmentally benign alternatives to chemical pesticides, but they are not entirely without risk (Jaronski *et al.* 2003). While the hazards posed to non-target organisms are typically low in comparison with broad spectrum pesticides, the potential environmental impacts following the release of a living biological agent must be carefully considered,

---

M. O'Callaghan
AgResearch Ltd., Biocontrol, Biosecurity & Bioprocessing, Lincoln Science Centre, Christchurch 8140, New Zealand
e-mail: maureen.ocallaghan@agresearch.co.nz

as this can be an irreversible process. In most cases, the greater the scale of use of a microbial agent, the greater the potential for impacts on non-target organisms (Cook et al. 1996). In certain eradication and control programmes, thousands of hectares may be treated, sometimes repeatedly. Here, environmental safety of the microbial control agent is paramount. This is also an area of growing public concern, as demonstrated by the increase in community groups opposing biopesticide spray programmes in many countries, including the USA, Canada, Europe and New Zealand. For these reasons, availability of robust and balanced data on environmental impacts of microbes is essential for both regulators and the general public.

Key characteristics of pathogens that determine their efficacy in microbial control vary widely between groups and even within each group of pathogen. In general, specific pathogens (many of which cause natural epizootics) have been used in classical biological control approaches, whereby a pathogen is introduced into a pest population with the goal of it becoming established and spreading within that population to exert a sustained suppressive effect over the long-term, e.g. entomophthoralean fungi such as *Entomophaga maimaiga* and *Neozyygites fresenii* (Steinkraus et al. 2002, see also Chapter 11). In such cases, high numbers of infective spores would be present in the environment only for short periods when host densities are high and weather conditions are suitable for sporulation events to occur. Pathogens that are more generalist in nature and have a wider host range, e.g. fungal biocontrol agents such as *Beauveria bassiana* and *Metarhizium anisopliae*, and bacteria such as *Bacillus thuringiensis* (*Bt*), may be used as biological insecticides, with repeat applications being made as necessary over a cropping cycle. While these organisms frequently do not persist over extended periods, they are present in the environment – for a short time at least – at high levels immediately after application.

Information on host specificity, mode of action, persistence in the environment, transmission and dispersal properties is necessary to allow a full evaluation of environmental safety (Glare & O'Callaghan 2003). A full environmental impact assessment as required for registration packages generally also includes data on mammalian safety, which will not be addressed in this chapter. The human health impacts of the widely used microbial control agent, *Bt*, have been reviewed by Levin (see Chapter 16).

17.1.1 Specificity

One of the key environmental benefits of some microbial control agents is their host specificity. The vast majority of insect viruses developed for pest control are baculoviruses. To date, no negative ecological effects following their wide scale release have been documented (Cory 2003). Host range studies of baculoviruses isolated from Lymantriidae indicate that their spectrum of activity is very narrow, and in most cases restricted to the species from which they were originally isolated; this explains the lack of observed effects on non-target organisms. A highly specific baculovirus (*Ld*MNPV) developed as a microbial pesticide for the gypsy moth (*Lymantria dispar*) also cycles naturally in the host population (see Chapter 11). Similarly, non-occluded viruses used to control rhinoceros beetle, *Oryctes rhinoceros*, in the western Pacific will recycle within the host population, but in spite of ecological persistence this virus does not infect other insects in the same environment (Cory 2003, see also Chapter 8). However, a high level of specificity may also limit the broader utilisation of these disease-causing agents and few highly specific pathogens have been widely used to control invasive pests. One example where this has been achieved is with the tylenchid nematode *Beddingia (Deladenus) siricidicola*, which is highly specific to its host, *Sirex noctilio* (see Chapter 12).

In contrast to these highly specific microbial agents, various strains and subspecies of *Bt* have relatively wide host spectra, and have been applied as biopesticides to economically-important crops and for control of vector and nuisance species for over 40 years (Roush 1994). Sprays consist of wettable and flowable formulations of spores and protein crystals derived from the bacterium (Llewellyn *et al.* 1994). In addition to the physiological and biochemical factors influencing activity, toxicity also depends on ingestion (insects have to be actively feeding to become intoxicated), exposure time, and the frequency with which the insect is in contact with the bacterium (Chowdhury *et al.* 2001). A major benefit of *Bt* spray formulations is their effectiveness at controlling over 150 problematic insects, including many invasive species, e.g. European corn borer (*Ostrinia nubilalis*), diamondback moth (*Plutella xylostella*), gypsy moth (*L. dispar*) and various mosquito species.

Bt kurstaki has been used extensively in eradication and control programmes in New Zealand, the USA and Canada (see Chapters 4 & 5). It is primarily effective against Lepidoptera, but activity against species of Coleoptera, Diptera, Hymenoptera, Hemiptera and several other insect orders has also been reported (Glare & O'Callaghan 2000). Therefore careful consideration of potential side-effects on non-target species is particularly important when intensive spray campaigns with *Bt kurstaki* are undertaken, as may be required for the eradication of invasive pests.

Under natural conditions, some entomopathogenic fungi occasionally cause spectacular epizootics, but in many crops these occurrences cannot be relied upon to prevent economic damage. Most attempts to exploit these agents have focused on their use through inundative releases. Few negative effects have been reported following the use of entomopathogenic fungi in commercial crops and forests. Comprehensive reviews on the safety of the most widely used species, *B. bassiana*, *Beauveria brongniartii* and *M. anisopliae*, and the relevance of current regulatory testing requirements are already available (Jaronski *et al.* 2003, Vestergaard *et al.* 2003, Zimmermann 2007a, b).

Another group of insect pathogens, the protists, are obligate pathogens; they only reproduce and mature within living cells, which limits their potential use as biopesticides. However, practices that promote the natural prevalence and spread of these microbes can enhance their impact on pest populations. The protists (formerly known as microsporidia) have been infrequently used to control invasive pests, most notably against gypsy moth (see Chapter 11).

17.1.2 Persistence and Transmission

Environmental safety of any pesticide is closely related to the extent to which the active ingredient persists in the environment. Microorganisms released into the environment as biopesticides have raised special concern because of their potential to establish, multiply and spread (Cook et al. 1996). Knowledge of the persistence of a microbial agent is important given the potential for spray drift, especially during pest eradication programmes. *Bt*-related mortality of susceptible species has been detected several kilometers from areas sprayed to kill pest Lepidoptera (Whaley et al. 1998).

There may be greater risk of non-target impacts from agents that persist in the environment as the likelihood of non-target exposure is increased. However, microorganisms applied to control a specific pest generally decline to a density that is naturally sustainable within the environment, often to undetectable levels. Following a soil application, inoculum generally falls to between 10^2–10^5 propagules/g soil. This trend has frequently been reported for non-sporeforming entomopathogenic bacteria (O'Callaghan 1998), and also for fungi where natural background levels of inoculum frequently reach 10^3 propagules/g soil, especially in the vicinity of diseased insect cadavers (Rath 2002, Brownbridge et al. 2006). *Bt* spores can persist for extended periods in soil although they do not appear to replicate outside of a host (Smith & Barry 1998). Following aerial application of the *Bt*-based biopesticide Foray 48B in a campaign to eradicate the white spotted tussock moth in New Zealand, viable *Bt kurstaki* spores persisted in soil with only limited reduction in numbers (Gribben et al. 2002). However, with *Bt*, persistence of the active toxin is the key factor in considerations of environmental impact, as it is the primary insecticidal component produced by the bacterium. The intact pro-toxins contained in *Bt* preparations are generally considered safe for most non-target organisms owing to the specific conditions required for their activation and activity in the gut of susceptible species, and their degradation in the soil environment (West 1984, Meadows 1993, Tabashnik 1994). *Bt* spores and toxins are more rapidly deactivated on foliage than in soil; a range of biotic and environmental conditions (temperature, ultra violet light) strongly influences persistence on the leaf surface.

Similarly, the viability of fungal inoculum quickly declines following application to foliage where damaging effects of ultraviolet light and high temperatures inactivate infective propagules (Hajek 1997a, Zimmermann 2007a, b). In soils, fungi are generally afforded more protection from environmental extremes, but persistence in the upper soil layers is strongly influenced by fungal strain and soil type (Hajek 1997a, Vänninen et al. 2000, Nelson et al. 2004, Brownbridge et al. 2006, Zimmermann 2007a, b).

Transmission and dispersal are additional key factors impacting the spatial range and extent of potential environmental impacts. Bacteria used as biopesticides typically have limited dispersal and, in the case of *Bt*, have limited ability to maintain themselves in the insect population. Other bacterial species, such as *Paenibacillus popilliae*, recycle in the pest population and can maintain inoculum in the environment for extended periods. The naturally-cycling fungus *E. maimaiga* is an obligate pathogen of larval stages of gypsy moth, and while not used as a typical biopesticide, the fungus is responsible for decimating large populations of this destructive, introduced pest (see Chapter 11). At least part of its success in gypsy moth suppression can be attributed to its ability to spread through the pest population; *E. maimaiga* has shown remarkable natural dispersal over a period of several years (or maybe even decades) in the northeastern USA (Hajek 1997b, 1999). The fungus displays an interesting adaptation to the host, predominantly producing infective conidia on early instar larvae, while resting spores for winter survival are mostly produced within larger larvae.

17.2 Issues in Environmental Impact Assessment

Broad guidelines for assessment of non-target effects of microbial control agents on non-target species have been reviewed by Hajek & Goettel (2000) and Jaronski *et al.* (2003). The quality and reliability of environmental impact assessment data depend on a number of factors, including the ecological relevance of the non-target species selected, the parameters that are measured, and the rigor with which laboratory and field studies are carried out.

17.2.1 Selection of Non-Target Species for Testing

It is essential that non-target species representative of the habitat in which the microbial control agent is to be applied are used. For example, much of the non-target testing of the mosquito control agents, *Bt israelensis* and *Bacillus sphaericus*, has focused on aquatic species. Because of the particularly sensitive nature of aquatic habitats in terms of essential resources for humans and the rich complexes of aquatic species, it is important that agents used for pest control cause little or no deleterious effects (Lacey & Merritt 2003).

Potentially significant amounts of microbial inoculum applied during large scale eradication programmes will reach the soil. Therefore, an evaluation of effects on components of the soil biota is an integral part of an environmental impact assessment. Because of their key role in soil ecosystem function, there are clear precedents and justifications for using Collembola as representative test species (Klironomos & Kendrick 1995, Frampton 1997). Collembola are now used in standardized ecotoxicological tests in Europe to assess effects of pesticides and other pollutants (Fountain & Hopki 2005). The collembolan species *Folsomia candida* is frequently used in laboratory evaluations. Cultures of *F. candida* are easy to maintain and its short reproductive cycle make this species ideal for ecotoxicological experiments (Fountain & Hopki 2005). While data from these "standardized" tests cannot be directly extrapolated to the field, they facilitate the assessment of direct toxic and chronic sub-lethal effects of test materials under controlled conditions. It should be noted, however, that differing sensitivity of Australian collembolan species to a variety of toxicants was demonstrated by Greenslade and Vaughan (2003), highlighting limitations of single-species testing and the importance of selecting species that are ecologically relevant to local soils.

Earthworms also play a key role in soil ecosystem function. While our understanding of many aspects of earthworm feeding ecology is still poor, earthworms are likely to ingest entomopathogenic fungi and bacteria as part of their normal diet, and may be topically exposed to them and their insecticidal toxins in soil (Strasser *et al.* 2000). Testing is frequently carried out against the compost worm, *Eisenia fetida*, which is not widely distributed in most environments. Evaluating effects against more ecologically relevant species, ideally under field conditions, would be of greater value in determining non-target impacts (Lowe & Butt 2005). As with *F. candida*, the ease with which *E. fetida* can be reared makes it a useful representative test species; however, the same caveats should be applied to the data obtained from trials.

17.2.2 Protocols for Non-Target Testing

One of the primary potential environmental impacts of applications of entomopathogens is toxicity to, or infection of, non-target species. While susceptibility of selected non-target species can be readily tested in laboratory bioassays, there are often striking differences between results of laboratory assays and field approaches and observations. Species that show sensitivity in the laboratory are not necessarily susceptible to the microbial agent in the field. In part, these differences may result from the use of maximum challenge bioassays in laboratory testing, which do not reflect the typically lower doses to which non-target organisms are exposed in the field. For example, guidelines generally require that earthworms and/or Collembola are exposed to artificially high levels of inoculum in laboratory soils and biological responses (e.g. live/dead, weight gain, egg production) to the pathogen challenge are monitored (Hokkanen *et al.* 2003b). These tests have frequently been adapted from pesticide testing regimes, although they are increasingly being refined for microbial pesticides in acknowledgement of the fundamental differences between the agents.

There are a number of other reasons for observed differences in the susceptibility of non-target organisms between laboratory and field tests and these illustrate the complexity of environmental impact assessment. For example, there may be spatial and temporal barriers separating the microbial agent from non-target organisms, or behavioural and ecological barriers (e.g. insect life cycles) that minimize direct exposure of non-target species, in comparison with the target pest. In most cases, the physiological host range is much wider than the ecological host range (Hajek & Goettel 2000). In contrast with the situation in maximum challenge laboratory bioassays, an insect may never come in contact with sufficient inoculum to succumb to infection in the field. For example, Hajek *et al.* (2000) examined the field prevalence of larvae infected with *E. maimaiga* in areas where infection rates in gypsy moth populations were high. Only two individual larvae from a total of 1,511 larvae from 52 species of Lepidoptera that were collected were infected with *E. maimaiga*. Despite *E. maimaiga* being a pathogen of the Lymantriidae in the laboratory, species other than gypsy moth are unlikely to be infected in the field unless they spend significant periods of time inhabiting leaf litter when infective spores of *E. maimaiga* are abundant in this habitat (see Chapter 11).

While there is disparity between laboratory bioassays and field experiments, bioassays remain the first step in determining the potential susceptibility of at-risk species, such as predators and parasites or rare indigenous species. However, if bioassay results are to be used to predict activity and non-target effects in the field, pertinent environmental and exposure parameters must be incorporated into the bioassay design (Butt & Goettel 2000). Bioassay design must include a consideration of factors such as standardization of dose and uniformity of test species, and an accurate estimation of dose actually consumed by or reaching the test species (Navon & Ascher 2000).

Another complicating factor is the variability in insecticidal activity of different isolates. Frequently, individual isolates of fungal entomopathogens are more host-specific than is suggested by the aggregate host range data from published reports on that pathogen (Goettel *et al*. 1990, Hajek & St. Leger 1994). For example, although *B. bassiana* has been reported as infecting over 700 species of arthropods, laboratory studies show that isolates are generally most virulent to the host from which they were originally isolated (Goettel *et al*. 1990). Conversely, some strains of insect pathogens can exhibit high pathogenicity to previously un-encountered hosts (Prior 1990).

17.2.3 Sublethal Effects

Ecological impact assessment and host range testing of pathogens has largely focused on direct lethal effects on non-target species but microbial agents can also have significant sublethal effects when species survive a pathogen challenge, or are exposed to low concentrations of the pathogen, for example as a result of spray drift following an aerial application (Whaley *et al*. 1998). Rothman and Myers (1996) reported that viruses caused sublethal effects, such as altered developmental rates, reduced fecundity, and lower pupal weights in Lepidoptera. A limited number of studies have reported sublethal effects of *Bt* strains on pests; effects include delayed development, reduction in larval and pupal weights, reductions in adult emergence, reduction in adult fecundity and reduced egg viability (reviewed in Glare & O'Callaghan 2000). Some side-effects of *Bt* toxins have been noted following conventional foliar applications, with sublethal effects reported on certain natural enemies (Salama *et al*. 1982, Flexner *et al*. 1986, Giroux *et al*. 1994). Sublethal effects could potentially be found in susceptible non-target species but information in this area is sparse, perhaps because of the greater difficulties inherent in measuring sublethal effects, in comparison with direct toxicity.

17.2.4 Effects in Combination with Other Control Measures

In eradication and control programmes, it is sometimes necessary to use a combination of control measures to ensure rapid kill and to limit the spread of invasive pests. For example, both *Bt israelensis* and the insect growth regulator methoprene were used in an eradication programme for the salt marsh mosquito (*Ochlerotatus camptorhynchus*, previously known as *Aedes camptorhynchus*) in New Zealand (New Zealand Ministry of Health 2008). Ground sprays with the organophosphate insecticide chlorpyrifos and/or Decis, a synthetic pyrethroid insecticide, were made in addition to intensive aerial spraying with Foray 48B (*Bt kurstaki*) to eradicate painted apple moth, *Teia anartoides*, in New Zealand (Chapter 4). Many studies of chemical insecticides used in combination with *Bt* sprays report enhanced insecticidal activity, although some studies reported that the interaction was neutral or antagonistic (reviewed in Glare & O'Callaghan 2000).

Co-infections of insects can also occur. Gypsy moth, for example, can be infected with both the fungus *E. maimaiga* and the nucleopolyhedrovirus, *Ld*MNPV. Laboratory studies showed that depending on dosage and timing, there was sometimes an apparent synergistic interaction between the two pathogens, thereby significantly increasing overall mortality (Malakar *et al.* 1999).

17.2.5 Effects of Formulation Components

Consideration of formulation components is also a factor in assessment of non-target impacts, although it is generally very difficult, if not impossible, to obtain information on formulation components, which are usually closely guarded trade secrets. Formulations are mostly developed to enhance persistence of the active agent. For example, some formulations of *Bt israelensis* are designed to enhance persistence of the toxin in the water column, which increases the likelihood of exposure of non-target species. Addition of feeding stimulants can increase ingestion by both the target pest and non-target species. Addition of synergists (e.g. acids, salts) can cause up to a 40x increase in efficacy over *Bt* used alone (Burges & Jones 1998), which could also be significant in terms of non-target toxicity. Addison and Holmes (1995, 1996) reported non-target impacts on a collembolan and earthworm when high doses of oil formulations of *Bt kurstaki* were used, but no effects when unformulated or aqueous suspensions were used.

Formulation of *B. bassiana* has been shown to alter parasitism and survival of the parasitoid *Eretmocerus eremicus* on whitefly nymphs (Armstrong 2000). This study compared BotaniGard 22-WP and BotaniGard ES (wettable powder and emusifiable suspension, respectively) and formulation blanks. When the fungal treatments and parasitoid releases were made on the same day, parasitism was significantly lower in the ES and ES blank treatments compared to the untreated control and WP sprays. Under this treatment regime, the WP formulation appeared to be compatible, the ES antagonistic. The oil may kill whiteflies or render them unsuitable as hosts for the parasitoid, host-searching may have been compromised by the oil coating which can interfere with the wasp's receptor organs, or volatiles from the distillate may have been repellent. When whiteflies were sprayed with *B. bassiana* 3 days after parasitoid release, parasitoid emergence was significantly lower in populations treated with the ES and formulation blank; at the time of treatment, parasitoid eggs would be hatching or early first instar parsitoids would be beneath the whitefly. During this period, the host may have been killed or compromised as a result of the insecticidal action of the oil used in the ES, denying the parasitoid a viable host in which to complete its development. The oil may also have permeated beneath the host and had a direct toxic effect on the young parasitoid larvae. Thus the sequence in which treatments are made, the timing of treatments, and the formulation used, can all affect compatibility.

17.3 Non-Target Effects

17.3.1 Non-Target Invertebrates

Bacterial, fungal and viral biopesticides have been used in the field for many years, and through non-target testing in the laboratory and field monitoring of selected species post-application, much can be learned about effects of biopesticides on non-target arthropods. Impacts of hypocrealean fungi (Vestergaard *et al.* 2003), viruses (Cory 2003), nematodes (Ehlers 2003) and microsporidia (Solter & Becnel 2003) have been reviewed previously. Undoubtedly the most well studied microbial control agent is *Bt kurstaki*, which is generally regarded as specific to Lepidoptera when used in the field; a large body of literature supports this (Wagner *et al.* 1996, Peacock *et al.* 1998, Whaley *et al.* 1998, Glare & O'Callaghan 2000). The large scale gypsy moth eradication programmes carried out in British Columbia provided an excellent opportunity to monitor the severity of effects on non-target species after application of *Bt kurstaki*. Boulton (2004) reported that uncommon species groups were significantly less abundant in treatment sites, as were most of the common species. Lepidopteran species richness was also reduced after spraying. A few species were unaffected because their life-history characteristics minimized exposure to the spray. Boulton *et al.* (2007) subsequently examined non-target lepidopteran populations 1 and 4 years after the gypsy moth eradication programme. Total numbers of caterpillars in the treated sites remained lower than the reference sites, indicating that at a guild level recovery of non-target Lepidoptera was not complete 4 years after *Bt kurstaki* spraying. There were some signs of partial or full recovery of most of the impacted species but overall recovery was incomplete.

While thorough studies of non-target impacts in the field are limited, largely because of the inherent difficulties in this work, these studies provide valuable data on effects on habitat-specific non-target species. An example is the study by Parker *et al.* (1997) examining non-target effects of an indigenous strain of *B. bassiana* that was being evaluated for control of pear thrips, *Taeniothrips inconsequens*, a pest of sugar maple (*Acer saccharum*) in the northeastern United States (Brownbridge et al. 1993). Originating from Europe, the pest was first recognized as a pest of maple in the late 1970s and was primarily responsible for the defoliation of ~ 400, 000 and 200,000 ha of maple in Pennsylvania and Vermont, respectively, in 1988. Various ground-dwelling arthropods play important roles in the forest ecosystem as scavengers, natural enemies or decomposers, and would be contacted by any treatments applied to the forest floor for pear thrips control. Consequently, their sensitivity to *B. bassiana* was assessed in a series of replicated trials. Invertebrates were collected in pitfall traps after treatment of field plots and held in the laboratory to determine infection levels. The majority of the invertebrates collected were spiders and beetles (especially ground beetles, Carabidae, and weevils, Curculionidae). Over 3600 arthropods were collected; of these 2.8% were infected but infected individuals were also recovered from the control plots (Parker et al. 1997). Considering the low level of infection, it was concluded that the fungus posed minimal risk to the forest-dwelling non-target arthropod population.

Clover root weevil (CRW), *Sitona lepidus*, is an exotic pest that has severely impacted white clover (*Trifolium repens*) pastures since it arrived in New Zealand (Barratt et al. 1996). Larvae destroy nitrogen-fixing nodules, thereby affecting the nitrogen-fixing capacity of plants. An exotic fungal isolate of *B. bassiana*, strain F418 derived from infected *Sitona* sp. collected in Romania, is being evaluated as a biopesticide for control of this pest. The fungus is being developed for application to soil as a granular formulation, targeting the damaging larval stages. This development has included an assessment of effects on selected soil- and ground-dwelling non-target invertebrates that are commonly found in clover fields. Using formulated granules, soil mesocosm assays were run against the collembolan *F. candida* and the earthworm *Aporrectodea caliginosa*. No adverse effects were observed on *F. candida* population development or earthworm survival, weight gain or time to sexual maturity over a 12 week study (Brownbridge et al. 2007). Non-target species were also monitored in a 2006 field trial in which *B. bassiana* F418 was drilled into three CRW-infested white clover/ryegrass pastures in the Waikato, New Zealand. Sites were sampled at regular intervals to determine numbers and diversity of above-ground coleopteran species and collected specimens were maintained in the laboratory to determine the prevalence of *B. bassiana* infections in individuals that died. Number and diversity of Coleoptera varied between sites and seasons. The most abundant species (and % of total) were: *S. lepidus* adults (52%), Argentine stem weevil, *Listronotus bonariensis* (38%), and coccinelid species (4%). Native weevils represented < 1% of the total Coleoptera collected. *B. bassiana* infections were prevalent at all sites and treatments, and *B. bassiana* was recovered from all three major groups of beetles. Of the total insects collected, from all sites and treatments, 21, 16 and 17% of the CRW, Argentine stem weevil and coccinellids were infected, respectively. However, there was a significant site effect on the prevalence of *B. bassiana* among beetles but no significant treatment effect, i.e. infection levels in beetles sampled from control plots were the same as those from treatment plots (McNeill et al. 2007). Safety of the pathogen to non-target Coleoptera and soil-dwelling invertebrates was thus demonstrated.

Similarly, extensive studies (reviewed in Chapter 11) on the host range and field infectivity of the gypsy moth pathogen *E. maimaiga* indicate that it is of very low risk to non-target Lepidoptera. Solter and Hajek (Chapter 11) speculate that *E. maimaiga* infection levels are low because few lepidopteran larvae, particularly native lymantriids, spend much time at the soil surface where fungal spores are most abundant, although this is a common habitat for late instar gypsy moth larvae. The environmental safety of another group of pathogens active in control of many insect species including gypsy moth, the protists, is reviewed in detail elsewhere (Solter & Becnel 2003, see also Chapter 11).

17.3.2 Beneficial Non-Target Species

Data on the direct toxicity of microbial agents/biopesticides to beneficial species is an essential requirement for product registration. Tests typically focus on commercially-important production species such as bees and silkworms, pest control species such as naturally occurring predators and parasites, and endangered species found in the environment (Hajek & Goettel 2000, Goettel *et al*. 1990, Sample *et al*. 1996, Wagner *et al*. 1996). No less important, though, are species involved in ecosystem processes such as pollinators, and earthworms and soil microarthropods, which are essential components of a healthy soil biota. Many species of earthworms and Collembola specifically feed on fungal spores and bacteria in the soil. Studies on their diet have focused on ingestion and dispersal of saprophytic species or plant pathogens, but few have addressed their susceptibility to bacterial or fungal entomopathogens.

17.3.2.1 Bees

The majority of studies on *Bt* have failed to show any adverse effects on honey bees (*Apis mellifera*) or other pollinators (Glare & O'Callaghan 2003).

Testing for non-target effects on bees has been an essential part of the search for control agents active against the ectoparasitic mite *Varroa destructor*. Originating in Asia, this pest is now almost cosmopolitan in its distribution, and has had a major impact on European honey bee colonies worldwide (see Chapter 15). Previously, laboratory studies showed that caged honey bees were susceptible to high concentrations of *M. anisopliae* and *B. bassiana* (Vandenberg 1990, Butt *et al*. 1994, Jaronski *et al*. 2003). Similarly, some of the pathogens that are active against *Varroa* are also pathogenic to bees (Shaw *et al*. 2002, Meikle *et al*. 2006). However, in these assays there were problems of high control mortality, largely due to overcrowding and high moisture levels in the assay containers, so the bees were likely compromised and more susceptible to infection under these conditions. The dangers of extrapolating such laboratory data to estimate potential side-effects in the field are clearly demonstrated when reviewing results of field trials utilizing the same isolates. No adverse effects on behaviour, worker mortality rates, or colony health have so far been observed, even when fungi have been directly applied to the hive (Alves *et al*.

1996, Butt *et al.* 1998, James *et al.* 2006, Jaronski *et al.* 2003, Kanga *et al.* 2003, 2006, Meikle *et al.* 2007, 2008). Honey bees have also been successfully used to disseminate a range of biological control fungi with no apparent adverse effects on the bees (Brownbridge *et al.* 2006); for example, honey bees delivered *M. anisopliae* to *Brassica* spp. flowers at levels that provided effective control of pollen beetles (*Meligethes aeneus*) (Butt *et al.* 1998), and Al-mazra'awi *et al.* (2006a) used honey bees to disseminate *B. bassiana* conidia for control of tarnished plant bug (*Lygus lineolaris*) on canola. The pathogen was also safe for bumble bees, *Bombus impatiens*, which were used to disseminate *B. bassiana* conidia to control tarnished plant bug and western flower thrips (*Frankliniella occidentalis*) on greenhouse sweet peppers in Canada (Al-mazra'awi *et al.* 2006b). However, a previous study on *M. anisopliae* and *B. bassiana* found that under favourable conditions, these two relatively broad-spectrum species, while apparently safe for honey bees, can infect bumble bees (Hokkanen *et al.* 2003a).

The relative safety to honey bees of the locust pathogen *M. anisopliae* var. *acridum* (previously *M. flavoviride*) in an oil formulation has also been demonstrated (Ball *et al.* 1994). When exposing honey bees to field doses that effectively killed the target hosts, 11% of bees died as a result of mycosis and only 8% of bees died when conidia were formulated in water. In comparison, honey bee mortality from the recommended chemical pesticide, fenitrothion, was 100%, even when applied at rates that were sub-lethal to locusts. Thus, the risks to bees posed by microbial control agents are extremely low, particularly when compared to alternative chemical controls.

17.3.2.2 Earthworms

Regulatory authorities generally require data on impacts of biopesticides on earthworms but there are surprisingly few studies published in this area. Most information remains confidential in registration data packages and is not in the public domain. This is unfortunate, as the greater the body of information readily available to regulators and the public that demonstrates the inherent safety of these microorganisms, the easier the regulatory pathway should become for more products. From the limited studies available, it can be seen that most microbial agents likely to be used for control of invasive pests have little or no impact on earthworms. Addison and Holmes (1996) reported that unformulated and aqueous *Bt kurstaki* at 1000× field concentration had no effect on the forest earthworm species *Dendrobanena octaedra*; however, an oil formulation of *Bt kurstaki* reduced survival, growth and cocoon production, highlighting the potential non-target effects of formulation components.

Hozzank *et al.* (2002) carried out laboratory trials on the earthworm *Lumbricus terrestris*, which was exposed to *M. anisopliae* conidia in soil mesocosms. The earthworms moved conidia within the system but none died as a direct result of fungal infection over the 2-month experimental period. Similarly, no adverse effects were observed when the earthworms were placed in soil inoculated with a commercial product of *B. brongniartii* (Keller *et al.* 2003). Soil concentrations of

B. brongniartii were significantly higher than those recommended for field use. Earthworm feeding was unaffected and served to concentrate conidia in worm casts without harmful effects to this non-target organism. In field sites where *B. brongniartii* was applied to control the cockchafer *Melolontha melolontha*, no differences in earthworm density or species composition were detected between treated and untreated plots (Keller *et al.* 2003).

Earthworm toxicity studies were required for the registration of the *B. bassiana* GHA strain (BotaniGard) in the EU (S Jaronski personal communication). Earthworms (*E. fetida*) were exposed to *B. bassiana* at 0 (negative control), 0 (attenuated control), 130, 216, 360, 600, and 1000 mg spore powder/kg dry soil, under normal conditions of pH, temperature, humidity and light for this species. There were no treatment-related effects in any group. The authors of the study concluded that the 14-day LC_{50} for *E. fetida* exposed to *B. bassiana* GHA strain was > 1000 mg dry spore powder/kg dry soil. The fact that no effects were detected at this level of inoculum, a level that would never be attained in the field, attests to the safety of this pathogen.

While the available data are limited, there is no direct evidence suggesting that entomopathogens are harmful to earthworms. No toxic or pathogenic effects have been reported, and avoidance behaviour has not been reported. In fact, earthworms may play a beneficial role in microbial control, potentially aiding the dispersal of biocontrol inoculum in the soil.

17.3.2.3 Collembola

Atlavinyte *et al.* (1982) reported that numbers of Collembola decreased in soil following treatment with a commercial formulation of *Bt galleriae*. However, no negative effects were observed when *F. candida* were fed exclusively on pure *Bt* cultures (*kurstaki, aizawai, entomocidus, israelensis*), demonstrating a potential role of Collembola as scavengers of biopesticides in the environment (Peterson & Luxton 1982, Broza *et al.* 2001). A wettable powder formulation of *Bt kurstaki* (Dipel) did not affect Collembola, but an oil-based formulation reduced soil populations, suggesting that the oil carrier was toxic to these insects rather than the pathogen (Addison 1993, Addison & Holmes 1995).

Studies with entomopathogenic fungi have reported few to no effects on Collembola (e.g. Broza *et al.* 2001, Dromph and Vestergaard 2002). Broza *et al.* (2001) found that although conidia germinated on the cuticle of *F. candida*, they failed to penetrate and infect exposed individuals. Dromph (2001, 2003) showed that *Folsomia fimetaria*, *Hypogastrura assimilis* and *Proisotoma minuta* were able to carry viable conidia of *B. bassiana*, *B. brongniartii* and *M. anisopliae* both on the cuticle and in the gut, effectively vectoring the fungi. Direct exposure of the Collembola (by dipping) to the three fungal species showed that they were not susceptible to infection, even when high concentrations of inoculum (10^8/ml) were used. When exposed to a substrate containing a high concentration of spores, though, one *B. brongniartii* isolate and one *M. anisopliae* isolate significantly increased the mortality of *F. fimetaria* over the control (Dromph & Vestergaard 2002). However, the test concentrations used were much higher than would be encountered following an application of entomopathogenic fungi for biological control and it was concluded that the fungi could be considered either non-virulent or of low virulence to Collembola. The studies demonstrated the importance of testing multiple species and conditions, which can lead to different interpretations on the relative pathogenicity of fungi for Collembola.

Reinecke et al. (1990) found that the developmental Bayer product Bio1020 (based on *M. anisopliae* originally isolated from the codling moth, *Cydia pomonella*, in Austria), did not infect Collembola. Brownbridge (2006) reported that *B. bassiana* GHA strain (commercialised as BotaniGard and Mycotrol for control of sweetpotato whitefly (*Bemisia tabaci*) and silverleaf whitefly (*B. argentifolii*) in North America) and *M. anisopliae* var. *acridum* (IMI strain; Green Muscle; for control of locusts) had no effect on survival and reproduction in *F. candida*.

In contrast to most of the studies, Ireson and Rath (1991) found that the lucerne flea, *Sminthurus viridis*, was susceptible to *M. anisopliae* DAT F-001, with up to 100% mortality being recorded after 17 days at 12 °C (Rath 1991). Although DAT F-001 was developed as a commercial biopesticide for control of indigenous soil scarabs in Tasmania, and was a locally isolated strain, this study demonstrates potential variation in sensitivity of different collembolan species.

17.3.2.4 Predators and Parasites

The selection of beneficial invertebrates to be assessed for non-target impacts depends on the setting in which the microbial control agent will be applied. Because *Bt kurstaki*-based products are typically used against caterpillar pests in crops and forests, potential for non-target impacts against predators (such as carabid beetles) and parasitoids (such as hymenopteran wasps) must be considered. Most direct and indirect tests of *Bt* products against predators and parasites have not shown adverse effects (Glare & O'Callaghan 2000). Enhanced parasitism after *Bt* application has occasionally been reported, possibly because of the altered behaviour of the treated pest made it more susceptible to parasitoids (Nealis et al. 1992). A sublethal *Bt* treatment caused exposed gypsy moth larvae to develop more slowly, which in turn provided parasitoids with a greater window of opportunity to parasitize early instars, thereby elevating the overall rate of parasitism (see Chapter 11).

Compatibility of two formulated commercial preparations of *B. bassiana* [Naturalis-L, an emulsifiable concentrate (Troy BioSciences, Inc.) and BotaniGard-WP, a wettable powder (ex. Mycotech Corp.)] with the predatory mites *Phytoseiulus persimilis* and *Neoseiulus* (*Amblyseius*) *cucumeris* has also been demonstrated in greenhouse trials carried out in the UK (Jacobson et al. 2001, Chandler et al. 2005). These treatments were being tested when considering use of *B. bassiana* for regulation of western flower thrips (originally from North America) and two-spotted spider mite (*Tetranychus urticae*).

Several studies have examined non-target impacts of *B. bassiana* on predators and parasites of whiteflies, *Bemisia* spp. As invasive species in many countries, *B. tabaci* and *B. argentifolii* have achieved near-global distributions and major pest status on a variety of economically-important crops. Several predators and parasitoids are commercially available for whitefly control in the US, or contribute to the natural regulation of these pests in field crops. Whitefly is also susceptible to a range of fungal entomopathogens, some of which regularly cause natural epizootics in crops (Lacey *et al.* 1996), so if these pathogens are to be used in conjunction with natural enemies for whitefly control, a good understanding of non-target effects is required. In laboratory trials, Jones and Poprawski (1996) showed that whitefly nymphs parasitized by *E. eremicus* were immune to infection when the *B. bassiana* GHA strain was applied 3 or more days after parasitism. However, parasitized nymphs became infected if applications were made within 48 hours of parasitism, before the parasite eggs, which are laid beneath the scale, had eclosed and the parasitoid larva had penetrated the whitefly. Once within the host, the parasitoid larva appears to produce antibiotic compounds that inhibit fungal infection, even though germination and pre-penetration events can occur on the whitefly cuticle. In field trials on infested cantaloupe, however, no significant differences in parasitism rates were detected between control plots and those treated with *B. bassiana*.

Similarly, Jaronski *et al.* (1998) reported that even when *B. bassiana* GHA was applied at greater than label rate in field-grown cotton, there was minimal impact on predaceous insects. Poprawski *et al.* (1998) found that when whitefly treated with *B. bassiana* GHA strain was fed to first instars of the coccinellid *Serangium parcestosum*, mortality levels in the predator progressively declined with a corresponding increase in time from treatment to prey consumption. There was a concurrent decline in inoculum levels on the whiteflies and Poprawski *et al.* (1998) concluded that the predators were thus progressively exposed to lower concentrations of inoculum over time, leading to the decline in infection levels. Under field conditions it is likely that the natural enemies could complement each other, especially as inoculum levels would quickly decline on treated foliage as a result of environmental degradation.

17.3.3 Microorganisms and Plants

Microbial control agents typically have no adverse effects on plants (e.g. EPA 1998; Zimmermann 2007a, b). From an environmental safety viewpoint, a concern is that microorganisms applied inundatively for biological control could potentially displace non-target microorganisms. In soil at least, microorganisms applied artificially do not generally interfere with the microbial equilibrium, e.g. monitoring of strains of *B. bassiana* applied inundatively against *Dendrolimus punctatus* in southwest China showed that within 1 year, indigenous strains predominated in the local environment, indicating that they were not displaced by exotic ones (Wang *et al.* 2004).

Castrillo *et al.* (2004) evaluated effects of mass releases of a commercial formulation of *B. bassiana* GHA on naturally-occurring conspecific strains by comparing prevalence of, and genetic diversity within, indigenous populations of *B. bassiana* in fields with no history of GHA treatment and in fields representing a range of GHA application histories. Genetic diversity in *B. bassiana* isolates collected from soil from *B. bassiana*-treated potato farms in Maine and New York, was examined using amplified fragment-length polymorphisms and random amplified polymorphic DNA markers. There was greater diversity among populations in untreated fields than in GHA-treated fields, with displacement of indigenous strains in treated fields by GHA or GHA-similar haplotypes. This displacement, however, appeared to be temporary with recovery of native strains over time.

The potential for genetic recombination among microorganisms is another important component of risk assessment studies of entomopathogens that are sprayed repeatedly in large quantities in agricultural and forestry settings. The likelihood of recombination and the impact of resulting recombinants on non-target organisms need to be considered because recombinants can potentially have different virulence and host range characteristics than the parent strain(s). Castrillo *et al.* (2005) assessed the likelihood of recombination between *B. bassiana* GHA and indigenous conspecific strains by determining vegetative compatibility groups present in potato fields in Maine and New York with various histories of GHA application. Thirty-seven strains out of 110 soil isolates characterized using AFLP and RAPD markers were selected to represent the different cluster groups of indigenous and GHA-similar genotypes observed. Results showed that strains from all three genetic clusters of indigenous populations found in both Maine and New York were vegetatively incompatible with GHA, indicating a low likelihood of parasexual recombination between the introduced and indigenous strains.

The effects of biopesticides on other microorganisms have rarely been studied with the exception of *Bt*. While Yudina and Burtseva (1997) reported some *in vitro* activity of *Bt* endotoxins against prokaryotes, other studies have failed to detect any effect of the toxins on soil microorganisms (e.g. Ferreira *et al.* 2003, Muchaonyera *et al.* 2004). Studies examining effects of *Bt*-based biopesticides Dipel (Visser *et al.* 1994) or Foray 48B (O'Callaghan *et al.* 2007) on soil microorganisms have not detected any significant effects on parameters such as soil microbial diversity and respiration.

17.3.4 Aquatic Species

There is potential for microbial control agents to enter waterways, for example through spray drift during an aerial spray programme or, in the case of invasive mosquito species, the microbial control agent must be applied directly to water. There is a large body of literature on aquatic non-target effects of the two bacterial species *Bt israelensis* and *B. sphaericus*, which have been used extensively in mosquito control programmes in West Africa and Germany. Neither species poses direct or indirect threats to non-target invertebrate species or fish (Boisvert & Boisvert 2000, Lacey & Merritt 2003). This positive safety data largely facilitated registration of *Bt israelensis* for use in New Zealand to combat several invasive species of mosquitoes capable of vectoring human diseases. *Bt israelensis* was used in intensive spray programmes in localized areas to contain the spread of the Australian southern saltmarsh mosquito, *Ochlerotatus camptorhynchus* (Glare & O'Callaghan 1998, O'Callaghan *et al.* 2002).

The few studies that have examined effects of *B. bassiana* and *M. anisopliae* on aquatic species were reviewed in Zimmermann (2007a, b), who concluded that these fungi were unlikely to pose any significant hazards. An earlier study by Milner *et al.* (2002) found that the level of *M. anisopliae* conidia likely to enter water during campaigns to control the Australian plague locust, *Austracris guttulosa*, was a small fraction of that required to kill cladocerans, which was the only non-target aquatic species tested that showed any sensitivity to the fungal control agent.

17.4 Conclusion

Bacterial biopesticides have been used extensively in eradication of invasive pests and the widespread use of products based on *Bt* and *B. sphaericus* has provided a wealth of data on environmental impacts of these species in a range of different settings. The laboratory testing and post-application monitoring indicate that, in comparison with other control methods, these microbial control agents have minimal ecological impact. Even when used in very intensive and prolonged eradication programmes, they have not resulted in major ecological disturbances. It would be imprudent to assume that this would be true for all bacterial biopesticides and extrapolation of results to other agents, which may have differing modes of action and disease cycles, is not possible. Even within one species of insect pathogen, there can be wide variability in the insecticidal activity between strains and this factor must be addressed in non-target safety testing. Vestergaard *et al.* (2003) considered it was not possible to generalize about the safety of all entomopathogenic hypocrealean fungi at this stage but products developed to date, mainly based on *Beauveria* and *Metarhizium* spp., have good non-target safety records as do the more specific agents, such as viruses and nematodes (reviewed in Hokkanen and Hajek 2003). In addition, the use of microbial control agents potentially increases opportunities to utilize other compatible biological control agents, or complement the activities of natural enemies in the field. As several studies have shown, risks can vary according to the natural enemy, the timing of the treatment, and the formulation used.

Even if products pass all regulatory requirements and do not demonstrate any non-target effects in field tests, it cannot be said that these products are completely safe to non-target organisms elsewhere. Vigilance is required when microbial control agents are used on a large scale and in new environments. Only through long-term and wide scale use will long-term effects on non-target organisms surface (Vestergaard *et al.* 2003). Ecosystem level studies are very difficult and expensive to conduct, but are essential in sound environmental impact assessment. In addition, improved understanding of the ecological interactions between the control agent, its target and relevant non-target species is required if we are to be able to predict potential adverse effects over the long term (Goettel & Hajek 2001).

While no pest control strategy is entirely without risk, the levels of risk associated with the use of microbial control agents must be considered in context, i.e. what are the risks compared to those posed by current control methods and the costs of taking no action at all? For some new pest incursions, the potential damage in terms of economic losses in productivity and/or impacts on indigenous ecosystems is enormous, and rapid action must be taken to control or hopefully eradicate invasive pests. In these situations, particularly when large scale control programmes must be implemented, microbial control agents can be regarded as the most environmentally acceptable pest control option.

References

Addison JA (1993) Persistence and nontarget effects of *Bacillus thuringiensis* in soil: a review. Can J Forest Res 23:2329–2342

Addison JA, Holmes SB (1995) Effect of two commercial formulations of *Bacillus thuringiensis* subsp. *kurstaki* (Dipel R 8L and Dipel R 8AF) on the collembolan species *Folsomia candida* in a soil microcosm study. Contam Toxic 55:771–778

Addison JA, Holmes SB (1996) Effect of two commercial formulations of *Bacillus thuringiensis* subsp. *kurstaki* on the forest earthworm *Dendrobaena octaedra*. Can J For Res 26:194–1601

Al-mazra'awi MS, Shipp L, Broadbent B, Kevan P (2006a) Dissemination of *Beauveria bassiana* by honey bees (Hymenoptra: Apidae) for control of tarnished plant bug (Hemiptera: Miridae) on canola. Environ Entomol 35:1569–1577

Al-mazra'awi MS, Shipp L, Broadbent B, Kevan P (2006b) Biological control of *Lygus lineolaris* (Hemiptera: Miridae) and *Frankliniella occidentatlis* (Thysanoptera: Thripidae) by *Bombus impatiens* (Hymenoptera: Apidae) vectored *Beauveria bassiana* in greenhouse sweet pepper. Biol Control 37:89–97

Alves SB, Marchin LC, Pereira RM, Baumgratz LL (1996) Effects of some insect pathogens on the africanized honey bee, *Apis mellifera* L. (Hym., Apidae). J Appl Entomol 120:559–564

Armstrong CM (2000) Compatibility of two biological control agents, *Eretmocerus eremicus* (Rose & Zolnerowich) and *Beauveria bassiana* (Balsamo) Vuillemin, for silverleaf whitefly, *Bemisia argentifolii* (Bellows & Perring), control on poinsettia. MS Thesis, University of Vermont, Burlington, Vermont, USA. 132 pp

Atlavinyte O, Galvelis A, Daciulyte J, Lugaukas A (1982) Effects of enterobacterin on earthworm's activity. Pedobiol 23:372–379

Ball BV, Pye BJ, Carreck NL, Moore D, Bateman RP (1994) Laboratory testing of a mycopesticide on non-target organisms: the effects of an oil formulation of *Metarhizium flavoviride* applied to *Apis mellifera*. Biocontr Sci Technol 4:289–296

Barratt BIP, Barker GM, Addison PJ (1996) *Sitona lepidus* Gyllenhal (Coleoptera: Cuculionidae), a potential clover pest new to New Zealand. NZ Entomologist 19: 23–30

Boisvert M, Boisvert J (2000) Effects of *Bacillus thuringiensis* var. *israelensis* on target and nontarget organisms: a review of laboratory and field experiments. Biocontr Sci Technol 10:517–561

Boulton TJ (2004) Responses of nontarget Lepidoptera to Foray 48B® *Bacillus thurinigensis* va. *Kurstaki* on Vancouver Island, British Columbia, Canada. Environ Toxicol Chem 23:1927–1304

Boulton TJ, Otvos IS, Halwas KL, Rohlfs DA (2007) Recovery of nontarget Lepidoptera on Vancouver Island, Canada: One and four years after a gypsy moth eradication program. Environ Toxicol Chem 26:738–748

Brownbridge M (2006) Entomopathogenic fungi: status and considerations for their development and use in integrated pest management. Recent Res Devel Entomol 5:27–58

Brownbridge M, Humber RA, Parker BL, Skinner M (1993) Fungal entomopathogens recovered from Vermont forest soils. Mycologia 85:358–361

Brownbridge M, McNeill M, Nelson TL (2007) Effects of the clover root weevil pathogen *Beauveria bassiana* F418 on soil invertebrates and above ground non-target Coleoptera in New Zealand pastures [abstract]. In: Proceedings, 40th Annual Meeting of the Society for Invertebrate Pathology, Quebec City, Quebec, Canada, 12–16 August 2007, p 70

Brownbridge M, Nelson TL, Hackell DL, Eden TM, Wilson DJ, Willoughby BE, Glare TR (2006) Field application of biopolymer-coated *Beauveria bassiana* F418 for clover root weevil (*Sitona lepidus*) control in Waikato and Manawatu. NZ Plant Prot 59:304–311

Broza M, Pereira RM, Stimac JL (2001) The nonsusceptibility of soil Collembola to insect pathogens and their potential as scavengers of microbial pesticides. Pedobiol 45:523–534

Burges HD, Jones KA (1998) Formulation of bacteria, viruses and protozoa to control insects. In: Burges HD (ed) Formulation of Microbial Biopesticides, Beneficial Microorganisms, Nematodes and Seed Treatments. Kluwer, Dordrecht, NL, pp 33–127

Butt TM, Goettel MS (2000) Bioassays of entomogenous fungi. In: Navon A, Ascher KRS (eds) Bioassays of Entomopathogenic Microbes and Nematodes. CAB International Press, Wallingford, UK, pp 141–195

Butt TM, Ibrahim L, Ball BV, Clark SJ (1994) Pathogenicity of the entomogenous fungi *Metarhizium anisopliae* and *Beauveria bassiana* against crucifer pests and the honey bee. Biocontr Sci Technol 4:207–214

Butt TM, Carreck NL, Ibrahim L, Williams IH (1998) Honey-bee-mediated infection of pollen beetle (*Meligethes aeneus* Fab.) by the insect-pathogenic fungus *Metarhizium anisopliae*. Biocontr Sci Technol 8:533–538

Castrillo LA, Mishra P, Annis SL, Groden E, Vandenberg JD (2004) Field releases of *Beauveria bassiana* strain GHA affect genetic diversity of indigenous conspecific populations [abstract] In: Proceedings, 37th Annual Meeting of the Society for Invertebrate Pathology, Helsinki, Finland, 1–6 August 2004. p112

Castrillo LA, Annis SL, Groden E, Mishra PK, Vandenberg JD (2005) Low likelihood of recombination between the introduced *Beauveria bassiana* strain GHA and indigenous conspecific strains based on vegetative compatibility groupings [abstract]. In: Proceedings, 38th Annual Meeting of the Society for Invertebrate Pathology, Anchorage, Alaska, USA, 7–11 August 2005. p 50

Chandler D, Davidson G, Jacobson RJ (2005) Laboratory and glasshouse evaluation of entomopathogenic fungi against the two-spotted spider mite, *Tetranychus urticae* (Acari: Tetranychidae), on tomato, *Lycopersicon esculentum*. Biocontr Sci Technol 15:37–54

Chowdhury AB, Jepson PC, Howse PE, Ford MG (2001) Leaf surfaces and the bioavailability of pesticide residues. Pest Manag Sci 57:403–412

Cook RJ, Bruckart WL, Coulson JR, Goettel MS, Humber RA, Lumsden RD, Maddox JV, McManus ML, Moore L, Meyer SF, Quimby PC, Stack JP, Vaughan JL (1996) Safety of microorganisms intended for pest and plant disease control: a framework for scientific evaluation. Biol Control 7:33–351

Cory JS (2003) Ecological impact of virus insecticides: host range and non-target organisms. In: Hokkanen HMT, Hajek AE (eds) Environmental Impacts of Microbial Insecticides. Need and Methods for Risk Assessment. Kluwer, Dordrecht NL, pp 73–92

Dromph KM (2001) Dispersal of entomopathogenic fungi by collembolans. Soil Biol Biochem 33:2047–2051

Dromph KM (2003) Collembolans as vectors of entomopathogenic fungi. Pedobiol 47:245–256

Dromph KM, Vestergaard S (2002) Pathogenicity and attractiveness of entomopathogenic hyphomycete fungi to collembolans. Appl Soil Ecol 21:197–210

Environmental Protection Agency (1998) EPA Reregistration Eligibility Decision (RED) *Bacillus thuringiensis*. US EPA, Prevention, Pesticides and Toxic Substances, EPA738-R-98-004, 91 pp

Ehlers R-U (2003) Biocontrol nematodes. In: Hokkanen HMT, Hajek AE (eds) Environmental Impacts of Microbial Insecticides. Need and Methods for Risk Assessment. Kluwer, Dordrecht NL. pp 177–220

Ferreira LHPL, Molina JC, Brasil C, Andrade G (2003) Evaluation of *Bacillus thuringiensis* bioinsecticidal proetin effects on soil microorganisms. Plant Soil 256:161–168

Flexner JL, Lighthart B, Croft CA (1986) The effects of microbial pesticides on non-target, beneficial arthropods. Agric Ecosys Environ 16:203–254

Fountain MT, Hopkin SP (2005) *Folsomia candida* (Collembola): a "standard" soil arthropod. Annu Rev Entomol 50:201–222

Frampton GK (1997) The potential of Collembola as indicators of pesticide usage: evidence and methods from the UK arable system. Pedobiol 41:179–184

Giroux S, Côté J-C, Vincent C, Martel P, Coderre D (1994) Bacteriological insecticide M-One effects on predation efficiency and mortality of adult *Coleomegilla maculata lengi* (Coleoptera: Coccinellidae). J Econ Entomol 87:39–43

Glare TR, O'Callaghan M (1998) Environmental and health impacts of *Bacillus thuringiensis israelensis*. Report for New Zealand Ministry of Health, 57 pp. http://www.moh.govt.nz/mosquito.html

Glare TR, O'Callaghan M (2000) *Bacillus thuringiensis*: Biology, Ecology and Safety. John Wiley and Sons, Chichester, UK. 100 pp

Glare TR, O'Callaghan M (2003). Environmental impacts of bacterial biopesticides. In: Hokkanen HMT, Hajek AE (eds) Environmental Impacts of Microbial Insecticides. Need and Methods for Risk Assessment. Kluwer, Dordrecht NL, pp 119–150

Goettel MS, Hajek AE (2001) Evaluation of nontarget effects of pathogens used for management of arthropods. In: Wajnberg, E, Scott JK, Quimby PC (eds) Evaluating Indirect Ecological Effects of Biological Control. CABI Press, pp 81–97

Goettel MS, Poprawski TJ, Vandenberg JD, Li Z, Roberts DW (1990) Safety to non-target invertebrates of fungal biocontrol agents. In: Laird M, Lacey LA, Davidson EW (eds) Safety of Microbial Insecticides. CRC Press, Boca Raton, Florida. pp 209–231

Greenslade P, Vaughan GT (2003) A comparison of Collembola species for toxicity testing of Australian soils. Pedobiol 47:171–179

Gribben JR, Lewis GD, Wigley PJ, Broadwell AH (2002) Environmental persistence and growth dynamics of the *Bacillus thuringiensis* Foray 48B biopesticide. In: Akhurst RJ, Beard CE, Hughes P (eds) Biotechnology of *Bacillus thuringiensis* and its environmental impact. Proceedings of the 4th Pacific Rim Conference. pp 200–204

Hajek AE (1997a) Ecology of terrestrial fungal entomopathogens. Adv Microbial Ecol 15:193–249

Hajek AE (1997b) Fungal and viral epizootics in gypsy moth (Lepidoptera: Lymantriidae) populations in Central New York. Biol Cont 10:58–68

Hajek AE (1999) Pathology and epizootiology of *Entomophaga maimaiga* infections in forest Lepidoptera. Microbiol Mol Biol Rev 63:814–835

Hajek AE, St. Leger RJ (1994) Interactions between fungal pathogens and insect hosts. Annu Rev Entomol 39:293–322

Hajek AE, Goetell MS (2000) Guidelines for evaluating effects of entomopathogens on nontarget organisms. In: Lacey LA, Kaya HK (eds). Manual of Field Techniques in Insect Pathology, Kluwer, Dordrecht NL. pp 847–868

Hajek AE, Butler L, Liebherr JK, Wheeler MM (2000) Risk of infection by the fungal pathogen *Entomophaga maimaiga* among Lepidoptera on the forest floor. Environ. Entomol 29:645–650

Hokkanen HMT, Hajek AE (eds) (2003) Environmental Impacts of Microbial Insecticides. Need and Methods for Risk Assessment, Kluwer, Dordrecht NL. 269 pp

Hokkanen HMT, Zeng Q-Q, Menzler-Hokkanen I (2003a) Assessing impacts of *Metarhizium* and *Beauveria* on bumblebees. In: Hokkanen HMT, Hajek AE (eds) Environmental Impacts of Microbial Insecticides. Need and Methods for Risk Assessment. Kluwer, Dordrecht NL. pp 63–72

Hokkanen HMT, Bigler F, Burgio G, van Lenteren JC, Thomas MB (2003b) Ecological risk assessment framework for biological control agents. In: Hokannen HMT, Hajek AE (eds) Environmental Impacts of Microbial Insecticides. Need and Methods for Risk Assessment. Kluwer, Dordrecht, NL. pp 1–14

Hozzank A, Keller S, Daniel O, Schweizer C (2002) Impact of *Beauveria brongniartii* and *Metarhizium anisopliae* (Hyphomycetes) on *Lumbricus terrestris* (Lumbricidae). IOBC WPRS Bull 26:31–34

Ireson JE, Rath AC (1991) Preliminary observations on the efficacy of entomopathogenic fungi for control of the lucerne flea, *Sminthurus viridis* (L) and the redlegged earth mite, *Halotydeus destructor* (Tucker). In: Proceedings of the First Redlegged Earth Mite Workshop, Perth September 1991

Jacobson RJ, Chandler D, Fenlon J, Russel KM (2001) Compatibility of *Beauveria bassiana* (Balsamo) Vuillemin with *Amblyseius cucumeris* Oudemans (Acarina: Phytoseiidae) to control *Frankliniella occidentalis* Pergande (Thysanoptera: Thripidae) on cucumber plants. Biocontr Sci Technol 11:391–400

James RR, Hayes G, Lelend JE (2006) Field trials on the microbial control of Varroa with the fungus *Metarhizium anisopliae*. American Bee Journal, November Issue: 968–972

Jaronski ST, Goettel MS, Lomer CJ (2003) Regulatory requirements for eco-toxicological assessments of microbial insecticides – how relevant are they? In: Hokkanen HMT, Hajek AE (eds) Environmental Impacts of Microbial Insecticides. Need and Methods for Risk Assessment. Kluwer, Dordrecht NL. pp 237–260

Jaronski ST, Hoelmer K, Osterlind R, Antilla L (1998) Effect of a *Beauveria bassiana*-based mycoinsecticide on beneficial insects under field conditions. In: Proc. 1998 BCPC-Pests and Diseases. BCPC, Farnham, Surrey. pp 651–657

Jones WA, Poprawski TJ (1996) Parasitized *Bemisia argentifolii* are immune to infection by *Beauveria bassiana*. In: Proc. 29th Annual Meeting of the Society for Invertebrate Pathology. 1–6 September, 1996. pp. 41

Kanga LHB, Jones WA, James RR (2003) Field trials using the fungal pathogen, *Metarhizium anisopliae* (Deuteromycetes: Hyphomycetes) to control the ectoparasitic mite, *Varroa destructor* (Acari: Varroidae) in honey bee, *Apis mellifera* (Hymenoptera: Apidae) colonies. J Econ Entomol 96:1091–1099

Kanga LHB, Jones WA, Gracia C (2006) Efficacy of strips coated with *Metarhizium anisopliae* for control of *Varroa destructor* (Acari: Varroidae) in honey bee colonies in Texas and Florida. Exp Appl Acarol 40:249–258

Keller S, Kessler P, Schweizer C (2003) Distribution of insect pathogenic soil fungi in Switzerland, with special reference to *Beauveria brongniartii* and *Metarhizium anisopliae*. BioControl 48:307–319

Klironomos JN, Kendrick B (1995) Relationships among microarthropods, fungi and their environment. Plant Soil 170:183–197

Lacey LA, Merritt, RW (2003) The safety of bacterial microbial agents used for black fly and mosquito control in aquatic environments. In: Hokkanen HMT, Hajek AE (eds) Environmental Impacts of Microbial Insecticides. Need and Methods for Risk Assessment. Kluwer, Dordrecht NL. pp 151–168

Lacey LA, Fransen JJ, Carruthers R (1996) Global distribution of naturally-occurring fungi of *Bemisia*, their biologies and use as biological control agents. In: Gerling CB, Mayer RT (eds) *Bemisia*: Taxonomy, Biology, Control and Management, Intercept, Andover, UK. pp 401–433

Llewellyn D, Cousins Y, Mathews A, Hartweck L, Lyon B (1994) Expression of *Bacillus thuringiensis* insecticidal protein genes in transgenic crop plants. Agri Ecosys Environ 49:85–93

Lowe CN, Butt KR (2005) Culture techniques for soil dwelling earthworms: a review. Pedobiologia 49:401–413

Malakar R, Elkington JS, Hajek AE, Burand JP (1999) Within-host interactions of *Lymantria dispar* (Lepidoptera: Lymantridae) nucelopolyhedrosis virus and *Entomophaga maimaiga* (Zygomycetes: Entomophthorales). J Invert Pathol 73:91–100

McNeill MR, Brownbridge M, Nelson TL (2007) Measuring the impacts of soil applied *Beauveria bassiana* on above ground non-target Coleoptera in Waikato pasture. NZ Plant Prot 60: 315.

Meadows MP (1993) *Bacillus thuringiensis* in the environment: ecology and risk assessment. In: Entwistle PF, Cory JS, Bailey MJ, Higgs S (eds) *Bacillus thuringiensis*, an environmental biopesticide: theory and practice. Wiley, Chichester, England. pp 193–200

Meikle WG, Mercandier G, Girod V, Derouané F, Jones WA (2006) Evaluation of *Beauveria bassiana* (Balsamo) Vuillemin (Deuteromycota: Hyphomycetes) strains isolated from varroa mites in southern France. J Apicultural Res 45:219–220

Meikle WG, Mercandier G, Host N, Nansen C, Girod V (2007) Duration and spread of an entomopathogenic fungus, *Beauveria bassiana* (Deuteromycota: Hyphomycetes), used to treat varroa mites (Acari: Varroidae) in honey bee (Hymenoptera: Apidae) hives. J Econ Entomol 100:1–10

Meikle WG, Mercandier G, Holst N, Nansen C, Girod V (2008) Impact of a treatment of *Beauveria bassiana* (Deuteromycota: Ahyphomycetes) on honeybee (*Apis* Nester*mellifera*) colony health and on *Varroa destructor* mites (Acari: Varroidae). Apidologie 39:1–13

Milner RJ, Lim RP, Hunter DM (2002) Risks to the aquatic ecosystem from the application of *Metarhizium anisopliae* for locust control in Australia. Pest Manag Sci 58:718–723

Muchaonyera P, Waladde S, Nyamugafata P, Mpepereki S, Ristori GG (2004) Persistence and impact on microorganisms of *Bacillus thuringiensis* in some Zimbabwean soils. Plant Soil 266:41–46

Navon A, Ascher KRS (eds) (2000) Bioassays of entomopathogenic microbes and nematodes. CABI Publishing, Wallingford UK

Nealis VG, van Frankenhuyzen K, Cadogan BL (1992) Conservation of spruce budworm parasitoids following application of *Bacillus thuringiensis* var. *kurstaki* Berliner. Can Entomol 124:1085–1092

Nelson TL, Willoughby BE, Wilson DJ, Eden T, Glare TR (2004) Establishing the fungus *Beauveria bassiana* in pasture for clover root weevil (*Sitona lepidus*) control. NZ Plant Prot 57:314–318

New Zealand Ministry of Health (2008) http://www.moh.govt.nz/mosquito.html. Accessed 29 February 2008

O'Callaghan M (1998) Establishment of microbial control agents in soil. In: O'Callaghan M, Jackson TA (eds). Proceedings of the 4th International Workshop on Microbial Control of Soil Dwelling Pests, February 17–19, 1998. AgResearch, ISSN 1174-653X. pp 83–89

O'Callaghan M, Glare TR, Jackson TA (2002) Biopesticides for control of insect pest incursions in New Zealand. Defending the Green Oasis: New Zealand Biosecurity and Science. In: Goldson SL, Suckling DM (eds) Proceedings of a New Zealand Plant Protection Society Symposium. ISBN 0-473-09386-3. pp 137–152

O'Callaghan M, Gerard E, Sarathchandra U (2007) Analysis of non-target impacts of Foray 48B on soil micro-organisms. In: Cote J-C, Otvos IS, Schwartz JL, Vincent C (eds) Proceedings of the 6th Pacific Rim Conference on the Biotechnology of *Bacillus thuringiensis* and its Environmental Impact. Erudit, Montreal. pp 133–134

Parker BL, Skinner M, Gouli V, Brownbridge M (1997) Impact of soil applications of *Beauveria bassiana* and *Mariannaea* sp. on nontarget forest arthropods. Biol Control 8:203–206

Peacock JW, Schweitzer DF, Carter JL, Dubois NR (1998) Laboratory assessment of effects of *Bacillus thuringiensis* on native Lepidoptera. Environ Entomol 27:255–258

Peterson H, Luxton M (1982) A comparative analysis of soil fauna populations and their role in decomposition processes. Oikos 39:287–388

Poprawski TJ, Legaspi JC, Parker PE (1998) Influence of entomopathogenic fungi on *Serangium parcestosum* (Coleoptera: Coccinellidae), an important predator of whiteflies (Homoptera: Aleyrodidae). Environ Entomol 27:785–795

Prior C (1990) The biological basis for regulating the release of microorganisms, with particular reference to the use of fungi for pest control. Ann Appl Biol 24:231–238

Rath AC (1991) Pathogens for the biological control of mites and collembolan pests. Plant Prot Quarterly 6:172–174

Rath AC (2002) Ecology of entomopathogenic fungi in field soils. In: Proc. 35th Annual Meeting of the Society for Invertebr Pathol, Iguassu Falls, Brazil, Aug. 18–23, 2002. pp 65–71

Reinecke P, Andersch W, Stenzel K, Hartwig J (1990) Bio1020, a new microbial insecticide for use in horticultural crops. In: Proceedings, BCPC Brighton Crop Protection Conference- Pests and Diseases, November 1990. pp 49–54

Rothman LD, Myers JH (1996) Debilitating effects of viral disease on host Lepidoptera. J Invert Pathol 67:1–10

Roush RT (1994) Managing pests and their resistance to *Bacillus thuringiensis*: can transgenic crops be better than sprays? Biocontr Sci Technol 4:501–516

Salama HS, Zaki FN, Sharaby AF (1982) Effect of *Bacillus thuringiensis* Berl. on parasites and predators of the cotton leafworm *Spodoptera littoralis* (Boisd.). Z. Angew Entomol 94:498–504

Sample BE, Butler L, Zivkovich C, Whitmore RC, Reardon R (1996) Effects of *Bacillus thuringiensis* Berliner var. *kurstaki* and defoliation by the gypsy moth (*Lymantria dispar* L.) (Lepidoptera: Lymantriidae) on native arthropods in West Virginia. Can Entomol 128:573–592

Shaw KE, Davidson G, Clark SJ, Ball BV, Pell JK, Chandler D, Sunderland KD (2002) Laboratory bioassays to assess the pathogenicity of mitosporic fungi to *Varroa destructor* (Acari: Mesostigmata), an ectoparasitic mite of the honeybee, *Apis mellifera*. Biol Control 24:266–276

Smith RA, Barry JW (1998) Envoronmental persistence of *Bacillus thuringiensis* spores following aerial application. J Invertebr Pathol 71:263–267

Solter LF, Becnel JJ (2003) Environmental safety of microsporidia. In: Hokkanen HMT, Hajek AE (eds) Environmental Impacts of Microbial Insecticides. Need and Methods for Risk Assessment. Kluwer, Dordrecht NL. pp 93–118

Steinkraus DC, Boys GO, Rosenheim JA (2002) Classical biological control of *Aphis gossypii* (Homoptera: Aphididae) with *Neozygites fresenii* (Entomophthorales: Neozygitaceae) in Californian cotton. Biol Control 25: 297–304

Strasser H, Vey A, Butt TM (2000) Are there any risks in using entomopathogenic fungi for pest control, with particular reference to the bioactive metabolites of *Metarhizium*, *Tolyplocadium*, and *Beauveria* species? Biocontr Sci Technol 10:717–735

Tabashnik BE (1994) Evolution of resistance to *Bacillus thuringiensis*. Annu Rev Entomol 39:47–49

Vandenberg JD (1990) Safety of four entomopathogens for caged adult honey bees (Hymenoptera: Apidae). J Econ Entomol 83:755–759

Vänninen I, Tyni-Juslin J, Hokkanen H (2000) Persistence of augmented *Metarhizium anisopliae* and *Beauveria bassiana* in Finnish agricultural soils. BioControl 45:201–222

Vestergaard S, Cherry A, Keller S, Goettel M (2003) Safety of Hyphomycete fungi as microbial control agents. In: Environmental Impacts of Microbial Insecticides. Need and methods for risk assessment. Kluwer, Dordrecht NL. pp 35–62

Visser TG, Addison JA, Holmes SB (1994) Effects of Dipel 176, a *Bacillus thuringiensis* subspecies against prokaryotes. Microbiol 66:17–22

Wagner DL, Peacock JW Carter JL Talley SE (1996) Field assessment of *Bacillus thuringiensis* on nontarget Lepidoptera. Environ Entomol 25:1444–1454

Wang C, Fan M, Li Z, Butt TM (2004) Molecular monitoring and evaluation of the insect-pathogenic fungus *Beauveria bassiana* in southeast China. J Appl Micro 96:861–870

West AW (1984) Fate of the insecticidal, proteinaceous parasporal crystal of *Bacillus thuringiensis* in soil. Soil Biol Biochem 16:357–360

Whaley WH, Anhold J, Schaalje GB (1998) Canyon drift and dispersion of *Bacillus thuringiensis* and its effects on selected nontarget lepidopterans in Utah. Environ Entomol 27:539–548

Yudina TG, Burtseva LI (1997) Activity of delta-endotoxins of four *Bacillus thuringiensis* subspecies against prokaryotes. Microbiol 66:17–22

Zimmermann G (2007a) Review on safety of the entomopathogenic fungus *Beauveria bassinana* and *Beauveria brongniartii*. Biocontr Sci Technol 17:553–596

Zimmermann G (2007b) Review on safety of the entomopathogenic fungus *Metarhizium anisopliae*. Biocontr Sci Technol 17:879–920

Part VI
Conclusions

Chapter 18
Considerations for the Practical Use of Pathogens for Control and Eradication of Arthropod Invasive Pests

Travis R. Glare, Maureen O'Callaghan and Ann E. Hajek

Abstract Exotic arthropod introductions into new regions are increasing in frequency every year, along with the ongoing expansion in international trade and travel. Broad spectrum chemical pesticides have been used to combat many invasive species, but use of these pesticides is increasingly difficult due to concerns over mammalian and environmental safety. Invertebrates succumb to diseases caused by specific microbial pathogens. These microbes have potential to be used to control or even eradicate invasive pests and have additional environmental and safety benefits because of their high specificity and low non-target effects. In this book, case studies on the use of microbial pathogens against a wide range of invasive species are documented, showcasing systems in which pathogens have been successful and describing some novel methods of use. This chapter examines the benefits and limitations of pathogens for control and eradication of invasive species. Apart from the biological and ecological aspects of using live microorganisms, control of invasive species requires consideration of many key issues, including risk assessment, regulatory requirements, community attitudes to control programmes and rapidity of response. Even deciding that a programme has been successful, such as declaring a pest eradicated, is not straightforward, as it involves determining the probability that an undetected low density population no longer exists. Finally, the future of use of pathogens against invasive species is briefly discussed.

18.1 Introduction

There is no doubt that with ongoing globalisation, and the increased availability and volume of international transportation, we will see more and more hitchhiking arthropods moving from areas where they are native to areas where they are exotic (see Chapter 1). In ever-increasing numbers, arthropods will continue to breach

T.R. Glare
AgResearch Ltd., Biocontrol, Biosecurity & Bioprocessing, Lincoln Science Centre, Christchurch 8140, New Zealand
e-mail: travis.glare@agresearch.co.nz

the biogeographical barriers that defined their original distributions. Modern rapid transportation systems now allow introductions of species that previously would not have survived slow journeys by sea over long distances (Nentwig 2007). While only a small fraction of species that are introduced become established over the long-term (Simberloff & Gibbons 2004) and only a small fraction of those becoming established cause serious problems (Williamson 1996), those invasive species that do become established have the potential to cause devastating losses in terms of productivity and environmental damage. Accidental introductions of new arthropod species most commonly occur near ports and airports, typically in heavily populated areas where invasive species cannot be easily controlled using chemicals with high non-target toxicity risks. Microbial control agents have been the control option of choice in these settings and, in many cases, have provided perhaps the only option for management of intractable invasive species. However, there are many additional issues to be considered when a new species lands on your doorstep, as detailed in the following sections.

18.2 Responding to a New Pest Invasion

After discovery of a new invasive species, a number of critical steps must be taken to determine the optimal response. Accurate pest identification is crucial, as specific identification of the species, possibly to subspecies level, is needed to ensure that appropriate control agents are selected. It is a peculiar facet of response to invasive arthropods that invasive species are often largely unknown in their native location/country so there is likely to be a lack of knowledge and expertise on which to call when assessing whether the new invasive poses a threat. Depending on the obscurity of the new invasive in its area of origin, it may take some time to find experts and/or to learn about how to control it. In the case of the white spotted tussock moth in New Zealand (see Chapter 4), no literature on control measures for this pest were found in standard databases when this insect was first discovered. Thus, determining the host range, ability to spread, potential as a pest and possible control measures was very difficult. In some cases, there will be no live specimens of the invasive species available for testing with potential control agents. This necessitates testing to be undertaken offshore but accurate identification of the invasive is essential to ensure that testing undertaken elsewhere is against the same species (or subspecies). Alternately, testing can be conducted using colonies of the pest established in containment in the invaded country; this is not always a desirable option for a high risk pest and in some cases regulators have not allowed colonies to be maintained because of risk.

As described above, it can be very difficult to assess the threat from a new invasive species, but this is a crucial first step in the decision to attempt eradication or employ control measures to limit establishment or damage and spread of the pest. Where the pest poses extreme threat to a country's productivity or human health, a rapid response is essential, which means decisions must sometimes be made relatively quickly based on incomplete information.

18.2.1 Level of Risk Posed by the Invasive Species

Many countries maintain lists of unwanted organisms, where the risk posed by key species has already been assessed, but in some cases it will be necessary to seek information about the invasive species from its country of origin, if this is known.

There can be multiple potential impacts from an invasive species, including severe economic loss, human and animal health impacts, loss of biodiversity, threats to natural areas, and even aesthetic impacts such as unsightly dead plants. Most responses focus primarily on impacts to human health and potential economic losses in agriculture or forestry, the latter because many invasive arthropods are plant feeders. Risks can be difficult to calculate as, on first arrival, the host range of the invertebrate may not be fully understood. Often the invasive species can pose a serious threat to more than one sector, complicating estimation of potential impacts and economic losses. The examples given in Chapter 4 for lymantriids invading New Zealand had potential predicted economic losses ranging from a few million to hundreds of millions of dollars for the same insect (Ombudsman 2007).

It is especially difficult to assess the potential threat posed by the invasive species on the native ecosystems and biodiversity. Response of indigenous biota, previously unchallenged by the invader, cannot be predicted accurately. Threats to indigenous biodiversity are difficult to quantify in economic terms and most regions have iconic species that are valued far beyond any real economic contribution. Studies on the preferences of Asian gypsy moth and painted apple moth for New Zealand native flora found some unexpected hosts (Kay *et al.* 2002, Stephens *et al.* 2007). Some of these studies were conducted using New Zealand plants growing in arboreta around the world, an effective method to obtain field data on ability of an arthropod to reproduce and cause damage on native host plants before the invasive is well established.

Invasive species that pose risks to human and animal health will usually receive the most rapid response from authorities and are generally better understood, since reactions of mammals to arthropods (i.e. as pests of mammals or vectors of disease) are similar around the world and are well studied. Experience and knowledge derived from the extensive use of *Bacillus thuringiensis israelensis* (*Bti*) for mosquito control in West Africa and Germany facilitated a rapid response by the New Zealand Ministry of Health, when several disease-vectoring exotic mosquitoes were detected in New Zealand from 1998 (O'Callaghan *et al.* 2002). One of these species, *Ochlerotatus* (= *Aedes*) *camptorhynchus*, was discovered originally in the North Island of New Zealand and eradication programmes have been conducted at several sites, using *Bti* and the insect growth regulator S-methoprene.

18.2.2 Size and Distribution of the Invasive Population at Time of Discovery

For an eradication attempt of an invasive species to be feasible, the introduced population needs to have remained localised. Unfortunately, there are many examples of situations where the invasive species had spread significantly before they were discovered and eradication became impossible, e.g. *Varroa* mite in New Zealand (MAF 2007) and fire ants in the USA (Chapter 13). In some cases, particular characteristics of the invasive species (e.g. having life stages that spread rapidly) make the chances of eradication extremely unlikely, as with many aphid species (see Chapter 6). As detailed below, knowledge of the life cycle and biology of the invasive species is crucial in early decision-making about possible control measures that could be employed. Many questions must be asked: Does the invasive species multiply rapidly and have a flight stage, or is it likely to spread by a slow moving front? What anthropogenic activities aid its spread and can spread be limited to allow an eradication attempt?

Detailed surveys are required to determine the size and distribution of the invasive population. These are, for some pests, conducted using pheromones (sex or aggregation) or specialized traps but often little is known about the invasive species and there is no available technology for efficient detection of low density populations.

The eradication programmes for gypsy moth in western North America provide examples of the difficulties of detecting populations and deciding on eradication or control strategies. Low population densities can be difficult to locate; woodborers (covered in Chapters 9, 10 and 12) are an example of pests that are not easily detected at low densities because they live in cryptic habitats (and no efficient pheromone traps are available), making decisions about eradication and control difficult. Low density populations are often below a threshold needed for establishment (Allee effect, as described in Chapter 3). If a population is below the threshold for establishment, then eradication efforts would be wasted. Chapter 5 describes the differential reaction of authorities in the USA to findings of European (EGM) versus the more damaging Asian subspecies of gypsy moth (AGM). In the case of EGM, several adults caught in pheromone traps are insufficient to trigger a response whereas detection of any AGM immediately triggers a response and appropriate treatment.

18.3 Control or Eradicate?

Eradication requires the eventual elimination of all individuals of the invasive species, whereas control recognises that a population is sufficiently established and/or has spread to such an extent that it would be very difficult to eradicate, so limiting damage and preventing further spread are the goals. Response strategies for invasive species are rarely planned in advance. There are exceptions, such as disease-vectoring insects and very high profile pests like gypsy moth where the threat is well recognised. However, often some unexpected potential pest has been discovered, and government authorities, industry and the research community struggle to find information to make informed decisions on its control or eradication. Because of the high risks, costs and potential adverse community response associated with an eradication campaign, the decision to eradicate will often require a political decision (on advice from regulators, science and community advisory groups), especially for invasive arthropod species in urban areas. Generally, the decision to eradicate follows a cost-benefit analysis, which balances the potential economic losses and harm to humans and the environment against the high costs of eradication. The decision-making process must also include an analysis of the likelihood that the eradication can succeed, which will depend on factors such as the particular characteristics of the invasive pests (e.g. whether it is cryptic or can be easily accessed by control agents) and how far the invasive species has spread before it was detected. In some cases, the likelihood of reinvasion is seen as high and it can be argued that eradication is a waste of time (Myers *et al.* 1998). In this book, Liebhold and Tobin (Chapter 3) have suggested that eradication is useful, even if reinvasion is likely to occur. Reduction of damage from removing an invasive pest can be worthwhile, despite the need to repeat pathogen application many times. Such a situation occurs with gypsy moth in the USA (Chapter 5) and has now occurred with the painted apple moth in New Zealand (Chapter 4).

As discussed by Liebhold and Tobin (Chapter 3), some authors consider that it is impossible or impractical to eliminate every individual in a population (e.g. Dahlsten *et al.* 1989). Liebhold and Tobin argue, however, that as low density populations are often driven to extinction (the Allee effect) then not every last individual needs to be killed by intervention to cause eradication. Certainly, the successful eradications described in Chapters 4 and 5 (gypsy and tussock moths) indicate that invasive arthropod species can be driven to extinction by applications of biopesticides. Attempting the ambitious goal of eradication also relies on the availability of a suitable control agent, as discussed below. Pathogens can be used in different strategies for eradication and in this book numerous examples of strategies are given. However, all approaches have issues that can affect the outcome, as discussed below.

18.3.1 Issues to Consider in Eradication of Invasive Pests

"Eradication" is a value-laden word; it implies that the technology, finances and willingness to accept side effects are sufficient to eliminate a species from a geographic area (Myers *et al.* 1998). In some cases, for example where a new invasion of a pest with serious potential to impact human health has been detected early, the decision to eradicate is relatively easily made. However, there are a number of other issues that must be considered early in the planning of an eradication campaign, as discussed below.

18.3.1.1 Properties of Microbial Agents for Eradication Programmes

In this book, we present very few examples of insect pathogens used in eradication. *Bacillus thuringiensis* (*Bt*) is the only biopesticide used extensively as a primary agent during eradication programmes, although examples of how other pathogens might be used are presented (e.g. Chapter 10 – wood-boring beetles) or have been used to support a *Bt*-based eradication (Chapter 5 – gypsy moth). Eradication requires either a highly efficacious pathogen and/or very high density applications. In most situations it is hard to guarantee that a high density of pathogen will reach the pest; however, this issue is not specific to pathogens and also applies to chemical insecticides.

Biopesticides are typically perceived as being less persistent than chemical pesticides, which results in more frequent applications of biopesticides than chemicals. A benefit of using *Bacillus thuringiensis kurstaki* (Chapters 4, 5 and 11) is that there is no requirement for the applied bacterium to remain viable for extended periods, as its associated endotoxins determine its efficacy, rather than infection of insects by live cells. Lack of persistence may not always be a factor; entomopathogenic fungi applied in non-woven material to the outside of trees for woodborer control remained active for over 100 days, which was sufficient to allow effective control (see Chapter 10).

In some cases, the pathogen can become less virulent over time and the control/eradication programme is less successful until this is recognised. In Chapter 8, the eventual attenuation of the *Oryctes* virus in some populations/regions is suggested to be the cause of failure of suppression of the rhinoceros beetle. Similarly, long-term, inappropriate culturing of the nematodes used for control of *Sirex* woodwasps caused loss of virulence (Chapter 12).

18.3.1.2 Availability of Biopesticides

Because microbial agents still only constitute around 1% of the pesticide market (AGROW report 2004) and are most frequently used on a relatively small scale for insect pest control, it can be difficult, if not impossible, to find a source for the often very large amounts of biopesticide required for broad scale application, especially at short notice. Currently commercial production of any microbial control agent other than *Bt kurstaki* is likely to be too low to supply a new market in the quantities required for eradication. For example, recognising the serious threat posed by the then hypothetical invasion of New Zealand by the Asian gypsy moth (AGM) (*Lymantria dispar*, Asian subspecies), a number of microbial agents for use in potential eradication efforts were investigated (Glare *et al*. 1998). The nucleopolyhedrovirus of *L. dispar* (LdNPV) was a candidate agent, as it had been used in conjunction with *Bt kurstaki* during an eradication campaign for AGM in North Carolina (Chapter 5), was efficacious and had high specificity for *Lymantria* spp. The virus was produced as a semi-commercial product in the USA by the USDA Forest Service as Gypchek (Reardon & Podgwaite 1992, 1994), and also in Canada, (Disparvirus), Czech Republic (Biola), and Russia (Virin-Ensh). The main benefit of LdNPV over *Bt kurstaki* and other control measures is increased specificity (e.g. Glare *et al*. 1995). The virus also had the ability to recycle in the population after application. However, investigations showed that production levels in the USA, Canada and the Czech Republic in the 1990s would not have been sufficient to supply the amounts required for an emergency response in New Zealand, unless there was an outbreak at the same time in the Northern Hemisphere and one of the production units had scaled up production and exceeded local demand. Other sources of virus, such as the Russian Virin-Ensh product had potential purity issues (Glare *et al.* 2003). When gypsy moth was eventually detected in New Zealand, only the *Bt kurstaki*-based product Foray 48B could be sourced in sufficient quantities to allow eradication efforts to commence.

An alternative approach is the localised production of sufficient quantities of the microbial agent. Some of the microbial control agents described in this book can be easily grown *in vitro* (e.g. the fungi, *Beauveria* and *Metarhizium*). In many countries cottage industries already produce pathogens for localised control of pests, or production is carried out by a government-funded agency (e.g. production of nematodes for *Sirex* control, see Chapter 12). Whether such production could be initiated and sufficient quantities produced for an eradication programme would depend on having suitable expertise and facilities and would result in a slower response time than sourcing of a commercially available product.

For microbial agents that can only be produced in the host (e.g. most viruses and protists), localised production will be more difficult as highly specialised laboratory equipment and skilled personnel will be required. Although there are examples of authorities maintaining large insect rearing colonies of invasive arthropods in eradication areas (e.g. for use in trapping of tussock moths for delimitation, Chapter 4), use of rearing facilities for production of microbial agents has not been reported to our knowledge. The *L. dispar* virus produced as Gypchek in the USA is produced in insects, but this is completed in a subspecies endemic in the eastern USA and not within eradication zones.

18.3.1.3 Behaviour or Biology of the Invasive Arthropod Makes Eradication Difficult

Some arthropods have characteristics that preclude an eradication attempt. Fire ant nests are apparent above ground, but entrances are all outside the obvious nest, as far as 10 m away. This can make detection and treatment difficult. Other insects that spend large parts of their life cycle underground are also notoriously difficult to control, let alone eradicate. Insects that fly can be very difficult to contain within a defined area, as new populations can establish at alarming rates and remain undiscovered for years. In some cases, anthropogenic effects result in extensive movement of the invasive, such as gypsy moth egg masses on cars (e.g. Chapters 4 and 5).

18.3.1.4 Cost of Eradication

Depending on the situation, eradication can be a very expensive option that will only be undertaken where there is a high level of threat to the economy, health and/or environment. Microbial pathogens may be a more expensive alternative than chemical insecticides, if only because few are readily available in quantity. However, as discussed in the section above, microbial agents are often preferred over less expensive chemical pesticides because of their good biosafety profiles. In fact, in a growing number of situations, microbial pathogens or other biological approaches will be the only feasible option as broad-spectrum agents are unlikely to be used in urban areas and outside limited (targeted) agricultural or forestry applications. This means that the cost of microbial agents is generally not a limiting factor if eradication is the decided course of action, so microbial pathogens could be applied at rates not generally considered for routine pest control.

There are additional hurdles when developing an eradication strategy for an invasive species. It may be necessary to maintain colonies of the invasive species, to supply live insects for trapping and surveillance and also to use as test organisms in bioassays against prospective control agents. Such work must be carried out under strict quarantine facilities and is expensive and labour intensive. As was found during eradication of Lymantriidae in New Zealand, laboratory colonies of insects are notoriously prone to disease (e.g. tussock moths – Chapter 4).

In some cases, eradication campaigns need to be repeated or extended, such as when populations emerge outside the delimited zone. This can significantly increase costs, as well as try the patience of affected communities.

18.3.1.5 Community Attitudes

Mass applications of live microbes are unlikely to be without drama. As was observed during programmes to eradicate invasive lepidopteran species in Auckland and Hamilton, New Zealand, and Vancouver, Canada (see Chapters 4, 5 and 16), local communities need to support eradication efforts. In these programmes, eradication took place in large cities and many people were exposed to sprays. In New Zealand, over the course of three eradication programmes, public opposition became more organised and implacable, despite the vast majority of New Zealanders recognising the importance of excluding pests and showing support for the programme. In Canada, opposition culminated in the production of specific documents such as "151 reasons not to spray" (Schroeter & Wharton 2000). Such attitudes can be expected to occur even when the eradication occurs in a rural setting.

Most eradication programmes will require environmental and health impact assessments before a programme commences, detailing expected effects based on what is known about the agent. For biopesticides such as *Bt kurstaki*-based agents, literature reviews of thousands of articles are available (e.g. Auckland District Health Board 2002, Jenner Consultants Ltd 1996, MAF 2003). Such thorough impact assessments are not possible when the microbial agent has not been used as frequently or non-target and health impacts are largely unstudied. However, a commercially available biopesticide such as Foray 48B (*Bt kurstaki*-based) can raise other issues of public concern. A major factor in public dissent during the tussock moth eradications in New Zealand was the lack of public declaration of the formulation ingredients of Foray 48B. The formulation is an industrial trade secret and, while government officials and health authorities were informed of the ingredients and assessed them as safe, community groups remained unconvinced of the safety of undeclared additives and often maintained that the formulation included specific harmful additives despite official denials. Lack of openness about commercial formulations will continue to cause issues with high profile microbial pesticide applications.

As discussed in Chapter 5, some authors (and the public) consider that eradication and control programmes are motivated by trade issues rather than environmental or health concerns and could be unnecessary (e.g. Myers 2003). Cost-benefit analyses for new pests are notoriously difficult to conduct, considering the effect on the productive sectors is usually based on conjecture in the early stages of invasion. It is difficult at any time to put monetary value on indigenous flora and fauna.

Experience with eradication and control campaigns has repeatedly demonstrated that clear communication remains the key to retaining public support.

18.3.2 Issues to Consider in Control of Invasive Pests

It is more common, as the chapters in this book attest, for insect pathogens to be used to control invasive pests, rather than to be used to attempt eradication. In many ways insect pathogens are more suited for use in control programmes rather than eradication attempts, as no pathogen has evolved to kill all available hosts. If the pest has become too well established and spread too far for eradication, what are the control approaches that yield the best results?

As outlined in Chapter 1, microbial pathogens can be used for an inundative augmentation (widespread application) approach or as a classical biological control (point release and natural spread). Determining which approach is the most applicable is dependent on the characteristics of the pest and availability of suitable microbial control agents. Inundative augmentation aims to contact as many hosts as possible through application(s) of high density pathogen formulations, while classical biological control relies on the pathogen's ability to spread naturally through a pest population from a limited or point source application and then persist.

Typically control programmes involve lower costs than eradication attempts but may still require significant research funding to select the best microbe for the job and to develop mass production and application techniques, if a commercial product is not available. The cost of ongoing control of inundative augmentation is often an issue. If the microbial agent is used as a biopesticide (e.g. Chapter 7), costs of production, formulation and application should be comparable to other agents (e.g. chemical insecticides). If the microbial agent is released to form a self-sustaining population (e.g. Chapters 6 – entomophthoralean fungi and aphids and 14 – entomophthoralean fungus and mites), costs involved in discovery, importation and release will need to be commensurate with damage risk.

Public reaction is generally more favourable to control programmes than eradication attempts, largely because there are no highly publicised large-scale spraying events that occur with little forewarning.

18.3.2.1 Selecting and Sourcing a Suitable Insect Pathogen

Studies have demonstrated that invasive species commonly travel without a suite of naturally occurring enemies (see Chapter 2). While some pathogens in the new region may be virulent or adapt quickly to the new pest, in most cases no such pathogen challenge exists.

Commonly, the approach taken for control of incursion pests is to search the area of origin for naturally occurring pathogens (and other natural enemies). The theory is that the most effective pathogens are those from the point of origin, especially if the invasive is not an important pest in its home region. This classical biological control approach has been used in the control of fire ants in the USA (Chapter 13), cassava mites in Africa (Chapter 14), and mole crickets in the USA and Puerto Rico (Chapter 7). With gypsy moth, a virulent fungal pathogen from its area of endemism was accidentally introduced to North America and programs are now in place to introduce European microsporidia against gypsy moth in North America (Chapter 11). This approach is generally based on use of highly host-specific pathogens.

However, a constraint on using this approach for inundative augmentation was highlighted by Hajek and Bauer (Chapter 10), where entomopathogenic fungi were effective against the wood borers in their area of origin, but USA laws would have made it difficult to use exotic strains for development as biopesticides. Consequently, testing concentrated on selecting strains of fungi from the invaded area as control agents. This led to only limited work on one of the more promising pathogens for *Anoplophora glabripennis*, the fungus *Beauveria brongniartii*, as it was not thought to occur naturally in the USA at the time that studies were conducted. The species has since been found in the USA (see Chapter 10).

For either control approach, screening and selection of a suitable microbial control agent will require bioassays against the target invasive species and it may be necessary to carry out these tests on other populations of the invasive species, outside of the invaded country or regions. This can lead to problems if the population used for testing purposes differs at all from the invasive subspecies. If the microbe of choice is an exotic species, there can be delays in quarantine clearance and permission to release in the environment. Environmental and mammalian safety data are usually required and the time taken to collect appropriate data (which often involves testing against indigenous species in the new area), can lead to delays in use of the pathogen in the field.

In some cases inexpensive production methods using low-level technology and distribution methods have been developed and can result in highly effective control agents. In most areas where *Oryctes* virus has been released for control of rhinoceros beetles (Chapter 8), introduction of the virus into the local beetle population has been achieved through simple infection and release of live beetles. The nematodes for control of *Sirex* woodwasps are easily grown in flasks with the fungal symbiont and then distributed (see Chapter 12).

18.3.2.2 Behaviour or Biology of the Arthropods Makes Control Difficult

The characteristics of the invasive pest play a key role in determining the outcome of a control programme. As discussed above, a number of arthropod behaviours and life styles make eradication difficult, for example, rapid spread or cryptic habitats. With ongoing control programmes, the same issues are relevant but other aspects of

18 Considerations for the Practical Use of Pathogens

the pest-pathogen dynamics also come into play. Thus, if a pathogen does not sufficiently suppress the pest population before the pest reproduces, the next generation of pests may not be adequately suppressed. For example, the density-dependent egg laying behaviour of some soil-dwelling pests, such as the Argentine stem weevil *Listronotus bonariensis*, can compensate for mortality. When larval and/or adult stages are decimated, surviving females lay more eggs than usual to compensate (McNeill *et al.* 1998).

Other examples of pest characteristics that hinder control efforts are discussed throughout the book. The cryptic habitats of wood boring insects (Chapters 9, 10 and 12), social insects in protected nests such as fire ants with huge underground colonies (Chapter 13) and insects that live in close relationships with beneficial insects, such as *Varroa* mite living with honey bees, all present challenges to use of pathogens (or any other control method).

18.3.3 Environmental Safety of the Microbial Agent Used for Eradication and Control

Microbial agents are generally considered as environmentally benign alternatives to chemical pesticides, but they are not entirely without risk. Issues around assessment of environmental safety of microbial control agents have been discussed in Chapter 17, but in brief, there are several key characteristics of the microbes that determine their potential to impact on non-target species, including host specificity, mode of action, persistence in the environment, transmission and dispersal properties (Glare & O'Callaghan 2003). Highly specific pathogens such as gypsy moth (*Lymantria dispar*) nucleopolyhedrovirus (*Ld*NPV) have very low risk to non-target species while other pathogens with a wider host range can impact on non-target organisms closely related to the target pest (e.g. Boulton 2004, Boulton *et al.* 2007). It is essential that biopesticides used in eradication programmes have minimal impacts on beneficial species such as pollinators and biological control agents. The biopesticide used most extensively in eradication of invasive species, *Bt kurstaki*, has few non-target effects on beneficial species and soil biota and is non-toxic to birds and mammals (Glare & O'Callaghan 2000).

18.4 Regulations Affecting Pathogens Used Against Invasive Arthropods

Use of pathogens against invasive invertebrate pests can be regulated by various forms of legislation, depending on the country and situation. If an exotic pathogen is to be used to control an invasive (exotic) pest, then importation requirements as well as pesticide regulation requirements will need to be met.

18.4.1 Registration of Microbial Control Agents/Biopesticides

There is significant variation around the world in how microbial-based pesticides are treated for registration purposes. Most countries regulate all pesticides (insecticides, herbicides and others) through legislation requiring that human, animal and environmental safety are assured. These regulations require that physical characteristics of the pesticide are described and environmental persistence (residues) is considered as seriously as efficacy. Unfortunately, such regulations are difficult to use when looking at products based on live organisms. Conventional pesticide regulations can also impose high costs for those seeking to register biopesticides; regulations enacted primarily for chemical pesticides often require testing and data collection inappropriate for microbes such as the weight of the active ingredient (for some biopesticides this translates as weight of an individual spore or cell) and aspects of residue analysis. Many chemical pesticides are persistent in the environment after application, which can cause issues if the pesticide is toxic in large quantities or has non-target issues. However, for most live pathogens it is the lack of persistence rather than extended persistence that causes problems (i.e. lack of field efficacy), so residue analysis is not necessary.

Led by the USA (EPA 2008) many countries are moving to specifically recognise the nature of biopesticides that are based on live organisms. The US Environmental Protection Agency (EPA) registration regime for microbial agents is based on testing at different risk levels. Tier 1 testing involves high dose challenges and is quite simple. Any organism that fails a Tier 1 test must then pass more precise (and expensive) Tier 2 tests. Most toxicological tests need to be completed by specialists and this makes the testing expensive and time-consuming. Generally, if microbial pathogens fail Tier 1 tests, then registration is not pursued due to the complexity and expense of Tier 2 testing. Estimates of Tier 1 level mammalian testing start at US $150,000 (Gryzwacz 2003). In the USA, there is a group (IR-4) that provides funding to assist with the cost of testing for registration, to support biopesticide development (Anonymous 2008).

The EPA provides guidelines specifically for microbial pesticide registrations (EPA 2008). Basically, data are needed to show that the "maximum challenge" approach (Tier 1 testing) shows no mammalian or unexpected non-target impacts. Elements of the registration process include specific identification of the microbe, demonstration that the microbe is responsible for the pesticide effect, information on the manufacturing, characterization of any formulation, and data on toxicology, environmental safety and biopesticide efficacy (Braverman *et al.* 2003).

The EU regulations seem to have been tailored specifically to microbes (e.g. Pesticides Safety Directorate 2007), and require a full toxicology and environmental package, which is difficult to assemble quickly. Data must be sufficiently detailed to allow a full risk assessment to be completed, but are not considered to be as extensive as data required for chemical pesticides (Pesticides Safety Directorate 2007).

Other factors affecting the ability to register microbial pathogens include the amount of information available about its prior use elsewhere, ability to provide data for risk assessment and any history of safe use as a biopesticide (Songa 2003).

As an example, registration of the *Lymantria dispar* NPV in New Zealand would require only limited new data to be collected, as the virus was previously registered with the EPA and had been used in many countries without problems (Glare *et al.* 1998).

Not all countries have such well established registration systems (e.g. Gryzwacz 2003). In India, for example, laws to cover biopesticide registration were only enacted in 1999. Regulatory agencies typically have staff with chemistry rather than microbiology backgrounds. Lack of specific microbiological expertise in regulatory teams results in some of the inappropriate testing requirements mentioned above, which in turn hampers registration of microbial pathogens.

18.4.2 Other Regulatory Concerns

Any process that adds to the cost or reduces the speed of response is detrimental to control or eradication success. In addition to the registration of specific pathogens, other laws and regulations will be involved in response to the invasives. In some countries, special regulatory facilities are available to fast-track registration or importation, such as under New Zealand's Biosecurity Act (see Chapter 4). Other countries do not have such explicit provisions. Without such legislation, it could be difficult to use microbial pathogens in rapid response to invasive pests especially for eradication. All countries have at least a minimal system in place for regulating alien species introductions, but these systems are not always sufficient to facilitate an appropriate response.

In some cases, there can be several layers of regulation that can impact the response, from national (and international) to state, regional and even local authorities. If national legislation has no overriding powers to mandate for rapid response, local bylaws can inhibit a response, especially in controversial situations, such as intensive spray programmes in urban areas.

18.5 Declaring Success

It is important to establish the overall goal of any control attempt, against which the success of the programme can be measured. In control programmes seeking to mitigate damage or limit spread, the goal may be economic (e.g. reduction of damage to beneath a certain threshold that can be tolerated without economic losses in production), social (i.e. lack of large visible populations of nuisance pests, such as stinging insects) or some other measure. In the case of invasive species with potential to cause severe economic losses to key productive sectors or which can vector human disease, the aim must be complete eradication if the invasion is found early, but reducing impacts through control can be considered a success.

In the case of eradication programmes, declaring success would appear to be quite straightforward: Eradication is achieved when the invasive species can no longer be found. However, lack of a detectable population for a short time is no certainty of eradication. Undetected pest populations can develop in cryptic habitats, and can lead to a sudden re-emergence of the pest. Declaring eradication successful requires knowledge of the probability that complete eradication has been achieved, which will require understanding of the biology and ecology of the pest in the new ecosystem.

The traditional approach has been to wait a specified period of time that has some relationship to the life cycle of the pest, for example, two generations for the Asian gypsy moth eradications in the northwestern USA (Chapter 5). More recently, new literature, largely based on weed science (e.g. Regan *et al.* 2006), has examined how to determine if eradication has been successful. One approach uses probability models to calculate the probability of negative trapping results if in fact insects were still present (Barclay & Hargrove 2005). Kean and Suckling (2005) used data from the known population biology and sterile insect re-capture results generated during the infestation of the painted apple moth, *Teia anartoides*, in Auckland, New Zealand, to determine when lack of trap captures could be interpreted as "no population present". They used a general probability model with specific data on painted apple moth to model the probability of eradication over time. The model could accommodate new finds and predict probabilities based on different trap densities. The authors considered that the model could be modified for use with other species.

The correlation between trap density and time until eradication can be declared is an obvious one. More traps are more likely to detect small wild populations. However, maintaining traps is expensive and can be very labour-intensive, so most programmes use fewer traps in the final stages of the programme, thereby prolonging the length of time before eradication can be declared. The efficiency of the lures is also important as it impacts the probability that traps will detect a population (e.g. Chapter 5 – gypsy moth).

Discovering a new population of the invasive species poses questions too. If a pest was introduced once, it can be introduced again and often this happens (Chapter 3). Alternatively, control or eradication may have failed and the new find may result from a population resurgence (Carey 1991). The decision on whether a new find is indicative of a new or resurgent population makes a difference to the response. If it is a new occurrence, then eradication was successful, suggesting that another programme could also be successful. If the population has resurged, then the eradication programme failed and decisions about future efforts would need to take that into account.

For these reasons, methods that allow distinction between populations of a species are needed. Molecular biology has offered some tools for subspecies identification, but technique development depends on there being some genotypic difference(s). Molecular methods include sequencing of specific regions, such as the "barcoding" approach (Ball & Armstrong 2006). These methods have been used in the eradication programmes for painted apple moth in New Zealand (Chapter 4) and gypsy moth in the USA (Chapter 5), where new populations were discovered and shown to be different from the eradicated population. This is important, as government and the public need to be assured that eradication was successful.

Other techniques for distinguishing between individuals within a species have some promise. For example, an approach based on stable isotope analysis can indicate in which climatic region an insect pupates, and has been used by the authorities in New Zealand to distinguish between painted apple moth populations. By examining isotope signatures from a newly discovered population after the eradication spray programme had finished, it was possible to show that the new population had not pupated in Auckland where it was found, i.e. it was a new introduction and not a resurgent population (Benbow *et al.* 2007).

18.6 Conclusions and Future Opportunities

It would be pleasing to have many chapters in this book detailing the extensive use of pathogens in eradication programmes, but in reality in many situations, microbial pathogens would not be the first choice of tool for eradication if environmental and health concerns were not recognised. Toxic chemical agents, which are inexpensive and widely available, have typically been selected before their full impacts were known. However, as this book demonstrates, some high profile eradication programmes have been undertaken using microbial agents, and their benign environmental and biosafety profiles should lead to increasing use of microbial agents in the future.

As some of the chapters demonstrate, there is a lack of published material on eradication programmes against invasive arthropods. Publications that do exist often lack detail, as the principal researchers involved rapidly move on to other projects after completion of the eradication campaign. Government agencies are reluctant to fund researchers to record the process of eradication (Hosking *et al.* 2003). Because of the urgent nature of eradication programmes, all funding and resources are focussed on practical aspects of the campaigns. This book is an effort to collate some of these stories. There is much better information available on control programmes than eradication. We urge researchers involved to better document all eradication and control programmes using microbes to inform future efforts.

Pathogens are specific control agents and, as such, often have specific applications. Chapter 13 described the situation of common entomopathogenic fungi applied against fire ant nests, where colony collapse did not occur although some of the ants would be killed by the fungi. Because the fungi failed to kill queens and did not move between colonies, it was difficult to see these as effective control options over the range of fire ant invasions. However, discovery of chronic virus infections that are transmitted between nests and can cause colony collapse presents a new opportunity for control through pathogens (Chapter 13).

Availability of application methods that ensure the microbial control agent comes into contact with the pest is an issue for any pesticide and some of the examples in this book highlight difficulties in this area. *Varroa* mites in bee hives (Chapter 15), fire ants (Chapter 13) and mole crickets (Chapter 7) in soil, and borers that spend large parts of their life cycle inside wood (Chapters 9, 10 and 12) are protected from broadcast spray applications and more inventive applications methods are required. Methods include inoculation of inert materials/substrates (e.g. in baits) with pathogens, and strategic placement of inoculated material where contact with the pest is assured (e.g. longhorn beetle and non-woven bands containing fungi, see Chapters 9 and 10), specific contamination of habitats targeting only one life stage (e.g. rhinoceros beetle breeding sites, see Chapter 8) or application methods that allow penetration in to cryptic habitats (e.g. injection of nematodes into trees, see Chapter 12, or application of fungi that contact larvae of borers through cracks in the bark, see Chapter 10).

In addition, chronic pathogens may in some cases hold the key to suppression of invasive species. Chronic pathogens, which kill slowly or have largely sub-lethal effects, have not received the same level of attention and research interest has been directed to pathogens that kill rapidly. However, chronic pathogens are often transmitted vertically and can be debilitating in combination with other stresses (e.g. other pathogens). Thus, chronic pathogens have potential to play a significant role in pest suppression (see Chapters 11 and 13 on microsporidia and viruses). Techniques that facilitate identification of these less obvious pathogens will greatly assist in their increased utilisation, as demonstrated by recent work looking at viruses for fire ant control (Chapter 13).

There is a lack of available products based on microbial pathogens. With the exception of *Bt*, very few pathogens are available immediately in quantities required for eradication or even large-scale control programmes. One issue hindering development of biopesticides is the inability of the developer to protect their investment from competitors. Most biopesticides are simple formulations based on a well-described microbial pathogen, over which only limited protection can be maintained since living organisms are difficult to patent in entirety. Formulations can be maintained as "trade secrets" to reduce competition. The nematode *Steinernema scapterisci* was patented for use against mole crickets. The history of development of *S. scapterisci* as a biopesticide described by Frank (see Chapter 7) shows the difficulties in progressing from discovery of a useful pathogen to recouping investment through product sales or licensing.

A method which has been more rarely used is modification of the environment to enhance natural infections. This approach, called conservation biological control, can be used to encourage indigenous pathogens to attack invasive species (Chapter 2) or to enhance activity of native or introduced pathogens against invasive species (Chapter 6). Most pathogens have specific requirements to achieve optimal efficacy and transmission, and there remains considerable potential to enhance the effectiveness of microbial control agents.

Environmental and mammalian safety of pest control technologies are growing issues in most countries around the world. Traditionally, microbial pathogens as eradication or control technologies have competed with chemical pesticides, and are often not as cost effective or had lower toxicity to the pest. However, the specificity, benign environmental impact profile (Chapter 17) and lack of mammalian toxicity (e.g. Chapter 16) are leading to greater focus on the use of pathogens in responses to invasive pests. In this book there are many examples where more effective chemical control methods exist, but use of chemicals is reducing due to concerns about environmental safety (e.g. Chapter 9 and 11) or chemicals cannot be used because of mammalian toxicity concerns (Chapters 4 and 5). Given that rates of exotic arthropod introductions into new regions are increasing every year, the future is bright for the use of microbial agents, but must be supported by significant on-going research, innovative approaches to pathogen discovery, formulation and application and improved registration and regulatory processes.

References

AGROW report (2004) Biopesticide Biocontrol and Semiochemical Markets. PJB Publications Ltd, London. 46 pp

Anon. (2008) The IR-4 Project. http://ir4.rutgers.edu/ [accessed March 2008]

Auckland District Health Board (2002) Health risk assessment of the 2002 aerial spray eradication programme for the painted apple moth in some western suburbs of Auckland. Report to the Ministry of Agriculture and Forestry. 66 pp + Appendices

Ball SL, Armstrong, KF (2006) DNA barcodes for insect pest identification: a test case with tussock moths (Lepidoptera: Lymantriidae). Can J Forest Res 36:337–350

Barclay HJ, Hargrove JW (2005) Probability models to facilitate a declaration of pest-free status, with special reference to tsetse (Diptera: Glossinidae). Bull Entomol Res 95:1–11

Benbow T, McMorran D, Harrison R (2007) painted apple moth, where did it come from? Alpha 132. www.rsnz.org/education/alpha/Alpha132.pdf. 4 pp [accessed March 2008]

Boulton TJ (2004) Responses of nontarget Lepidoptera to Foray 48B® *Bacillus thuringiensis* va. *Kurstaki* on Vancouver Island, British Columbia, Canada. Environ Toxicol Chem 23:1927–1304

Boulton TJ, Otvos IS, Halwas KL, Rohlfs DA (2007) Recovery of nontarget Lepidoptera on Vancouver Island, Canada: One and four years after a gypsy moth eradication program. Environ Toxicol Chem 26:738–748

Braverman M, Nelson W, Torla B, Mendelshon M, Matten S, Jones Steinwand B, Roberts A (2003) Understanding the biopesticide registration process at EPA. http://ir4.rutgers.edu/biopesticides/RWP/PowerPoint/Understanding%20the%20Registration%20Process%20at%20EPA.pdf [accessed March 2008]

Carey JR (1991) Establishment of the Mediterranean fruit fly in California. Science 253:1369–1373

Dahlsten DL, Garcia R, Lorraine H (1989) Eradication as a pest management tool: concepts and contexts. In Dahlsten DL, Garcia R (eds) Eradication of exotic pests. Yale University Press, New Haven. pp 3–15

EPA (United States Environmental Protection Agency) (2008) Regulating biopesticides. http://www.epa.gov/pesticides/biopesticides/index.htm [accessed March 2008]

Glare TR, O'Callaghan M (2000) *Bacillus thuringiensis*: Biology, Ecology and Safety. John Wiley and Sons, Chichester, UK. 100 pp

Glare TR, O'Callaghan M (2003). Environmental impacts of bacterial biopesticides. In: Hokkanen HMT, Hajek AE (eds) Environmental Impacts of Microbial Insecticides. Need and Methods for Risk Assessment. Kluwer, Dordrecht NL, pp 119–150

Glare TR, Barlow ND, Walsh PJ (1998) Possible agents for use in New Zealand for the eradication or control of gypsy moth, *Lymantria dispar*. Proc 51th NZ Plant Prot Conf. pp 224–229

Glare TR, Newby EM, Nelson TL (1995) Safety testing of a nuclear polyhedrosis virus for use against gypsy moth, *Lymantria dispar*, in New Zealand. Proc 48th NZ Plant Prot Conf. pp 264–269

Glare TR, Walsh PJ, Kay M, Barlow ND (2003) Strategies for the eradication or control of gypsy moth in New Zealand. Report for the Forest Health Research Collaborative of New Zealand. 178 pp

Gryzwacz D (2003) Development and registration of biopesticides in Asia. In: Wabule MN, Ngaruiya PN, Kimmins FK, Silverside PJ (eds) Proc PCPB/KARI/DFID/CPP workshop, 14–16 May 2003, Nakura, Kenya. Natural Resources International Ltd., Aylesford, UK. pp 101–109

Hosking GJ, Clearwater J, Handiside J, Kay M, Ray J, Simmons, N (2003) Tussock moth eradication: a success story from New Zealand. Int J Pest Mgt 49:17–24

Jenner Consultants Ltd (1996) Health risk assessment of *Btk* (*Bacillus thuringiensis* var. *kurstaki*) spraying in Auckland's eastern suburbs to eradicate white spotted tussock moth. (*Orgyia thyellina*). Report to the Ministry of Health and the Ministry of Forestry New Zealand, commissioned by the Northern Regional Health Authority, North Health, Jenner Consultants, Auckland, New Zealand 4 September 1996

Kay M, Matsuki M, Serin J, Scott JK (2002) A risk assessment of the Asian gypsy moth to key elements of the New Zealand flora. Report for the Forest Health Collaborative, New Zealand. 32 pp

Kean JM, Suckling DM (2005) Estimating the probability of eradication of painted apple moth from Auckland. NZ Plant Prot 58: 7–11.

MAF (2003) Environmental Impact Assessment of Aerial Spraying *Btk* in NZ for painted apple moth. February 2003. New Zealand Ministry of Agriculture and Forestry. http://www.biosecurity.govt.nz/pests-diseases/forests/painted-apple-moth/environmental-impact.html [accessed March 2008]

MAF (2007) Varroa Mite. New Zealand Ministry of Agriculture and Forestry. http://www.biosecurity.govt.nz/pests-diseases/animals/varroa.htm [accessed March 2008]

McNeill MR, Baird DB, Goldson SL (1998) Evidence of density-dependent oviposition behaviour by *Listronotus bonariensis* (Coleoptera: Curculionidae) in Canterbury pasture. Bull Entomol Res 88:527–536

Myers JH, Savoie A, van Randen E (1998) Eradication and pest management. Annu Rev Entomol 43:471–491

Myers JH (2003) Eradication: Is it ecologically, financially, environmentally, and realistically possible? In: Rapport DJ, Lasley WL, Rolston DE, Nielsen NO, Qualset CO, Damania AB (eds) Managing for healthy ecosystems. Lewis Publ., Boca Raton. pp 533–539

Nentwig W (2007) Pathways in animal invasions. In: Nentwig W (ed) Biological invasions. Springer, Berlin. pp 11–27

O'Callaghan M, Glare TR, Jackson TA (2002) Biopesticides for control of insect pest incursions in New Zealand. Defending the Green Oasis: New Zealand Biosecurity and Science. In: Goldson SL, Suckling DM (eds) Proc New Zealand Plant Protection Society Symposium. ISBN 0-473-09386-3, pp 137–152

Ombudsman (2007) Report of the opinion of Ombudsman Mel Smith on complaints arising from aerial spraying of the biological insecticide Foray 48B on populations of parts of Auckland and Hamilton to destroy incursions of painted apple moths and Asian gypsy moths respectively during 2002–2004. Office of the Ombudsman, Wellington, New Zealand. 108 pp

Pesticides Safety Directorate (2007). Biopesticides. http://www.pesticides.gov.uk/biopesticides_home.asp [accessed March 2008]

Reardon R, Podgwaite J (1992) The gypsy moth nuclear polyhedrosis virus product, Gypchek. AIPM technology transfer. USDA Forest Service, Northeast area. 9 pp

Reardon RC, Podgwaite JD (1994) Summary of efficacy evaluations using aerially applied Gypchek against gypsy moth in the USA. J Environ Sci Health B29(4):739–756

Regan TJ, McCarthy MA, Baxter, Peter WJ, Panetta DF, Possingham HP (2006) Optimal eradication: when to stop looking for an invasive plant. Ecol Lett 9:759–766

Schroeter BS, Wharton D (2000) 151 Reasons NOT to Spray: Aerial Spraying of *Btk* for Gypsy Moth in Ballard/Magnolia. The Society Targeting Overuse of Pesticides, (STOP). http://www.nosprayzone.org/btk_151.html [accessed March 2008]

Simberloff D, Gibbons L (2004) Now you see them, now you don't!—population crashes of established introduced species. Biol Invasions 6:161–172

Songa W (2003) Kenyan regulations for importation of biological control agents. In: Wabule MN, Ngaruiya PN, Kimmins FK, Silverside PJ (eds) Proc PCPB/KARI/DFID/CPP workshop, 14–16 May 2003, Nakura, Kenya. Natural Resources International Ltd., Aylesford, UK. pp 75–78

Stephens AEA, Suckling DM, Burnip GM, Richmond J, Flynn A (2007) Field records of painted apple moth (*Teia anartoides* Walker: Lepidoptera: Lymantriidae) on plants and inanimate objects in Auckland, New Zealand. Aust J Entomol 46: 152–159

Williamson M (1996) Biological invasions. Chapman & Hall, London

Index

A

Abies firma, 150
Abies fraseri, 5, 6
Abiotic influences, 34
Abscesses, 295
Acacia spp, 60
Acarapis woodi, 273, 274
Acari, 260, 264, 272, 273
Accidental introduction, 24, 181, 203, 214, 332
Acephate, 123
Acer saccharum, 314
Acer spp, 159, 161
Acheta domesticus, 120
Actinomycetes, 150
Acyrthosiphon kondoi, 96
Acyrthosiphon pisum, 95, 96
Additional release, 105
Adelges piceae, 5
Adult bees, 273, 274, 279, 282, 284, 285
Adult emergence, 154, 167, 169, 171, 175, 310
Aedes camptorhynchus, 312, 333
Aerial application, 30, 53, 54, 59, 64, 66, 71, 75, 76, 78, 80, 85, 159, 168, 174, 175, 176, 291, 292, 296, 297–300, 308, 311
Aerial spray, 49, 59, 61, 62, 64, 66, 77, 161, 173, 190, 191, 202, 292, 296–297, 298, 299, 312, 320
Aerosol, 296, 297
Aesthetic impacts, 333
Aethina tumida, 271, 274, 276, 286
Africa, 134, 214, 259–267, 286, 320, 340
Aggregation, 26, 35, 133, 138, 334
Aggregation pheromone, 133
Agricultural crops, 7, 150, 153
Agricultural pest, 7
Agriculture, 4, 5, 6, 7, 11, 50, 57, 93, 95, 183, 200, 239, 264, 272, 273, 299, 333
Agrilus planipennis, 7, 159, 160, 165–175
Air samples, 296, 297

Airborne spores, 297
Aircraft, 53, 55, 56, 59, 61, 296
Airport, 7, 8, 10, 332
Alabama, 119, 123, 127, 240, 249
Aleurothrixus aepim, 264
Aleyrodidae, 264
Alfalfa snout beetle, 4
Alginate, 106, 241
Alien species, 4, 34, 343
Allee effects, 35, 36, 38, 39, 41, 42, 94
Allee population threshold, 94
Allergies, 299
American elm, 165
American Samoa, 137
Amplified fragment length polymorphism (AFLP), 193, 320
Amylostereum areolatum, 215, 218
Anamorphic stages, 97
Andersen samplers, 297
Anisoplia austriaca, 11
Anoplophora chinensis, 162
Anoplophora glabripennis, 160, 161–165, 173, 340
Anoplophora malasiaca, 147
Ant colony, 241
Antagonist, 126, 128, 202, 241, 242, 312
Antagonistic effects between pathogens, 202
Anthelminthics, 143
Anthocoridae, 150
Anthonomus grandis, 41, 42
Anthrax, 292
Anthropogenic effects, 337
Anthropogenic movement, 41, 72, 76, 80
Antifeedant, 162
Aphelinus asychis, 102
Aphid, 21, 22, 26, 93–113, 334, 339
Aphididae, 21, 22, 26, 93–109, 334–339
Aphis glycines, 21, 95, 99
Aphis gossypii, 94, 95, 101–103

351

Apis cerana, 273
Apis mellifera, 150, 271–286, 315, 316
Apoanagyrus diversicornis, 264
Aporrectodea caliginosa, 314
Application, 8, 11, 39, 53, 54, 56, 57, 59, 61, 62, 64, 66, 75, 80, 103, 106, 107, 108, 120, 121, 122, 125, 127, 129, 144, 146, 149, 150, 153, 154, 160, 163, 167, 171, 173, 174, 176, 183, 192, 241, 246, 278
Application strategy, 278, 284
Apriona germari, 173
Aquatic non-target effects, 320
Aquatic species, 309, 320–321
Aquatic systems, 186
Aqueous suspension, 279, 312
Area of establishment, 9
Area-wide biological control, 126, 127
Argentina, 117, 119, 214, 229, 239, 242, 244, 245, 246, 247, 249, 250
Argentine ant, 41
Argentine stem weevil, 314, 341
Arsenic, 117
Arthropod behaviour, 340
Artificial media, 97, 144, 277
Artificially infested, 137
Ash tree, 165, 166, 167, 168, 169, 170, 171, 172, 176
Asia, 22–27, 50, 51, 62–65, 71–86, 98, 99, 133, 134, 139, 141, 147, 160, 161, 162, 165, 173, 188, 192, 199, 214, 273, 286, 299, 315, 333, 334, 336, 344
Asian gypsy moth, 50, 51, 62–65, 72, 76–77, 78–80, 84–86, 188, 199, 299, 333, 334, 336, 344
Asian honey bee, 273
Asian longhorned beetle, 160, 161–165, 173, 340
Aspen, 182, 186, 191
Aspergillus flavus, 144
Asthma, 298, 299
Asthmatic children, 296
Asynchronous development, 176
Atta mexicana, 241
Attractant, 40, 137, 138, 139, 160, 175
Attraction to light, 77
Attraction traps, 54
Auckland, 50, 51, 52, 53, 57, 59, 61, 62, 66, 299, 300, 338, 344, 345
Augmentation, 8, 9, 11, 13, 40, 42, 97, 106, 339, 340
Augmentative release, 196, 200
Austracris guttulosa, 321

Australia, 5, 21, 41, 50, 57, 60, 96, 97, 101, 116, 117, 124, 125, 128, 214, 215, 220, 221, 223, 224, 225–229, 230, 231, 232, 238, 309, 321
Australian plague locust, 321
Australian possum, 50
Austria, 109, 318
Autographa californica MNPV, 188, 190
Azadirachtin, 191
Azygospore, 96, 194

B
Bacillus anthracis, 292
Bacillus cereus, 292
Bacillus cereus-group, 292
Bacillus mycoides, 292, 293
Bacillus sphaericus, 292, 309, 320, 321
Bacillus subtilis, 292
Bacillus thuringiensis, 10, 12, 13, 39, 161, 173–175, 187, 291–300, 306, 335
Bacillus thuringiensis aizawai, 317
Bacillus thuringiensis darmstadiensis, 173
Bacillus thuringiensis entomocidus, 173, 317
Bacillus thuringiensis galleriae, 173–175, 317
Bacillus thuringiensis israelensis, 173, 309, 312, 317, 320, 333
Bacillus thuringiensis japonensis, 173
Bacillus thuringiensis konkukian, 295
Bacillus thuringiensis kurstaki, 51, 75, 183–188, 291, 307, 308, 312, 313, 316, 317, 318, 336, 337, 338, 341
Bacillus thuringiensis morrisoni, 173
Bacillus thuringiensis tenebrionis, 173
Bacillus thuringiensis thuringiensis, 187
Backpack hydraulic sprayer, 107
Bacteria, 10, 12, 13, 24, 73, 144, 150, 183, 185, 240, 277, 292, 293, 294, 295, 296, 306, 308, 310, 313, 315, 320
Bactimos, 295
Baculoviridae, 20, 135, 188, 306
Bahamas, 116
Baits, 117, 238, 241, 247, 346
Ballistospore, 96
Ballooning neonates, 80
Barbados, 118
Barcoding, 344
Bark beetles, 35, 36, 145, 146, 151, 173, 229, 231
Barrier zones, 9, 34
Batkoa, 95
Bayer, 279, 318
Beauveria bassiana, 12, 22, 25, 40, 60, 96, 141–154, 160, 162, 167–172, 239, 241–242, 262, 278, 306

Beauveria brongniartii, 24, 51, 147, 162, 307, 340
Beddingia siricidicola, 213, 214, 216–233
Beddingia siricidicola – Kamona strain, 228, 229, 231, 232
Beddingia wilsoni, 220
Bee brood, 274, 276, 282, 285
Bee colony, 278, 286
Beekeeping, 272–273, 275–276, 286
Beetle, 4, 11, 12, 13, 21, 22, 24, 25, 26, 133, 134, 135, 136, 137, 138, 142, 143, 145, 150, 159, 160, 161, 165, 167, 173, 174, 175, 229, 264, 274, 286, 291, 307, 314, 336, 340, 346
Behavior, 4, 25, 35, 38, 108, 116, 118, 120, 160, 240, 241, 242, 272, 276, 286
Bemisia argentifolii, 318
Bemisia tabaci, 318
Bendiocarb, 123
Beneficial species, 315, 341
Benefit-to-cost analysis, 86
Binding affinities, 184
Binding domains, 184
Bio 1020, 279
Bioaerosol, 296
Bioassay, 54, 60, 64, 101, 138, 144, 149, 150, 162, 164, 168, 174–175, 195, 240, 261, 262, 263, 277, 278, 310, 311, 338, 340
Bioassay design, 311
Bioblast, 277, 279
Biogeographical barriers, 332
Biological control, 3, 7, 9, 10–13, 21, 23, 24, 28, 51, 97, 98, 101, 102, 103, 104, 105, 106, 107, 108, 118, 119, 121, 123, 124, 125, 126, 127, 128, 172, 173, 183, 186, 200, 201, 203, 219, 220, 238, 239, 242, 243, 246, 247, 252, 260, 261, 262, 263, 272, 275, 276, 277, 286, 306, 316, 318, 319, 321, 339, 340, 341, 346
Biological control agents, 3, 28, 118, 119, 123, 124, 127, 142, 172, 173, 183, 203, 238, 239, 243, 247, 276, 286, 321, 341
Biological invasions, 4, 13, 35, 38, 43
Biopesticide, 8, 9, 12, 73, 74–76, 85, 97, 106, 107, 123–124, 201, 295, 306, 307, 308, 309, 313, 314, 315, 316, 318, 320, 335, 336–337, 338, 339, 340, 341, 342–343, 346
Biorational, 24
Biosecurity, 6, 50, 55, 61, 64, 343
Biosecurity Act, 55, 61, 64, 343
Black ash, 166
Black imported fire ant, 237–252

Blue green aphid, 96
Boll weevil, 41, 42
Bombus impatiens, 316
Bombycidae, 150, 264
Bombyx mori, 150, 151, 183, 264
Boston, 182, 193
Bostrichidae, 173
BotaniGard, 168, 169, 170, 171, 172, 312, 317, 318
Bothynoderes (= Cleonus) punctiventris, 11
Boverin, 12
Braconidae, 25, 60
Brassica sp., 316
Brazil, 116, 117, 118, 119, 214, 215, 225, 227, 229, 232, 239, 241, 242, 246, 247, 260, 262, 263, 264, 265, 267
Breeding sites, 135, 136, 137, 346
British Columbia, 73, 76, 77, 78, 85, 296, 298, 313
Broad-spectrum, 316, 338
Bt, 10, 161, 173, 174, 176, 183, 184, 185, 291, 292, 293, 294, 295, 298, 300, 306, 307, 308, 312
Bt-based products, 174, 188, 292
Bt-positive patients, 298
Bufo marinus, 118
Bulgaria, 196, 198, 199, 201
Buprestidae, 160, 167
Burning, 142, 143, 238
Bursaphelenchus xylophilus, 142

C

Caged field trials, 169
California, 24, 40, 73, 76, 101, 102, 238, 250
Callophrys (= Incisalia) fotis, 186
Callophrys sheridanii, 186
Calvia quatuordecimguttata, 24
Canada, 11, 55, 60, 77, 78, 85, 98, 99, 161, 166, 186, 187, 191, 214, 220, 231, 232, 271, 292, 296, 298, 300, 306, 307, 316, 336, 338
Cane toad, 118
Canola, 316
Capilliconidia, 96, 261
Carabidae, 186, 314, 318
Carbamates, 117
Cargo containers, 79
Carrier, 145, 146, 147, 191, 192, 317
Carya illinoinensis, 25, 26
Cassava, 13, 259–267, 340
Cassava green mite, 13, 259–267
Cassava mealybug, 264
Cerambycidae, 142, 160, 173

Ceratitis capitata, 40
Chalcidoidea, 60
Chemical-mediated behavior, 276
Chemical pesticides, 8, 10, 65, 117, 122, 125, 135, 305, 336, 337, 341, 342, 346
Chestnut blight, 141
Chilocorus bipustulatus, 24
China, 72, 96, 99, 144, 161, 162, 165, 167, 175, 176, 193, 214, 238, 245, 319
Chipping, 143, 166, 224
Chlordane, 117, 118
Chloropicrin, 118
Chlorpyrifos, 57, 58, 59, 312
Chronic pathogens, 346
Chrysomelidae, 173
Cicadellidae, 101
Cinnabar moth, 186
Citrus, 11, 21, 24, 167, 272, 276, 277
Citrus rust mite, 272, 276, 277
Classical biological control, 7, 9, 10–11, 13, 23, 100–103, 118–119, 123, 125, 126, 127, 128, 200–201, 219, 239, 260, 262, 286, 306, 339, 340
Climate models, 195
Clinical disease, 295
Clinical illness, 187, 298
Clover root weevil, 50, 314
Coccinelidae, 22, 24, 314
Coccinella septempunctata, 24
Cockchafer, 317
Cockroach, 4, 120
Coconut palm, 134, 139
Cocos nucifera, 134, 139
Co-infection, 312
Coleomegilla maculata, 25, 26
Coleoptera, 4, 11, 12, 13, 21, 22, 24, 25, 26, 133, 134, 135, 136, 137, 138, 142, 143, 145, 150, 159, 160, 161, 165, 167, 173, 174, 175, 229, 264, 274, 286, 291, 307, 314, 336, 340, 346
Collembola, 309, 312, 314, 315, 317, 318
Colonization, 7, 50, 57, 63, 94, 108, 117, 171
Colorado potato beetle, 12
Colydiidae, 142, 150
Commercial formulations, 186, 338
Common blue ash, 166
Community advisory group, 53, 58, 59, 61, 335
Community attitudes, 338
Competitive displacement, 201
Competitor, 26, 27, 346
Compost worm, 310, 317
Compsilura concinnata, 202
Conidiobolus obscurus, 100

Conidiobolus sp., 22, 240
Conidiobolus thromboides, 99
Conifer, 72, 214, 220
Conifer-feeding sawflies, 11
Conservation biological control, 103–104, 108, 346
Conspecific, 35, 36, 201, 319, 320
Contaminant, 116, 247, 248, 295, 298
Conventional control method, 143
Copra, 134
Coptotermes formosanus, 6
Corn, 11, 12, 22, 307
Corneal ulcer, 295
Cost-benefit analysis, 335
Cost of control efforts, 183
Cost of microbial agents, 338
Cost of preventing new introductions, 6
Cotesia melanoscelus, 202
Cottage industries, 337
Cotton aphid, 94, 95, 101–103
Cottony-cushion scale, 24
Crambidae, 22
Crickets, 120
Cronartium ribicola, 141
Crown dieback, 171, 172
Cryphalus fulvus, 145
Cryphonectria parasitica, 141
Cryptic habitats, 334, 340, 341, 344, 346
Crystal proteins, 173, 291, 292
Crystal solubilization, 174
Cry toxins, 173, 174, 185
Cuba, 116
Cultural control, 135, 223, 276
Culture medium, 147, 278
Curculionidae, 21, 41, 42, 50, 163, 173, 314, 341
Cyanide, 117
Cycloneda munda, 26
Cypovirus, 55, 60
Cytoplasmic membrane, 188, 292
Cytotoxic, 295
Czechoslovakia, 271, 272
Czech Republic, 65, 191, 199, 336

D
Damage, 3, 6, 94, 98, 117, 118, 122, 127, 134, 135, 137, 138, 139, 141, 142, 153, 181, 182, 214, 232, 307, 332, 333, 334, 343
Dastarcus helophoroides, 142, 150
DDT, 275
Decis, 57, 58, 312
Decision theory methods, 122
Declaring success, 62, 343–345

Decomposers, 314
Defoliation, 183, 186, 190, 194, 203, 314
Deladenus siricidicola, 213, 214, 216–233
Delayed-density-dependent activity, 137
Delimitation survey, 52, 58
Delimitation trapping, 72, 74
Delta-endotoxins, 184
Dendrobaena octaedra, 316
Dendrolimus punctatus, 319
Dendrolimus spectabilis, 153, 154
Denmark, 99
Density dependence, 137, 189, 198, 214, 341
Desiccation tolerance, 106
Detection, 4, 7, 8, 34, 51, 57–58, 63–64, 74–76, 84, 85, 119, 161, 166, 191, 223, 261, 263, 286, 334, 337
Detection of small arthropods, 4
Developmental rates, 311
Diagnosis, 138
Diamondback moth, 12, 17, 307
Diapause, 74, 145, 191
Diapausing eggs, 51
Diaprepes abbreviatus, 21, 27
Diarrhoeal enterotoxin, 295
Diffusion coefficient, 37
Diffusion models, 37
Diflubenzuron, 186
Dimilin, 186, 187, 192
Dinocampus coccinellae, 25
Dipel, 295, 317, 320
Diptera, 40, 118, 127, 161, 163, 239, 291, 307
Direct inoculation, 105, 150
Disease prevalence, 22, 28
Disease transmission, 101, 240
Disparlure, 64, 65, 74
Dispersal, 3, 4, 9, 25, 37, 38, 41, 42, 52, 72, 80, 82, 94, 151, 162, 165, 166, 195, 286, 306, 309, 315, 317, 341
Dispersal rate, 52
Diuraphis noxia, 21, 96, 98, 101–108
Dot-blot-DNA hybridisation technique, 296
Dried mycelial pellets, 106
Drift, 172, 186, 226, 297, 308, 311, 320
Droplet size, 53, 175, 296
Dutch elm disease, 141, 165
Dynastinae, 133

E

Earthworm, 120, 310, 312, 314, 315, 316, 317
Eastern Europe, 11, 197, 202
Ecological disturbances, 321
Ecological host range, 21, 28, 310
Ecological hypotheses, 5

Ecological impact, 311, 321
Economic analysis, 85, 122
Economic consequences, 34
Economic damage threshold, 152
Economic impact, 5, 6, 58, 63, 72, 238
Economic impact assessment, 58, 63
Economic thresholds, 182
EcoScience, 279
Ecosystem level studies, 321
Efficacy testing, 185–186
Egg laying, 138, 167, 218, 279, 341
Egg masses, 50, 63, 72, 74, 76, 77, 78, 80, 81, 83, 85, 189, 199, 337
Egg production, 310
Eggs, 6, 51, 72, 74, 94, 116, 120, 127, 134, 143, 145, 152, 154, 161, 164, 167, 169, 171, 174, 175, 215, 218, 219, 221, 228, 231, 233, 247, 273, 274, 313, 319, 341
Egg viability, 311
Eisenia fetida, 310, 317
Elaeis guineensis, 134, 135
Elatobium abietinum, 94, 99–100, 102, 103
Electrophoresis, 292
Emerald ash borer, 7, 159, 160, 165–175
Empoasca vitis, 101
Emulsifiable concentrate, 318
Emulsifiable suspension, 312
Encapsulation, 293
Encyrtidae, 264
Endangered species, 94, 315
Endemic pathogens, 12, 19–28, 97–100
Endemism, 11, 239, 340
Endophyte, 22
Endoreticulatus schubergi, 198
Endoreticulatus sp., 199
Enemy release hypothesis, 5, 19, 27, 28
England, 4
Enhancer, 185–186, 191
Enhancin genes, 191
Enterotoxin, 295
Entomopathogenic fungus, 142, 147, 239, 241
Entomopathogenic nematodes, 21, 22, 27, 117, 119, 121, 124, 227
Entomophaga maimaiga, 10, 11, 13, 20, 65, 183, 192–197, 203, 306
Entomophaga pyriformis, 100
Entomophaga sp., 95, 192, 193
Entomophthora chromaphidis, 98
Entomophthora planchoniana, 100
Entomophthora sp., 96
Entomophthorales, 22, 95, 96, 97, 98, 99, 100, 104, 106, 108, 193, 196, 240, 260, 277, 306, 339

Environmental damage, 332
Environmental degradation, 319
Environmental factors, 21, 22
Environmental human samples, 54, 173
Environmental impact assessment, 58, 59, 64, 306, 309–313
Environmental impacts, 42, 305–322
Environmentally-sensitive habitats, 82
Environmental safety, 65, 186–187, 195–196, 201, 306, 308, 315, 319, 341, 342, 347
Environmental stewardship, 182
Enzootic levels, 198, 200
Ephestia kuehniella, 183
Epizootic, 21, 98, 99, 101, 102, 104, 105, 106, 107, 137, 144, 189, 193, 202
Epizootiology, 28, 188–190, 194–195, 263
Eradication, 6, 7, 8, 9, 10, 36, 39, 40, 42, 49–66, 71–86, 153, 161, 163, 166, 174, 176, 192, 238, 239, 252, 292, 296, 305, 306, 307, 308, 312, 313, 321, 332–339, 343, 345
Eradication feasibility, 60
Eradication program, 73, 78, 79, 80, 166, 174, 175, 238
Eretmocerus eremicus, 312
Eriophyidae, 272
Erynia, 22, 95
Establishment, 7, 9, 11, 20, 21, 25, 27, 28, 34–36, 39, 40, 65, 72, 94, 100, 103, 196, 201, 214, 220, 232, 238, 243, 260, 261, 266, 267, 334
Eucalyptus nitens, 55
Eucharitidae, 239
Europe, 11, 12, 71, 100, 103, 141, 162, 197, 198, 200, 202, 203, 220, 273, 286, 306, 314
European balsam woolly adelgid, 5
European corn borer, 11, 22, 307
European gypsy moth, 63, 71–86, 181–204
European pine sawfly, 11
European spruce sawfly, 11
European wasp, 5
Euseius citrifolius, 264
Euseius concordis, 264
Evergreen ash tree, 169
Exotic, 4, 5, 9, 20, 21, 22, 23, 24, 25, 26, 27, 28, 51, 52, 57, 96, 97, 98, 100, 159, 239, 264, 265, 275, 314, 319, 331, 340, 341
Exotic aphids, 93–109
Exotic arthropods, 5, 20, 28
Exotic insects, 20, 51, 52, 94
Exotic strains, 340

Exposure parameters, 311
Exposure time, 307
Extinction, 34, 35, 36, 38, 39, 40, 335

F

Farm workers, 294
Fast-track registration, 343
Fecundity, 25, 105, 136, 139, 164, 169, 311
Feeding stimulants, 312
Feeding studies, 55
Fermentation, 174, 185, 278
Field, 24, 25, 26, 43, 52, 55, 56, 98, 100, 102, 103, 105, 106, 109, 116, 120, 121, 127, 128, 129, 136, 144, 145, 149, 152, 162, 163, 165, 167, 169, 175, 186, 190, 191, 195, 198, 199, 200, 201, 203, 228, 229, 233, 241, 242, 246, 248, 249, 262, 265, 267, 272, 276, 278, 279, 309, 310, 311, 314, 316, 319, 321, 340, 342
Field-cage studies, 196
Field efficacy, 342
Field experiments, 145, 149, 190, 191, 201, 242, 279, 311
Field prevalence, 310
Field trials, 121, 163, 168, 169–172
Fiji, 134, 136
Fire ant, 6, 39, 237–252, 345, 346
Fire ant baits, 238
Fire ant parasitoid, 239, 243
Flat fan nozzle, 168, 169, 171
Flies, 40, 118, 127, 161, 163, 239, 291, 307
Flight patterns, 64
Flights, 116, 243
Florbac FC, 295
Florida, 116, 117, 118, 121, 123, 124, 126, 127, 128, 239, 240, 244, 245, 249, 250, 272, 274, 277, 284, 286
Fluvalinate, 279
Folsomia candida, 309
Folsomia fimetaria, 317
Foray 48B, 53, 56, 59, 61, 64, 66, 75, 77, 79, 80, 174, 293, 295, 296, 297, 298, 299, 300, 308, 312, 320, 337
Foray 76B, 75
Forest, 5, 33, 50, 141, 150, 152, 153, 161, 166, 173, 176, 181, 182, 183, 186, 195, 201, 203, 223, 225, 233, 276, 314, 316
Forest defoliator, 50
Forest earthworm species, 316
Forest ecosystem, 166, 182, 314
Forest floor, 314
Formosan termite, 6
Formulated granules, 314

Index 357

Formulation, 75, 80, 106, 108, 147, 151, 152, 153, 168, 169, 185, 190, 192, 220, 222, 246, 277, 278, 279, 295, 312, 313, 314, 316, 317, 319, 321, 338, 339, 347
Formulation components, 312–313, 316
Founder populations, 34, 35, 36
France, 51, 71, 182, 184, 191
Frankliniella occidentalis, 316, 318
Fraser fir, 5, 6
Fraxinus americana, 165, 166, 170, 171
Fraxinus nigra, 166
Fraxinus pennsylvanica, 166, 168, 169, 170, 171
Fraxinus profunda, 166
Fraxinus quadrangulata, 166
Fraxinus sp., 165, 166
Fraxinus uhdei, 169
Fumigants, 142, 153
Fumigation, 79, 80, 143, 153
Fungal bands, 147, 149, 150, 151, 152, 153, 162, 163, 164, 165, 170, 175, 282, 283, 284
Fungal infections, 99, 101, 102, 171, 172
Fungal symbiont, 220, 340
Fungal-treated trees, 170

G
Gastroenteritis, 295
Generalist, 5, 306
Generation time, 94
Genetic variant, 191
Geographical distributions, 100
Georgia, 116, 119, 123, 128
Germany, 79, 135, 183, 199, 320, 333
Gilpinia hercyniae, 11
Globalization, 3–6, 28, 74
Global regulations, 4
Global regulatory organizations, 4
Glyptapanteles liparidis, 203
Gnats, 163
Golf courses, 117, 118, 122, 123, 125, 126, 127, 143
Gonads, 200
Government, 12, 50, 51, 52, 58, 61, 62, 66, 73, 151, 183, 286, 334, 337, 338, 344
Government agencies, 66, 73, 85, 286, 292, 345
Government officials, 338
Government response, 56, 60–62
Grain ships, 77
Granulovirus, 191
Graphognathus leucoloma, 21
Green ash, 166, 168, 169, 170, 171

Green spruce aphid, 94, 99–100, 102, 103
Greenguard, 143
Greenhouse, 108, 167, 168–169, 285, 294, 295, 316, 318
Gregarine protozoa, 24
Ground application, 59, 192
Ground beetles, 186, 314, 318
Ground-dwelling arthropods, 314
Ground-nesting birds, 6
Ground searches, 76
Ground spray, 53, 58, 59, 62, 185, 192, 202, 312
Gryllotalpidae, 115–129, 340, 345, 346
Gryllus sp., 120
Guadeloupe, 116
Guam, 135
Guinea pigs, 294
Gypchek, 76, 80, 82, 190, 191, 192, 202, 336, 337
Gypsy moth, 10, 11, 12, 13, 20, 38, 39, 40, 42, 49, 50, 51, 53, 58, 62, 63, 65, 71, 72, 73, 74, 76, 80, 83, 182, 183, 185, 187, 188, 190, 191, 193, 194, 196, 197, 198, 199, 200, 201, 203, 307, 309, 311, 313, 315, 333, 335, 336, 340, 341, 344
Gypsy moth hybrids, 79
Gypsy moth outbreaks, 75

H
Hamilton NZ, 50, 63, 64, 65, 299, 300, 338
Hardwood, 55, 72, 161
Harmonia axyridis, 20, 22, 23, 24, 26, 150
Hawaii, 4, 95, 249, 250
Health assessments, 64
Health authorities, 338
Health impact assessment, 298, 300, 338
Health impacts, 54, 56, 66, 296, 297, 306, 333, 338
Health monitoring, 59
Health perceptions, 299
Helicopter, 53, 59, 61, 78
Hemiptera, 10, 12, 21, 93, 95, 264, 307
Hesperomyces virescens, 24
Heteroptera, 150
Heterorhabditis bacteriophora, 27
Heterorhabditis sp., 21
High dose challenges, 342
Hinoki cypress, 150
Hippodamia convergens, 26
Hippodamia tredecimpunctata, 24
Hirsutella, 272, 277
Hirsutella thompsonii, 272

Hispaniola, 116
Hive, 272, 274, 276, 278, 279, 280, 282, 285, 286, 314
Hive temperature, 286
Home lawns, 117
Homoptera, 6, 24
Honey bee, 150, 271–286, 315, 316
Honey bee pests, 272, 273–286
Honeydew, 6, 94
Hong Kong, 238
Horizontal transmission, 164, 172, 194, 200, 244
Host density, 21, 105, 185, 189, 198
Host mortality, 40, 199, 202
Host plant, 27, 58, 72, 94, 183, 333
Host plant range, 72, 182
Host plant removal, 58
Host range, 20, 21, 26, 28, 63, 94, 100, 153, 190, 195, 244, 247, 264, 306, 311, 315, 320, 332, 341
Host specificity, 184, 190, 195, 199, 201, 247, 249–251, 264, 306, 341
House crickets, 120
Human exposure, 292, 294, 296, 297, 298, 300
Human health concerns, 187
Human health effects, 13, 291–300
Human health impacts, 296, 297, 306
Human illness, 297–300
Human infection, 298
Hungary, 199, 221
Hymenoptera, 13, 25, 60, 150, 264, 307
   Honey bee, 150, 271–286, 315, 316
   Bumble bees, 316
   Wasp, 5, 6, 50, 118, 119, 123, 124, 126, 128, 239, 313
Hyperparasites, 23
Hypocreales, 96, 97, 98, 99, 160, 192, 262
Hypogastrura assimilis, 317

I

Ibalia leucospoides, 214
Iceland, 94, 99–100, 102, 103
Icerya purchasi, 24
Ichneumonidae, 60
Iconic species, 333
Idaho, 73, 101, 105, 106, 107, 108
IgE antibodies, 294
IgG antibodies, 294
Imidacloprid, 162, 167
Immature life stages, 74
Immunocompromised mice, 294
Impacts of invasive species, 5, 6, 40
Impacts on non-target species, 292, 341

Importation, 51, 264–265, 339, 341, 343
Incineration, 153
India, 116, 124, 128, 134, 137, 220, 343
Indian Ocean, 134
Indigenous ecosystems, 322
Indigenous species, 3, 311, 340
Indonesia, 99
Inert ingredient, 280, 281
Infective stage juveniles, 120
Infectivity, 171, 225–226, 227–228, 264, 315
Infestations, 6, 52, 53, 75, 76, 81, 82, 84, 101, 125, 145, 163, 166, 172, 220, 226, 238, 242, 243, 250, 273
Ingestion, 135, 185, 197, 200, 307, 312, 315
Inhalation, 297
Inoculation biological control, 105–107, 108
Inoculative augmentation, 106
Inoculum levels, 319
Insect colony, 93, 94, 95, 160, 195, 337
Insect diet, 55
Insect growth regulator, 65, 312, 333
Insecticidal crystal proteins, 173
Insecticide, 23, 24, 49, 57, 58, 76, 103, 142, 143, 149, 153, 160, 162, 167, 170, 175, 183, 195, 243, 291, 312
Integrated pest management, 182, 183, 238, 275–276
International toxin units - insecticidal activity, 187
International trade, 173
International transportation, 331
International travel, 79, 86
Intracellular parasite, 246
Introduced species, 3, 4, 5, 7, 22, 23, 34, 35, 95
Introductions, 3, 4, 6, 10, 20, 24, 25, 33–34, 50, 72, 76, 94, 100, 102, 138, 193, 196, 203, 227, 243, 244, 263, 265, 332, 343
Inundation biological control, 107, 108
Inundative application, 13
Inundative augmentation, 8, 9, 11–12, 13, 339, 340
Invasion biology, 20, 28, 43
Invasion pathways, 34, 79, 86
Invasive arthropods, 3–13, 19–28, 333, 337, 345
Invasive hive pests, 273–286
Invasive species, 3–9, 12, 19, 40, 52, 73, 117, 160, 182, 307, 318, 331–347
Invasive weeds, 9
Irradiation of females, 60
Irrigation, 103, 106, 107, 108, 117, 122, 125, 126

Isaria (= Paecilomyces) farinosa, 143, 144, 162, 167, 192
Isaria fumosorosea, 96, 108, 167
Isaria sp, 106
Isotope, 62, 345
Isotope analysis, 62, 345
Israel, 98, 101, 173, 252, 309, 312, 317, 320, 333
Israeli Acute Paralysis virus of bees, 252

J
Jamaica, 116
Japan, 41, 51, 56, 63, 64, 72, 99, 142, 143, 144, 149, 151, 153, 160, 162, 163, 183, 193, 220, 274
Japanese beetle, 11, 12, 21
Japanese fir, 150
Japanese gypsy moth, 72
Japanese pine sawyer, 141–154, 160, 175
Jewel beetles, 160, 167

K
Karner blue butterfly, 76, 186
Korea, 51, 72, 161, 273

L
Laboratory bioassays, 101, 162, 174–175, 195, 278, 310, 311
Laboulbeniales, 24
Lady beetle, 22, 24, 314
Large scale application, 296
Larra bicolor, 118, 127
Larvae, 22, 24, 35, 53, 54, 55, 60, 74, 75, 77, 118, 127, 128, 135, 136, 137, 138, 144, 145, 146, 147, 149, 150, 151, 153, 154, 160, 161, 162, 164, 167, 170, 171, 172, 173, 174, 175, 182, 184, 185, 188, 189, 190, 191, 192, 193, 194, 196, 200, 201, 202, 218, 219, 223, 230, 244, 247, 249, 274, 291, 309, 311, 318, 346
Larval movement, 202
Lasiocampidae, 153
*Ld*MNPV, 65, 76, 85, 183, 188–192, 202, 307, 312, 336
Leading edge, 7, 73, 75, 80, 83, 187
Leaf droplet bioassay, 174
Lecanicillium lecanii, 167
Lecanicillium sp, 97
Legislation, 62, 261, 265, 341, 343
Lepidoptera, 10, 12, 22, 49–66, 76, 119, 150, 153, 161, 174, 182, 184, 185, 186, 187, 188, 190, 195, 196, 201, 264, 292, 307, 308, 311, 313, 315, 338
Lepidopteran-active, 185

Leptinotarsa decemlineata, 12
Lesser Antilles, 116
Life cycle, 23, 57, 94, 96, 120, 128, 166, 171, 197, 213, 216, 221, 228, 232, 246, 263, 310, 334, 337, 344, 345
Life stages, 36, 48, 63, 72, 74, 76, 80, 81, 334
Listronotus bonariensis, 314, 341
Live females, 40, 52, 55, 58, 60
Lizards, 6
Local authorities, 343
Local bylaws, 343
Long-distance dispersal, 37, 38, 41, 42
Long-horned beetles, 131, 142, 160, 161, 164, 173
Longicorn beetles, 131, 142, 160, 161, 164, 173
Louisiana, 24, 119, 123, 127, 244
Low density populations, 9, 33, 34, 35, 42, 73, 74, 334, 335
Lucerne flea, 318
Lumbricus terrestris, 316
Lure-and-infect, 139
Lures, 56, 64, 160, 224, 344
Lycaeides melissa samuelis, 76, 186
Lygus lineolaris, 316
Lymantria dispar, 10, 13, 20, 38, 50, 51, 62, 63, 71–86, 181–204, 296, 299, 307, 334, 336, 341, 343
Lymantria dispar asiatica, 72, 76–77, 78–80, 84–86, 188, 199, 299, 334, 336, 344
Lymantria dispar japonica, 72
Lymantria dispar praeterea, 62–65
Lymantria umbrosa, 62–65, 333, 336
Lymantriidae, 50, 51, 53, 55, 56, 60, 182, 195, 308, 332, 335, 337, 338

M
Maine, 144, 319, 320
Malaysia, 135, 138
Maldives, 137, 138
Male moth density, 81
Malpighian tubules, 200
Mammalian safety, 51, 295, 306, 340, 346
Mammalian safety data, 340
Mammals, 150, 166, 189, 275, 292, 300, 333, 341
Management tool, 166, 167, 176
Managing establishment, 38–41
Managing spread, 41–42
Manihot esculenta, 13, 259–267, 340
Maples, 159, 161
Martinique, 118
Mass application, 8, 338

Mass culture, 227, 277
Mass production, 11, 108, 145, 146, 185, 191, 252, 339
Mass rearing, 221–222
Mathematical models, 109, 129
Mating disruption, 39, 65, 81, 183
Mattesia geminata, 239
Mattesia sp., 240
Maturation feeding, 142, 143, 152, 166, 173, 176
Maximum challenge laboratory testing, 310
Mediterranean flour moth, 183
Mediterranean fruit fly, 40
Meligethes aeneus, 316
Melolontha melolontha, 317
Metarhizium anisopliae, 11, 22, 144, 159, 160, 162–165, 167, 239, 240, 277, 279–284, 306
Metarhizium anisopliae var. *acridum*, 316, 318
Metarhizium flavoviride, 316
Metchnikoff, 11, 22
Methods for detection, 8
Methyl bromide, 79, 118
Mexico, 214, 238, 249, 250, 273
Microbe, 10–12, 141–154, 272, 276, 286, 306, 307, 338, 339, 341, 342, 345
Microbial control agent, 13, 95, 144, 150, 159, 160, 161, 188, 190, 197, 241, 251, 272, 305–322, 332, 336, 337, 339, 340, 342–343, 344
Microhabitat, 103, 104, 120, 194
Microsporidia, 13, 20, 40, 183, 197–201, 202, 203, 246, 252, 277, 307, 340, 346
Midgut epithelial cells, 135, 184, 188
Military cargo, 79
Milky disease, 12, 308
Millipede, 21
Ministry of Agriculture and Forestry, New Zealand, 51, 57, 299
Mississippi, 119, 123, 127, 240, 244, 245
Mistblowers, 53
Miticides, 275
Mitochondrial DNA, 74
Mode of action, 184, 306, 341
Modelling approach, 62
Moisture conditions, 103
Mole crickets, 115–129, 340, 345, 346
Molecular analysis, 63
Molecular identification systems, 65
Molecular probes, 233, 265–266, 267
Molecular typing, 62
Monarch butterflies, 56

Monitoring, 8, 9, 52, 54, 57, 59, 72, 73, 75, 122, 127, 138, 186, 201, 224, 229, 232, 266–267, 277, 298, 313, 319, 321
Monitoring programs, 72
Monochamus alternatus, 141–154, 160, 175
Monochamus galloprovincialis, 144
Monochamus sp., 144
Monocultures, 5
Mononychellus tanajoa, 13, 259–267
Mosquitoes, 291, 333, 400
MtDNA, 62
Mulberry longicorn beetle, 173
Mulch, 118
Multiparasitized hosts, 127
Mummification, 262, 263
Mycar, 271, 272, 277
Mycelial formulations, 106, 108
Mycopesticides, 107
Mycosis, 21, 25, 26, 99, 169, 170, 316
Mycotrol O, 168
Myrmicinosporidium durum, 240

N
Nasal lavage cultures, 294
Nasal swabs, 296, 297
Native beech, 6
Native biodiversity, 5
Native bird, 5, 6
Native ecosystems, 6, 7, 333
Native invertebrates, 6
Native lymantriids, 196, 315
Natural enemies of fire ant, 237–252
Natural enemy, 20, 23, 100, 102, 104, 118, 127, 201, 203, 238–241, 319
Natural enemy complex, 102, 201
Naturalis-L, 318
Natural regulator, 105
Natural spread, 72, 339
Nectar, 117
Nematac S, 124, 125
Nematode, 115–129, 141–154, 213–233, 239
Nematode liberation, 220, 221, 223
Nematode transmission, 152, 154
Neocurtilla hexadactyla, 120, 121
Neodiprion sertifer, 11
Neonates, 35, 72, 77, 80, 138, 167
Neonicotinoid, 162, 167
Neozygites floridana, 262, 263, 264, 266
Neozygites fresenii, 96, 99, 100, 101, 102, 103
Neozygites sp., 260, 265
Neozygites tanajoae, 13, 259–267
New introduction, 4, 6, 345
New York, 4, 161, 162, 163, 193, 231, 319, 320

New Zealand, 6, 7, 40, 49–66, 96, 214, 216, 221, 231, 292, 296, 299, 300, 306, 307, 308, 312, 314, 332, 333, 334, 336, 337, 338, 343, 344, 345
Nitidulidae, 274
Noctuidae, 190
Non-target, 10, 13, 24, 42, 59, 65, 76, 85, 86, 168, 195, 196, 198, 201, 264, 292, 306, 307, 308, 309, 310, 311, 313, 314, 315, 317, 318, 319, 320, 321, 332, 341, 342
Non-target effects, 76, 85, 86, 172, 264, 309, 311, 313–321, 341
Non-target impacts, 10, 13, 24, 59, 308, 310, 312, 313, 316, 318, 342
Non-target Lepidoptera, 195, 201, 313, 315
Non-target microorganisms, 319
Non-target organisms, 168, 264, 305, 306, 308, 310, 320, 321, 341
Non-woven cellulose fabric strips, 147
Non-woven fiber bands, 147
North America, 11, 12, 20, 22, 24, 25, 26, 27, 55, 63, 72, 73, 75, 78, 95, 96, 98, 99, 142, 159–176, 180, 183, 188, 190, 192–194, 195, 200, 203, 213, 214, 223, 231–232, 245, 247, 250, 272, 273, 274, 318, 334, 340
North Carolina, 76, 79–80, 81, 117, 123, 238, 336
Norwalk virus, 295
Norway spruce, 99
Nosema lymantriae, 198, 199
Nosema portugal, 198, 199, 202
Nosema serbica, 198
Nothofagus fusca, 55
Nothofagus spp., 6
Novodor, 174, 295
Novo Nordisk, 191
Nucleopolyhedrovirus (= NPV), 11, 13, 55, 60, 65, 76, 183, 188–192, 193, 202, 312, 336, 341, 343

O

Oak, 51, 63, 147, 182, 183, 185, 189, 190, 201, 203, 265
Oak tannins, 185
Obligate pathogens, 194, 197, 307
Observation hives, 279
Occlusion bodies, 188, 189, 191
Occupation-related respiratory symptoms, 294
Ochlerotatus camptorhynchus, 312, 320
Octospores, 198, 246
Oil formulation, 312, 316
Oil palm, 134, 135
Oligonychus gossypii, 264
Olla v–nigrum, 25
Ommatoiulus moreletii, 21
Operation Evergreen, 52, 53, 54, 55, 56, 57, 61, 300
Ophiostoma ulmi, 141, 165
Optical brighteners, 185, 191
Orasema spp., 239
Orchards, 7, 26, 57, 147, 160, 162
Oregon, 73, 76, 78, 83, 85, 187, 273, 298
Organic, 126, 134, 138, 153, 168
Organic matter, 134, 138
Organophosphates, 117
Orgyia leucostigma, 55
Orgyia pseudotsugata, 55
Orgyia thyellina, 50, 51–57, 308, 332
Orius sauteri, 150
Ormia depleta, 118, 127
Orthoptera, 12, 115, 121
Oryctes rhinoceros, 11, 133–139, 307, 336, 340, 346
Oryctes rhinoceros nudivirus (OrNV), 133–139, 336, 340
Ostrinia nubilalis, 11, 22, 307
Otiorhynchus ligustici, 4
Outbreak, 7, 11, 63, 71, 72, 75, 77, 79, 85, 98, 134, 135, 138, 142, 182, 183, 187, 189, 190, 194, 195, 197, 199, 223, 275, 295, 336
Overwinter, 22, 24, 25, 26, 51, 153, 154, 167, 280, 282
Oviposition, 116, 134, 152, 162, 163, 164, 169, 174, 175, 215, 216, 219, 225, 242

P

Pacific Islands, 133–139
Paecilomyces, 96, 106, 144, 167, 192
Paenibacillus popilliae, 12, 308
Painted apple moth, 40, 50, 55, 57–62, 299, 312, 333, 335, 344, 345
Palau, 134, 137
Palm trees, 133
Pandora kondoiensis, 96
Pandora neoaphidis, 95
Pandora sp., 99
Papua New Guinea, 134, 137
Parasetigena silvestris, 202
Parasite, 11, 19, 22, 23, 25, 51, 60, 98, 135, 142, 198, 199, 202, 203, 214, 217, 238, 239, 246, 273, 274, 276, 277, 311, 315, 318, 319

Parasitoid, 9, 11, 24, 25, 28, 60, 102, 104, 117, 127, 176, 188, 189, 201–203, 214, 220, 239, 243, 264, 276, 312, 313, 318, 319
Parthenogenetic reproduction, 94
Passive surveillance, 297
Passive vectors, 104
Pasture, 101, 117, 118, 121, 122, 124, 125, 126, 129, 286, 314
Patent, 121, 124, 174, 239, 346
Pathobiology, 138
Pathogen challenge, 310, 311, 340
Pathogen identification, 265–266
Pathogen life cycle, 21
Pathogen load, 19
Pathogen spread, 106
PCR, 138, 244, 248, 249, 251, 265, 294, 295, 296
Pea aphid, 95, 96
Pear thrips, 313, 314
Pecan, 25, 26
Pectinophora gossypiella, 184
Pellets, 106, 146, 147, 241
Pennsylvania, 165, 314
Permethrin, 125
Persistence, 21, 23, 38, 53, 54, 117, 125, 168, 172, 185, 190, 199, 200, 201, 246, 272, 275, 306, 307, 308–309, 312, 336, 341, 342
Pest, 7, 8, 9, 10, 11, 19, 23, 24, 34, 43, 51, 52, 57, 63, 64, 73, 95, 98, 99, 101, 102, 104, 105, 107, 117, 118, 121, 122, 123, 124, 134, 135, 137, 141, 160, 162, 165, 176, 182, 183, 184, 187, 190, 197, 203, 214, 233, 238, 251, 260, 263, 264, 273, 274, 275, 276, 285, 286, 306, 308, 309, 312, 314, 315, 318, 332, 339, 344, 346
Pesticide regulations, 341, 342
Pest identification, 332
Pest incursions, 322
Pests of legume, 101
Pest outbreaks, 275
Pest of pasture, 101
Pest-pathogen dynamics, 341
Pest of wheat, 98
Phagostimulants, 185
Phenacoccus herreni, 264
Phenacoccus manihoti, 264
Phenolic glycosides, 186, 191
Phenylpyrazole, 117
Pheromone, 8, 35, 41, 42, 52, 54, 55, 56, 57, 58, 63, 64, 65, 72, 74, 75, 76, 77, 78, 80, 81, 133, 160, 183, 275, 334

Pheromone-baited traps, 40, 41, 42, 72, 74, 75, 76, 77, 78, 80, 81
Pheromone trapping programme, 57
Philippines, 137, 139
Phloem, 92, 94, 160, 162, 167, 168, 171
Phoridae, 239, 243
Phototaxis, 151
Phyllocoptruta oleivora, 272, 276, 277
Phylogenetic relationship, 198, 250
Physical health scores, 298
Physiological host range, 21, 100, 195, 310
Phytoseiidae, 264
Phytoseiulus persimilis, 318
Phytotoxicity, 150
Picea, 94, 99
Picea abies, 99
Picea sitchensis, 99, 100
Pinaceae, 150
Pine bark beetle, 145
Pine plantations, 214, 215, 232
Pine sawyer, 141–154, 160, 175
Pine tree pest, 147
Pine trees, 142, 143, 144, 145, 148, 149, 151, 152, 153, 154, 214, 215, 223, 233
Pine wilt disease, 141–154
Pinewood nematode, 141–154, 160
Pink bollworm, 184
Pinus radiata, 55, 225
Pinus thunbergii, 145
Pitfall traps, 122, 125, 126, 249, 315
Plasmid, 293
Plasmid transfer, 293
Pleistophora schubergi, 198
Plutella xylostella, 12, 17, 307
Point release, 339
Point source, 339
Poland, 160, 161, 162, 165, 166, 167, 199
Pollen beetles, 316
Pollination, 3, 272, 273
Pollinators, 315, 341
Popillia japonica, 11, 12, 21
Poplar, 161
Population dynamics, 7, 25, 149, 183, 195
Population ecology, 33–43
Population growth, 7, 35, 36, 37, 38, 39, 40, 41, 42, 105, 263, 273
Population resurgence, 344
Populus spp., 161
Ports, 7, 8, 10, 40, 50, 63, 72, 73, 77, 78, 86, 116, 332
Ports of egress, 72
Ports of entry, 7, 8, 40, 72
Portugal, 144, 198, 199, 201, 202

Post-establishment, 28
Post-release monitoring, 266
Potato crops, 117
Poultry chicks, 6
Predators, 9, 22, 23, 24, 35, 36, 94, 98, 104, 120, 135, 203, 264, 276, 311, 315, 318, 319
Predatory birds, 275
Predatory mites, 260, 264, 318
Pressurized sprayer, 284
Preventing establishment, 8–9, 28
Preventive spraying, 153
Primary conidia, 96
Primary infections, 294–295
Proactant, 124
Probability models, 344
Productivity losses, 6
Product licensing, 346
Proisotoma minuta, 317
Prophylactic applications, 41
Protists, 10, 13, 305, 307, 315, 337
Psacothea hilaris, 147, 164
Pseudacteon spp., 239
Pseudacteon tricuspis, 243
Public attitude, 8, 64
Public concern, 56, 292, 306, 338
Public disclosure, 56
Public health, 3, 56, 64, 291–300
Public opposition, 338
Public relations, 50
Public response regulation, 60–62
Public self-reports, 298
Puerto Rico, 116, 118, 124, 125, 126, 128, 340
Pulse-field electrophoresis, 292
Pumpkin ash, 166
Pupal parasites, 60
Pupal weights, 311
Pupation cells, 167
Pyrethroid insecticide, 58, 312

Q

Quarantine, 6, 8, 9, 28, 33, 34, 41, 50, 54, 84, 119, 134, 135, 162, 164, 166, 167, 220, 223, 264–265, 338, 340
Quebec, 76
Queen bee, 274
Queensland, 57, 101, 124, 228, 229
Quercus spp., 51, 63, 147, 182, 183, 185, 189, 190, 201, 203, 265

R

Rabbits, 50
Rainfall, 165, 167, 193, 195, 230, 263

Random amplified polymorphic DNA (RAPD), 227, 228, 265, 295, 296, 320
Rapidity of response, 332, 333, 343
Rapid response, 54, 73, 332, 333, 343
Rare species, 35
Rats, 50
Raven, 174
Recombination, 250, 293, 320
Red beech, 55
Red imported fire ant, 6, 39, 237–252
Red mite, 264
Red spider mite, 264
Reduced fecundity, 311
Reference sites, 313
Registration, 11, 61, 162, 191, 315, 317, 320, 342–343, 347
Regulations, 4, 54, 64, 341, 342, 343
Regulatory agencies, 166, 174, 176, 343
Regulatory issues, 54–55
Regulatory processes, 347
Regulatory requirements, 321
Reinfection, 137
Reintroduction, 74, 138
Reinvasion, 40, 73, 335
Release, 5, 9, 11, 19, 20, 23, 28, 39, 60, 61, 100, 101, 102, 105, 119, 120, 121, 123, 128, 135, 136, 137, 138, 151, 183, 191, 193, 196, 200, 201, 238, 247, 262, 266, 267, 305, 313, 339, 340
Release point, 101, 151
Repeated introductions, 138
Representative test species, 309, 310
Residential population, 79
Residue, 185, 342
Residue analysis, 342
Resistance, 22, 27, 189, 215, 275, 276
Response strategies, 334
Resting spore, 96, 101, 102, 193, 194, 195, 196, 266, 267, 309
Resurgent population, 344, 345
Rhabditis necromena, 21
Rhinoceros beetle, 11, 133–139, 307, 336, 340, 346
Ribosomal RNA (16S), 292
Riparian strips, 59
Risk analysis, 264
Risk assessment, 59, 61, 64, 320, 342
RNA viruses, 247, 248, 249, 250, 251
Rodolia cardinalis, 24
Romania, 196, 199, 314
Root weevil, 21, 27, 50, 163, 314
Rosaceae, 55
Rubber plant, 134

Russian Far East, 77, 78
Russian wheat aphid, 21, 96, 98, 101–102, 103–104, 105–106, 107–108

S
Safety, 51, 54, 65, 150–152, 173, 186–187, 190, 195, 201, 264, 306, 307, 308, 315, 316, 317, 337, 340, 341, 342, 347
Safety profile, 337, 345
Salacin, 191
Salamander, 187
Salix spp., 51, 161, 182
Salt marsh mosquito, 312, 320
Samoa, 134, 136, 137, 138
Sandy soils, 116
Scale insects, 11
Scapanes australis, 134
Scapteriscus abbreviatus, 115–129
Scapteriscus borellii, 115–129
Scapteriscus didactylus, 115–129
Scapteriscus vicinus, 115–129
Scarabaeidae, 21, 124, 133, 140, 173
Scolytinae, 35, 36, 145, 146, 151, 173, 229, 231
Secondary conidia, 96
Sensitization, 294
Septicemia, 173, 185
Serangium parcestosum, 319
Serratia marcescens, 144
Sex pheromone, 8, 35, 74
Seychelles, 137
Ship ballast, 116
Silkworm, 150, 151, 183, 264, 315
Silverleaf whitefly, 318
Simulation models, 194
Sirex noctilio, 5, 11, 213–233, 307, 336, 337, 340
Sitka spruce, 99, 100
Sitona lepidus, 50, 314
Sitona sp., 314
Skin tests, 294
Slovakia, 199, 201, 271, 272
Slowing spread, 9, 41
Slow the Spread Program, 75, 80–83, 86, 183, 187
Small hive beetle, 271, 274, 276, 286
S-methoprene, 333
Sminthurus viridis, 318
Snakes, 6
Social behavior of bees, 272, 286
Socioeconomic impacts, 72
Soil, 4, 10, 21, 22, 27, 104, 116, 117, 118, 120, 123, 125, 126, 129, 150, 166, 185, 189, 194, 196, 241, 242, 259, 286, 292, 308, 310, 316, 320, 341, 345
Soil antagonists, 126, 242
Soil application, 308
Soil biota, 309, 315, 341
Soil-dwelling pests, 341
Soil ecosystem, 309, 310
Soil fumigant, 118
Soil fungi, 150
Soil mesocosm, 314, 316
Soil microarthropods, 315
Soil as pathogen reservoir, 104, 194, 201
Soil type, 21, 308
Solenopsis blumi [= *S. quinquecuspis*], 244
Solenopsis (= *Labauchena*) *daguerrei*, 239
Solenopsis geminata, 244, 245, 247, 250
Solenopsis interrupta, 244, 245
Solenopsis invicta, 6, 39, 237–252
Solenopsis invicta virus 1, 247–252
Solenopsis macdonaghi, 244, 245, 247
Solenopsis quinquecuspis, 244, 245
Solenopsis richteri, 237–252
Solenopsis saevissima, 238, 239, 244, 245, 247
Solenopsis saevissima species complex, 238
South Africa, 96, 98, 103–104, 107–108, 213, 214, 221, 223, 227, 229–231, 232
South America, 115–116, 117, 118, 119, 124, 214, 221, 223, 226, 229, 232, 237, 239, 240, 242, 244, 250, 252, 260, 262
South Carolina, 73, 123
South East Asia, 99
Soybean aphid, 21, 95, 99
Spawn chips, 146, 147, 148, 149
Species composition, 317
Species richness, 186, 313
Specificity, 184, 190, 195, 199, 201, 247, 249, 252, 264, 306–307, 336, 341, 346
Sphecidae, 115
Spore powder, 12, 279, 280, 282, 317
Spotted alfalfa aphid, 96, 97, 101, 102
Spray applications, 53, 64, 107, 279, 300, 346
Spray deposition, 54
Spray distribution, 160, 185
Spray drift, 308, 311, 320
Spray nozzle types, 185
Spray programs, 51, 54, 56, 57, 59, 61, 64, 66, 77, 161, 292, 296, 298, 299
Spray rigs, 122
Spray zone, 53, 54, 59, 64, 77, 79, 296, 297, 298, 299
Spread of pest, 7, 37–38, 52, 135, 203, 309, 332
Spread rates, 41

Spruce, 94, 99
Stabilized mycelium, 106
Stable isotope analysis, 345
Stable isotope testing, 62
Steinernema carpocapsae, 27
Steinernema diaprepesi, 21
Steinernema feltiae, 119
Steinernema glaseri, 12
Steinernema neocurtillae, 121
Steinernema riobrave, 124
Steinernema scapterisci, 115–129, 344
Sterile insects, 39
Sterile male moths, 60
Sternorrhyncha, 93
Stethorus sp., 264
Stickers, 185
Stoats, 50
Stored combs, 274
Stratified dispersal, 37, 38, 80, 82
Sublethal doses, 162, 202
Sublethal effects, 311
Subsurface application, 122
Suburban, 10, 52, 73, 85, 160, 175, 176
Sugarcane, 117, 118, 124
Sugar maple, 314
Surveillance, 4, 9, 54, 296, 297, 298, 338
Surveillance methods, 4
Survey data, 40
Survey methods, 40, 42
Survey tools, 40, 298
Survival rates, 278
Sweden, 11
Sweetpotato whitefly, 318
Symbiotic bacterium, 120
Symbiotic fungus, 213, 215, 216, 218, 221
Symptom checklist, 299
Symptom reports, 299
Synergistic materials, 191
Synergists, 191, 312
Synthetic pheromone, 52, 55, 60
Systemic, 160, 162, 167, 175

T

Taeniothrips inconsequens, 313, 314
Taiwan, 51, 238
Talc powder, 280
Tansy ragwort, 186
Tanzania, 137, 214
Tarnished plant bug, 316
Tarsonemidae, 273
Tecra VIA, 295
Teia anartoides, 40, 50, 55, 57–62, 299, 312, 333, 335, 344, 345
Teleomorphic stage, 97

Temperature, 60, 105, 107, 128, 150, 163, 194, 195, 228, 232, 233, 249, 277, 278, 280, 281, 286, 308, 309, 317
Temperature effects, 228
Tenebrionidae, 173
Tetradonema solenopsis, 239
Tetranychidae, 260
Tetranychus bastosi, 264
Tetranychus urticae, 264, 318
Texas, 73, 117, 119, 127, 240, 271, 276, 279
Thailand, 274
Thelohania solenopsae, 239, 242–246
Therioaphis trifolii f. *maculata*, 96, 97, 101–102
Thresholds, 42, 164, 167, 182, 275
Thrips, 163, 313, 314, 316, 318
Thuricide, 188
Ticks, 163, 277, 279
Tier 1 testing, 342
Tier 2 testing, 342
Timber production, 182
Tokelau Islands, 134
Tonga, 134, 135, 137
Toxin production, 293
Trachea, 142, 143, 273, 274
Tracheal mite, 273, 274
Trade issues, 85, 339
Trade secret, 66, 312, 338, 346
Transmission, 101, 104, 143, 152, 154, 164, 172, 189, 190, 193, 199, 200, 240, 241, 244, 251, 252, 262, 306, 308–309, 341, 346
Transovarial transmission, 200, 242
Trap catches, 58, 65, 83
Trap data, 60, 62
Trap densities, 344
Trap trees, 224, 229, 231
Trapping male moths, 52
Trapping grid, 80
Tree-killing bark beetles, 35
Tree mortality, 5, 85, 161, 165, 167
Trifolium repens, 314
Trunk injection, 143
Turex, 295
Turf grass, 117
Tussock moth, 50, 51, 53, 55, 56, 60, 182, 308, 332, 335, 337, 338
Twospotted spider mite, 264, 318
Typhlodromalus aripo, 260
Tyria jacobaeae, 186

U

Ulmus americana, 165
Underground colonies, 341

Uniformity of test species, 311
United Kingdom, 272
United States, 4, 6, 24, 39, 40, 41, 84, 86, 159, 168, 183, 199, 200, 237, 292, 314
Unwanted organisms, 333
Urban areas, 10, 13, 55, 76, 85, 242, 292, 335, 338, 342
Urban environment, 49, 52, 66
Urticating hairs, 56, 85
Uruguay, 118, 119, 214, 229, 239
US Environmental Protection Agency, 118, 186, 190, 300, 342
US Food and Drug Administration, 292, 300
USSR, 12

V

Vairimorpha disparis, 198, 202
Vairimorpha invictae, 239, 246–247
Vancouver, 53, 77, 78, 186, 298, 300, 338
Varroa destructor, 50, 271, 273–274, 276–285, 315
Varroidae, 272
Vectobac, 295
Vector, 13, 94, 107, 121, 124, 142, 143, 144, 153, 307, 317, 333, 343
Vector MC, 123, 124
Vegetable oil, 168, 279, 281, 282
Vegetable seedlings, 117
Vegetative compatibility groups, 320
Venezuela, 117, 214, 261
Vermont, 314
Vertical transmission, 200, 246
Verticillium lecanii, 96, 294
Verticillium sp., 144
Vespidae, 6
Vespula germanica, 6
Vespula vulgaris, 6
Victoria, 77, 187, 221, 225, 226, 228, 296, 298, 300
Viral ESTs, 247
Virin-Ensh, 191, 336, 337
Virtuss, 55, 60
Virulence, 20, 26, 28, 60, 98, 138, 139, 183, 184, 187, 197, 200, 226, 252, 262, 278, 311, 318, 336, 340
Virus, 10, 11, 13, 20, 55, 60, 65, 76, 80, 135, 136, 137, 138, 139, 183, 188, 189, 190, 191, 202, 239, 247, 248, 249, 250, 252, 277, 295, 305, 306, 307, 313, 321, 336, 337, 341, 343, 346

Visual inspections, 8
In vitro production, 191, 261
In vivo production, 12, 76, 191, 196, 260, 265

W

Washington, 73, 75, 78, 79, 86
Water column, 312
Waterways, 4, 320
Weather conditions, 102, 105, 195, 306
Weevils, 21, 41, 42, 50, 163, 173, 314, 341
Weight gain, 310, 314
Western flower thrips, 316, 318
West Indies, 116, 117
Wetlands, 7
Wettable powder, 107, 312, 317, 318
Wheat bran pellets, 146, 147
White ash, 165, 166, 170, 171
White clover, 314
White pine blister rust, 141
White spotted tussock moth, 50, 51–57, 308, 332
Whitefly, 11, 264, 312, 313, 318, 319
Whitefly nymphs, 312, 319
Whitefly pest of cassava, 264
Whitefringed beetle, 21
Wildflowers, 117
Wildlife, 6, 166, 238
Willow, 51, 161, 182
Wine grapes, 24
Winged morphs, 93, 94
Winter survival, 281, 309
Wolbachia, 240
Wood borers, 8, 160, 173, 340
Wood-boring beetles, 13, 159–176, 335
Woodpecker, 166, 171

X

Xenorhabdus innexi, 120
XenTari, 174, 295

Y

Yeast, 274, 285

Z

Zoophthora anhuiensis, 95
Zoophthora occidentalis, 99
Zoophthora radicans, 95
Zoophthora sp., 98
Zwittermicin A, 174
Zygospores, 96